Endorsements for *Design for Environment*

"With all the recent talk about green technology and green jobs, Joseph Fiksel's book could not come at a better time. If your goal is to design and develop environmentally sustainable products that also drive shareholder value, then this book is a must read."

—Stuart Hart, S.C. Johnson Chair in
Sustainable Global Enterprise, Cornell University
Author, *Capitalism at the Crossroads*

"*Design for Environment* shows clearly that aligning green design with innovation not only is possible, but is essential to remaining competitive in our brave new economy. This should be required reading for anyone who wants to understand how to design for today's consumer—and tomorrow's."

—Joel Makower, Executive Editor, GreenBiz.com
Author, *Strategies for the Green Economy*

"A must read for all practitioners of a Design for Environment approach. This book makes the most compelling case yet for taking a more integrated and holistic approach to DFE—the bottom line! Green initiatives must increase profitability to be truly sustainable, and Dr. Fiksel provides the blueprint for how global companies are enhancing profits and winning in the marketplace by designing their way to competitive advantage."

—Jim Lime, Vice President, Environment, Health & Safety
ConAgra Foods

"Dr. Fiksel has developed a comprehensive and inspiring guide that provides a powerful case for integration of environmental principles into product development. *Design for Environment* is essential reading for any organization putting DFE into practice."

—Ken Strassner, Vice-President, Global Environment,
Safety, Regulatory and Scientific Affairs
Kimberly-Clark Corporation

"Dr. Fiksel 'hits the nail on the head' with this practical and informative approach to life-cycle design and management. A must read to help you achieve both exceptional environmental stewardship and exceptional business results."

—Joe Allen, General Manager and Director of Sustainable Development
Caterpillar Remanufacturing Division

"The book offers important industry perspectives on how companies develop and design innovative solutions to complex environmental and societal challenges. It goes well beyond theory, offering case studies with quantifiable results that illustrate how companies can save money while improving the environment and helping local communities. It shows how companies of all types are using resources more efficiently, sometimes by teaming up with other industries, to achieve results that balance the triple bottom line of people, planet, and prosperity. This richly detailed study should be of great interest to industry leaders, policymakers, scholars, and students alike. We are all fortunate to have Joseph Fiksel working on sustainable development."

—Andy Mangan, Executive Director
U.S. Business Council for Sustainable Development

Design for Environment

A Guide to Sustainable Product Development

Joseph Fiksel

Second Edition

New York   Chicago   San Francisco
Lisbon   London   Madrid   Mexico City
Milan   New Delhi   San Juan
Seoul   Singapore   Sydney   Toronto

ISBN 978-0-07-177622-6
MHID 0-07-177622-2

Sponsoring Editor	**Copy Editor**
Michael Penn	Nancy W. Dimitry
Editing Supervisor	**Proofreader**
Stephen M. Smith	Donald Paul Pomeranz
Production Supervisor	**Indexer**
Pamela A. Pelton	Seth Maislin
Project Managers	**Art Director, Cover**
Joanna V. Pomeranz and	Jeff Weeks
Nancy W. Dimitry,	**Composition**
D&P Editorial Services, LLC	D&P Editorial Services, LLC

Printed and bound by RR Donnelley.

"To the people of poor nations, we pledge to work alongside you to make your farms flourish and let clean waters flow; to nourish starved bodies and feed hungry minds. And to those nations like ours that enjoy relative plenty, we say we can no longer afford indifference to suffering outside our borders; nor can we consume the world's resources without regard to effect. For the world has changed, and we must change with it."

President Barack H. Obama
Inaugural Address
January 20, 2009

About the Author

 Dr. Joseph Fiksel is Executive Director of the Center for Resilience at The Ohio State University, and Principal and Co-Founder of the consulting firm Eco-Nomics LLC. He is a recognized authority on sustainable business practices, with over 25 years of research and consulting experience for multi-national companies, government agencies, and consortia such as the World Business Council for Sustainable Development. A native of Montreal, Dr. Fiksel began his career at DuPont of Canada, and subsequently was Director of Decision and Risk Management at Arthur D. Little, Inc., and Vice President for Life Cycle Management at Battelle. He has published over 70 refereed articles and several books, and is a frequent invited speaker at professional conferences. Dr. Fiksel holds a B.Sc. from M.I.T., an M.Sc. and Ph.D. in Operations Research from Stanford University, and an advanced degree from La Sorbonne. He can be reached at Fiksel.2@osu.edu.

Contents at a Glance

Contents

Preface

Since the second edition of this book was published in hardcover, the world has experienced a continuing stream of disruptive events, including a volcanic eruption in Iceland, an oil spill in the Gulf of Mexico, a disastrous tsunami in Japan, floods and tornadoes in the midwestern U.S., and dramatic political upheavals in the Middle East. However, my basic premise has stayed constant: environmental sustainability is compatible with economic growth—in fact, growth is a "natural" process. The alternative seems unacceptable, namely, stagnation and decline of human civilizations. Yet today the challenges of climate change and natural resource degradation are more daunting than ever. Personally, I remain an optimist, believing that humans are ingenious and resilient enough to overcome these challenges. However, we need to get going, sooner rather than later!

Although sustainability has become fashionable, it seems in danger of being trivialized. While recycling containers, eating local foods, and using energy-efficient light bulbs are positive steps, the magnitude of the problem is much larger. Our economic systems are generating a mountain of hidden wastes and emissions that are gradually threatening the ecological goods and services—energy, water, land, and biological resources—needed to support our lifestyles and enable worldwide prosperity. These environmental impacts are not just "issues" that can be easily corrected by making products "greener" and more "friendly." We will need to work collectively to reinvent the supply chain systems and infrastructure that form the basis of production and consumption.

This book describes an important part of the solution, a field of engineering that I have been involved with for many years, Design for Environment (DFE). The evolution of DFE has followed the same path as my own professional career—from analyzing environmental risks to preventing waste and pollution at the source, from complying with regulations to integrating environmental sustainability into enterprise strategy and decision making. You will find in these pages a business rationale for developing sustainable products and processes, as well as a comprehensive toolkit for corporations to practice DFE in the context of product life-cycle management. Perhaps most interesting, you will find a series of examples drawn from actual case histories of innovative companies that have applied these tools to their businesses. The book should be equally useful to managers, product developers, environmental specialists, government officials, and academics.

The global economic collapse of 2008 demonstrated that sustainability does not guarantee short-term enterprise resilience—many corporate sustainability leaders stumbled into crisis mode. And, as the recovery gains momentum, social responsibility will not be enough. DFE is essential to the revitalization and continuity of our global economy. Yesterday's technologies and business models will not be adequate to achieve the quantum changes that are needed. Therefore, creative design and breakthrough innovation are necessary for a rapid shift to an economically sustainable path. Regulatory intervention is inevitable, but governments absolutely need to collaborate with corporations in developing and implementing new policies. Principled and enlightened leadership from the private sector will be a critical factor in helping to address the complexities of global warming and resource depletion, not to mention poverty, corruption, and armed conflict. Companies are already moving beyond enterprise boundaries to establish industry coalitions and strategic partnerships with governments and environmental advocacy groups.

As I said above, DFE is only part of the solution, and the scope of this book is restricted deliberately in several respects:

- The focus is on how *environmental* excellence creates business value, although we touch on other aspects of sustainability, such as human capital and stakeholder engagement.
- The focus is mainly on *United States* institutions and enterprises, although we describe broader international developments that have influenced U.S. practices.
- The focus is on design *engineering*, although we discuss the roles of other functions, including R&D, marketing, logistics, communications, and information technology.
- The focus is on the design of *products* and associated processes, although we mention the major strides that have occurred in the sustainable design of the built environment.

The ultimate purpose of DFE is to protect the capacity for humans and other organisms to flourish together on Earth. We now recognize that, because of unsustainable industrial patterns, natural systems may be destabilized and quality of life for future generations may be endangered. Fortunately, Nature is extremely resilient, and living systems can adapt to change. If we are agile enough, then DFE can become a cornerstone of our own successful adaptation strategy.

Joseph Fiksel

Acknowledgments

This book is the culmination of over a quarter century of research and practice during which I have been influenced, inspired, and supported by many talented people. The following is only a partial list of collaborators and contributors, my fellow travelers on this excellent journey.

First, I would like to thank my editors, Michael Penn and Tai Soda; Stephen Smith and the rest of the capable production team at McGraw-Hill; and Joanna Pomeranz and Nancy Dimitry of D&P Editorial Services. Thanks also to Warner North, Bill Balson, David Cohan, and Ken Wapman of Decision Focus, and the many other pioneers who contributed to the first edition of *Design for Environment*. Second, I salute the vision of my former colleagues at Battelle—Wayne Simmons, Bruce Vigon, Jeff McDaniel, Ben Parker, Dave Spitzley, Joyce Cooper, Vinay Gadkari, Tiffany Brunetti, Ken Humphreys, Marylynn Placet, Bob Quinn, and others who helped me to develop the Life Cycle Management practice. Third, I am indebted to the many clients and consulting partners of Eco-Nomics LLC, including John Stowell, Bonnie Nixon, John Harris, Bert Share, Les Artman, Alan Dudek, Jim Thomas, Jon Low, Steve Poltorzycki, Bob Axelrod, Steve Hellman, and Amy Goldman, who helped give me the opportunity to practice what I preached.

In recent years I have been fortunate to work with an outstanding group of faculty members, students, researchers, and administrators at The Ohio State University, who have helped to establish and grow the Center for Resilience. These include Marc Posner, Suvrajeet Sen, Dave Woods, Julie Higle, Rajiv Ramnath, Doug Lambert, Keely Croxton, Tim Pettit, Emrah Cimren, Kieran Sikdar, Joe Bolinger, Anil Baral, Shweta Singh, Cullen Naumoff, Ravishankar Rajagopalan, Maria Aguilera, Fernando Bernal, James Balch, Aparna Dial, Greg Washington, Randy Moses, and Bud Baeslack, among others. The most profound influence in my recent career has been Bhavik Bakshi, co-founder of the Center for Resilience, and an international thought leader in the field of life-cycle assessment. Contributions from Bhavik and his research group appear in many chapters of this book.

There were a number of key individuals who took time out of their busy schedules to help me incorporate some great stories into this book about their Design for Environment initiatives:

- Joe Allen, Caterpillar
- Patty Calkins and Wendi Latko, Xerox
- Terry Cullum, General Motors
- Heidi Glunz, McDonald's
- Erika Guerra, Holcim
- John Harris and John Kindervater, Eli Lilly
- Kathy Hart and Clive Davies, U.S. EPA
- Al Ianuzzi, Johnson & Johnson
- Jim Lime, ConAgra Foods
- Elissa Loughman, Patagonia
- Lisa Manley and Jeff Seabright, Coca-Cola
- Ken Martchek, Alcoa
- Debbie Mielewski and Jon Newton, Ford Motor
- Keith Miller, 3M
- Sandy Nessing and Bruce Braine, American Electric Power
- Bonnie Nixon, Hewlett Packard
- Scott Noesen and Cliff Gerwick, Dow Chemical
- Ted Reichelt, Intel
- Dawn Rittenhouse, DuPont
- Bert Share, Anheuser-Busch
- David Spitzley and Ken Strassner, Kimberly-Clark
- Jim Thomas, JCPenney
- Stan Wolfersberger and Frank O'Brien-Bernini, Owens Corning

Among my broader circle of friends and colleagues, I want to acknowledge long-standing relationships with a number of innovators and leading practitioners in the field of environmental sustainability who have helped to shape my knowledge and beliefs. These include Derry Allen, Brad Allenby, Paul Anastas, Vince Covello, John Ehrenfeld, Tom Gladwin, Glenn Hammer, Stuart Hart, Alan Hecht, Tom Hellman, Mike Long, Andy Mangan, Bill McDonough, Adrian Roberts, Jed Shilling, Paul Tebo, Warren Wolf, and the late

Joe Breen. I also wish to honor my dear uncle and first true mentor, Fred Fiksel, who launched me on the path of intellectual discovery.

Finally, I want to thank the most important person in my life— my spouse and soul-mate, my friend and business partner, my toughest critic and most ardent advocate, Diane Guyse Fiksel. She not only contributed substantive content to this book, but also care-fully reviewed and helped to refine the entire manuscript. Together, we have raised three admirable young men, Justin, Cameron, and Brandon, who are now entering their college years. This book is really about their future—the first generation of the new millennium.

CHAPTER 1

Introduction

The challenge is to build a new economy and to do it at wartime speed before we miss so many of Nature's deadlines that the economic system begins to unravel. —LESTER BROWN [1]

A Sense of Urgency

Human prosperity and environmental integrity are closely intertwined because the fulfillment of basic human needs—food, clothing, materials, energy—ultimately depends upon the availability of natural resources. Since the dawn of *homo sapiens,* we have recognized this fact, and in most ancient cultures nature was respected and revered. Yet, over the last several hundred years, during a period of dramatic industrialization, innovation, and global expansion, we humans have not only taken the natural environment for granted, but we have literally plundered and abused nature to serve our growing appetites. With no natural enemies, we conquered the planet, only to realize that we may be our own worst enemy.

Thankfully, over the last fifty-odd years, we gradually rediscovered the importance of protecting vital resources, such as soil, air, water, trees, and other organisms. What began as a fringe movement in the 1960s has evolved into a mainstream concern, as economists and politicians have gradually recognized that we are depleting fossil fuel resources and pumping greenhouse gases into the atmosphere at an alarming rate. Yet even today, many people do not understand the magnitude of these problems and tend to trivialize the solutions. Environmental awareness has become chic and is embraced by celebrities and brand marketers, while our major industrial systems continue to operate as before, with superficial changes. The good news is that we are no longer in denial, but the bad news is that we can't seem to break our old habits.

Perhaps we need another wake-up call. Global warming is only one of many disturbing trends identified by the scientific community—

1

sea level is rising, fresh water is growing scarce, we are running out of arable land, our forests are disappearing, and biodiversity is threatened due to changes in natural habitats. Meanwhile, global population continues to increase; more countries are developing into resource hogs, while over three billion people still live in abject poverty on less than $2.50 per day, many deprived of even basic sanitation and fresh water.

Humans are nothing if not ingenious problem-solvers. Can we cleverly escape, like Indiana Jones, from this predicament in which we find ourselves? We are faced with the most daunting set of problems in our recorded history. Once we squarely confront these problems, we have two paths to choose—hopeless or hopeful:

1. The hopeless path is to resign ourselves—accept that the world will soon run out of resources; that there will be an inevitable collapse of civilization; and that our best alternative is to hunker down, break our dependencies, become self-sufficient, live off the land, and pray that we are not destroyed in a global conflict or natural catastrophe.

2. The hopeful path is to reinvent our way of life—not to retreat but to join forces in an unprecedented sustainability campaign. We are past the point where better housekeeping will solve the problem. We need to completely redesign industrial products, processes, and supply chains to dramatically lower their resource intensity, while assuring that developing nations do not replicate the old designs.

Pursuing the hopeful path will require extraordinary collaboration. While breakthrough innovation can be accomplished by the private sector, governments and private citizens also have an important role to play. First, we need a governance framework for sustainability that provides the right incentives while minimizing the barriers to change. Second, we need to redesign the publicly managed physical infrastructure systems that support economic activity—transportation, energy, water, and waste management. Finally we need to accomplish a massive behavior change in order to shift toward sustainable consumption patterns.

This journey will be an enormous challenge—a moonshot for the twenty-first century. We have already waited too long, and we need to act with a sense of urgency. Achieving sustainability will draw upon the best thinking in every discipline—management and engineering, science and policy. This book does not provide a complete blueprint for the journey to sustainability, but it does provide some important tools for slowing and perhaps even reversing environmental degradation through innovative product and process design.

The Basic Premise

The underlying premise of this book is that environmental sustainability is compatible with economic development. Just as growth is essential to living organisms, growth is also essential to a healthy society, especially in developing nations afflicted by widespread poverty [2]. But to achieve global sustainability, *we need to radically redesign our industrial systems to create more value with fewer resources.* The idea of sustainable growth is a departure from traditional world views, in which environmentalists considered industry to be the enemy and businesses considered environmental protection to be a burden. But it is not pie-in-the-sky idealism. Global manufacturers in virtually every industry category are adopting sustainability goals, and environmental activist groups have begun to collaborate with them. Sustainable development through "eco-efficiency" has become the rallying cry of companies who see competitive advantage in conservation of resources and environmental stewardship. This book is filled with examples of leading companies who have embraced these concepts. Here are a few notable examples:

- Dow Chemical has adopted sustainability as a strategic imperative throughout its global businesses. Over the decade from 1995 to 2005, Dow spent about $1 billion on environmental, health, and safety improvements and realized about $5 billion in value (see Chapter 13).

- General Electric's ecomagination^{SM} portfolio of energy efficient and environmentally advantageous products and services exceeded $17 billion in 2008 and has been the company's fastest growing segment, even during the economic downturn.

- Wal-Mart, the retail giant, has adopted aggressive environmental goals for energy and waste reduction, and has urged its suppliers to help meet these goals through product development, packaging, and transportation (see Chapter 19).

- The World Business Council for Sustainable Development, based in Geneva, is a consortium of over 150 global companies that have been working together since 1990 on global sustainability issues such as climate change and clean water.

While large corporations speak enthusiastically about sustainability, some critics have assailed the "win-win" philosophy as misleading, claiming that environmental management costs will continue to outweigh the benefits and that regulatory controls are the only reliable way to reduce industrial wastes. As early as 1970, the eminent economist Milton Friedman declared that corporate social and environmental responsibility was a distraction from the fundamental purpose of a corporation—to create value for shareholders. In other words, "the business of business should be business." It is certainly true that environmental initiatives do not automatically produce financial benefits; they need to be evaluated in the same light as any other investments. However, it is also true that a full consideration of environmental factors and trade-offs will reveal opportunities to simultaneously enhance customer satisfaction, profitability, and competitiveness. This will be amply demonstrated in later chapters. In the long run, businesses can only remain competitive if they are attuned to the broader needs of society. In the words of management theorists Michael Porter and Mark Kramer [3]:

WE NEED TO RADICALLY REDESIGN INDUSTRIAL SYSTEMS TO CREATE MORE VALUE WITH FEWER RESOURCES.

> The mutual dependence of corporations and society implies that both business decisions and social policies must follow the principle of *shared value*...a company must integrate a social perspective into the core frameworks it already uses to understand competition and guide its business strategy.

The Hidden Mountain

The commitment of major corporations to environmental sustainability is certainly a hopeful sign. Yet, even with widespread adoption of corporate environmental responsibility, worldwide levels of energy and material use continue to rise. Paradoxically, the more efficient companies become in terms of resource utilization, the more rapidly the global economy grows; this is known to economists as the "rebound effect." It is becoming apparent that voluntary, incremental environmental improvements by individual companies will be inadequate to significantly offset the growth of the global economy. The rapid industrialization of China, India, and other emerging economies will likely exacerbate this problem. Ecological footprint analysis suggests that humanity's ecological demands already exceed what nature can supply, and we are now eroding our "natural capital" rather than living off the interest (see Chapter 9). Clearly, we need to dig deeper into the source of the problem.

Few of us in the developed world are aware of the enormous environmental impacts of everyday lifestyles. As we pursue our habitual patterns—mealtimes, commuting to work, occasional recreation—we have no clue about the hidden flow of resources needed to support these seemingly innocent activities. Each of us is actually living on top of a mountain of resources, including energy and materials, all of which originated from the natural environment. Ecological goods and services are embedded in everything that we consume (see Figure 1.1). It has been estimated that the average American citizen accounts for about 30 metric tons of material per

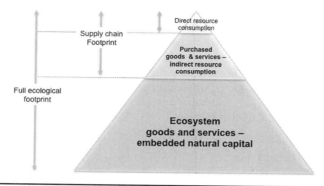

FIGURE 1.1 The hidden mountain of resource consumption.

year [4]. We never see most of that material because it is released to the environment in the form of trash, wastewater, and airborne emissions—mainly carbon dioxide. Only about 5% finds its way to the customer in the form of consumable products such as food, and durable goods such as furniture. A small fraction is recycled, but the rest is simply thrown away.

For example, assuming that an individual consumes 1 plastic quart bottle per day, her annual consumption generates about 5 kg of solid waste per year (11 lb). Recycling may reduce that total, although it still requires resources to bring the bottles back to the point where they reenter the manufacturing stream. However, if we consider the resources—both energy and materials—required in the full *life cycle* for those plastic bottles, including resource extraction, manufacturing, and transportation, it requires a total of about 250 kg of materials per year, 98% of which end up as solid waste. In addition, the supply chain associated with plastic bottle manufacturing consumes about 20 liters of water per bottle, amounting to about 230 million metric tons of water per year for the U.S. as a whole. This enormous water footprint is mainly due to thermo-electric power generation and crop irrigation [5].

Plastic bottles are not unusual in this respect. The processes involved in manufacturing and supporting most products have significant impacts on the environment, including the generation of waste, the disruption of ecosystems, and the depletion of natural resources. About 20 billion tons of industrial wastes are generated annually in the United States, and over a third of these are hazardous wastes. As a result, U.S. communities are rapidly filling up available landfill space. In the European Union, where unused land is scarce, a number of strict directives have been issued that require manufacturers to recover and recycle discarded products and packaging (see Chapter 3). But still the mountain grows.

The sobering message is that our current patterns of industrial development threaten to exceed the capacity of ecosystems in terms of resource utilization and waste absorption, and also pose potential threats to global climate, vegetation, and agriculture. Who is responsible for this escalating crisis—complacent consumers or profit-driven producers? Unfortunately, contentious debates are only a distraction from developing solutions. Playing the blame game is not going to help—we need to accept our collective guilt and move on. Making real progress will require disruptive innovation and fundamental redesign of industrial systems.

The Emergence of DFE

The concept of Design for Environment (DFE) originated in the early 1990s, largely through the efforts of a handful of private firms that were attempting to build environmental awareness into their product development efforts.* The strategic importance of DFE and examples of DFE practice were first described in an innovative 1992 report by the U.S. Congress Office of Technology Assessment [6], and in the same year the American Electronics Association produced a ground-breaking primer for the benefit of member companies [7]. Since that time, the level of interest has mushroomed, and DFE has become a common theme in corporate environmental stewardship and pollution prevention programs. Typically, the scope of DFE includes the following objectives:

- Environmental protection—assurance that air, water, soil, and ecological systems are not adversely affected due to the release of pollutants or toxic substances.

- Human health and safety—assurance that people are not exposed to safety hazards or chronic disease agents in their workplace environments or personal lives.

- Sustainability† of natural resources—assurance that human consumption or use of natural resources does not threaten the availability of these resources for future generations.

For purposes of this book, we view DFE as a collection of design practices aimed at creating products and processes that address the above objectives. Hence the following definition:

Design for Environment is the systematic consideration of design performance with respect to environmental, health, safety, and sustainability objectives over the full product and process life cycle.

---

*DFE is often referred to by other names, including Eco-Design, Life Cycle Design, Design for Eco-efficiency, and Sustainable Design.
†This is "sustainability" in the narrow sense originally intended. The term has evolved into a popular buzzword used to encompass environmental, health and safety, economic, social and ethical issues.

The practice of DFE has spread quickly in today's business environment, as major firms have recognized the importance of environmental and social responsibility to their long-term success. Even the U.S. Environmental Protection Agency has established a DFE program to encourage reduction of pollution at the source (see Chapter 3). The company case studies in Part 3 of this book illustrate how DFE provides competitive advantage by reducing the costs of production and waste management, driving product innovation, speeding time to market, and attracting new customers. The business benefits of DFE are summarized in Figure 1.2, and are further described in Chapter 4.

Example   Here is a recent example of DFE applied to sustainable packaging: Sam's Club redesigned its milk jugs with a new cubical shape that is easier to transport. The company estimates that this kind of shipping has cut labor by 50% and water use for cleaning by 60 to 70%. More gallons fit on a truck and in Sam's Club coolers, and no empty crates need to be picked up, reducing trips to each Sam's Club store to two a week from five—a substantial fuel savings. Also, Sam's Club can now store 224 gallons of milk in its coolers in the same space that used to hold 80. The only drawback was that consumers had a hard time getting used to pouring milk from the new jugs. However, the company was able to pass on savings of 10 to 20 cents per gallon to the consumer.

DFE originated in the early 1990s due to the convergence of several driving forces that made global manufacturers more aware of the environmental implications of their product and process designs. For one thing, consumers were becoming increasingly concerned about the environmental "friendliness" of the products that they purchase. The International Organization for Standardization (ISO) was developing the 14000 series of standards for environmental

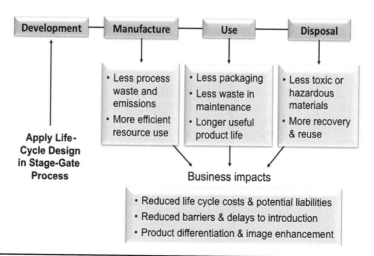

FIGURE 1.2   Benefits of Design for Environment.

management systems, analogous to the ISO 9000 series for quality management systems. Many government agencies, notably in the European Union, were taking aggressive steps to assure that manufacturers are responsible for recovery of products and materials at the end of their useful lives. At the same time, there was a growing voluntary commitment on the part of major manufacturing firms to assure environmental responsibility for both their internal operations and their suppliers. This led to the flourishing of consortia such as the Global Environmental Management Initiative and the World Business Council for Sustainable Development, as well as government-sponsored programs such as ENERGY STAR. All of these historic changes are described in Chapter 3.

DFE can be seen as a conceptual crossroads between two major thrusts that began in the 1980s and transformed the nature of manufacturing throughout the world. As illustrated in Figure 1.3, these two thrusts are *enterprise integration* and *sustainable development*.

Enterprise integration is the reengineering of business processes and information systems to improve teamwork and coordination across organizational boundaries, thereby increasing the effectiveness of the enterprise as a whole. The total quality management (TQM) movement provided a strong motivation for enterprise integration, and *integrated product development* (IPD) has been widely adopted as a strategy for agile manufacturing, allowing companies to release higher-quality products while reducing time to market. As described in Chapter 5, IPD involves cross-functional design teams who consider the entire spectrum of quality factors, including safety, testability, manufacturability, reliability, and maintainability, throughout the

Figure 1.3   DFE is at the crossroads between enterprise integration and sustainable development.

product life cycle. Since environmental performance is an important aspect of total quality, DFE fits naturally into this process—in fact, *life-cycle thinking* is at the core of DFE.

Sustainable development is industrial progress that "meets the needs of the present without compromising the ability of future generations to meet their own needs" [8]. The implied challenge is how to assure continued industrial growth without adverse ecological and social impacts. To address this challenge, the traditional economic concept of exchange value must be extended to include both man-made and natural capital (see Figure 1.4). Instead of accounting purely for the labor and man-made materials that are inputs to production, the broader approach of *ecological economics* takes into account the value of natural resource inputs and the effect of waste outputs [9]. Sustainability has recently emerged as a prominent global issue due to concerns about shrinking fossil fuel reserves and evidence of global warming. As mentioned above, climate change is just one of many sustainability issues that need to be considered in a broader systems view of economic development.

Historically, the issues addressed in DFE were managed by environmental, health and safety (EH&S) groups that tended to be isolated from the mainstream in terms of both the strategic and the operational aspects of a business. However, the role and positioning of EH&S groups have shifted, and sustainability issues that were previously considered esoteric have been incorporated into a comprehensive business strategy. The emergence of DFE has resulted in active collaboration among company groups that rarely had contact in the past—environmental managers and product development managers.

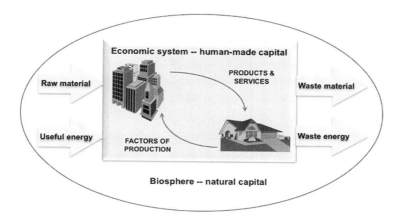

Figure 1.4   Broadening traditional economics to include natural capital (adapted from [9]).

Moreover, companies are increasingly working with suppliers, customers, and external stakeholders to design sustainable solutions. This book describes the exciting opportunities that are made possible by this collaborative, enterprise-wide approach.

DFE Implementation Challenges

A newfound passion for sustainability has led many companies to set aggressive environmental performance goals and issue glossy reports extolling their commitment to environmental and social responsibility. But veteran companies that have walked this path for years realize that putting these concepts into practice in a way that genuinely transforms a business is not as simple as it may appear. It is relatively easy to "pick the low-hanging fruit" by adopting best practices such as low-energy lighting systems and recovery of post-industrial scrap. With a bit more effort, companies can implement ISO 14001-style environmental management systems and measure their continuous improvement in performance. A much greater challenge is to integrate sustainability thinking into a company's business processes to achieve significant, lasting change. In particular, to perform Design for Environment consistently and effectively is challenging for several reasons:

- The necessary environmental expertise is not widely available among product development teams, including marketers and engineers.
- The complex and open-ended nature of environmental phenomena makes it difficult to analyze the effects of design improvement.
- The economic systems in which products are produced, used, and recycled are much more difficult to understand and control than the products themselves.

Despite the many examples of successful DFE efforts described in this book, the current state of DFE practice can be characterized as mainly opportunistic. Well-motivated and well-informed teams may be able to identify product improvements that are environmentally beneficial or that reduce life-cycle costs. However, in order to implement DFE fully as a component of the new product development process, systematic methods and processes must be introduced and integrated into the daily work of development teams. Much of this book is devoted to describing these methods, including stage-gate processes, design guidelines, performance metrics, and analytical tools.

Finally, it is important to position DFE correctly in the broader context of corporate innovation and social responsibility. As described in Chapter 4, those companies that consistently deliver shareholder value combine a relentless drive for technological superiority with an acute awareness of the expectations of stakeholders. DFE is part of a broader landscape in which companies continuously reexamine emerging challenges and opportunities, identify critical drivers of superior performance, evaluate the competitive position of their products and processes, and pursue purposeful innovation to achieve sustained excellence. Environmental, health and safety performance is one of many considerations that are part of this strategic feedback loop. Other key characteristics of a successful and resilient company include engagement with its key stakeholder groups, transparency of its communications, diversity of its workforce and suppliers, assurance of ethical practices such as avoidance of child labor, and contribution to economic development in the regions in which it operates (see Chapter 20).

Using This Book

This book describes the basic principles of DFE and outlines the steps necessary to make DFE an integral component in the design and development of new products and processes. For those companies who see the wisdom of DFE and wish to embark on this path, this book provides a guide to the design and development of environmentally responsible products and processes. Consider it a roadmap to sustainable product development.

Part 1, *Answering the Call*, describes the emergence of corporate environmental responsibility in the United States and abroad, including the transition from regulatory compliance to corporate responsibility and the broad range of voluntary initiatives sponsored by government, industry, and nonprofit organizations. It focuses specifically on the external forces and business drivers that motivate adoption of sustainability and DFE practices.

Part 2, *Charting the Course*, explains how DFE fits within the broader paradigm of concurrent engineering for integrated product development and life-cycle management. It sets forth the principles and methodology for implementation of DFE, including life-cycle thinking, performance metrics, design rules and guidelines, and supporting analysis tools.

Part 3, *Walking the Talk*, consists of a series of chapters that describe DFE practices in a variety of industries, ranging from basic commodities to consumer products and services. Each chapter contains case studies from several progressive companies in the United States and abroad, describing their approach to DFE and the lessons that they have learned.

Part 4 concludes with an examination of the global challenges involved in the journey to environmental sustainability and discusses how we can extend current DFE practices to meet these challenges. The final chapter provides a concise summary of the entire book.

Checklist for Getting Started with DFE

Business Motivations

- Do we have one or more business units for whom DFE appears to be a competitive factor?
- Have our customers expressed strong concerns about the environmental performance of our products or manufacturing operations?
- Do we envision regulatory changes or new standards that will influence our ability to profitably produce, distribute and support our products?
- Do environmental excellence and sustainability have a strong influence upon our company reputation or brand image?

Environmental Posture

- Do our corporate responsibility policy and mission statement support the practice of DFE?
- Are we prepared to shift from a compliance-driven to a pro-active environmental management strategy (or have we already done so)?
- Have we established corporate or divisional environmental, health and safety, and/or sustainability improvement goals?
- How might our DFE initiative relate to existing environmental responsibility programs (e.g., Responsible Care®)?

Organization Characteristics

- Are we planning (or have we begun) to implement an environmental management system (EMS) that is integrated with our existing management systems?
- Do we practice concurrent or simultaneous engineering in our new product development, using cross-functional teams?
- Do we have a system for managing product and process quality that can be extended to incorporate environmental attributes?
- Do we have the right organizational resources in line positions to support environmental and product stewardship?

Checklist for Getting Started with DFE *(continued)*

- Do we have appropriate accountability and reward systems to provide incentives for meeting environmental improvement goals?

Existing Experience

- What accomplishments have we made in "green" design, and what practical issues and barriers have we encountered?

- Have we performed any life-cycle assessments for products and / or facilities?

- Have we established programs and expertise in material recycling, resource conservation, waste reduction, or asset recovery?

- Have we implemented any initiatives in pollution prevention and environmentally conscious manufacturing?

- Have we attempted to introduce environmental performance measurement and management systems into our operations?

- Have we developed any useful enabling technologies for DFE, such as computer-based modeling or decision support tools?

Strategic Opportunities

- Can we describe a business case that indicates DFE will contribute to our profitability or business development?

- Can we identify desirable environmental improvements in specific products or processes?

- Have we considered key partnerships or alliances with suppliers or customers that are needed to pursue DFE opportunities?

- Is it valuable for us to enhance environmental awareness among our employees, customers, suppliers, communities, or other stakeholders?

- Are we prepared to move toward life-cycle environmental accounting systems that use an activity-based structure to reveal "true" costs and benefits?

References

1. Lester R. Brown, Earth Policy Institute, *Plan B 3.0: Mobilizing to Save Civilization* (New York: W. W. Norton and Company, 2008), p. 22.
2. J. E. Stiglitz, *Globalization and Its Discontents* (London: Norton, 2002).
3. M. E. Porter and M. R. Kramer, "Strategy and Society: The Link Between Competitive Advantage and Corporate Social Responsibility" (Cambridge: *Harvard Business Review*, December 2006), p. 84.
4. J. Fiksel, "A Framework for Sustainable Materials Management," *Journal of Materials*, August 2006, p. 16.
5. Environmental footprint estimates are based on the Eco-LCA™ tool from The Ohio State University, described in Chapter 9.
6. Office of Technology Assessment, *Green Products by Design: Choices for a Cleaner Environment*, OTA-E-541, October 1992.
7. American Electronics Association, *The Hows and Whys of Design for Environment*, Washington, D.C., November 1992.
8. World Commission on Environment and Development, *Our Common Future*, Oxford University, 1987.
9. H. E. Daly, "The Perils of Free Trade," *Scientific American*, November 1993, pp. 50–57.

PART 1

Answering the Call: The Green Movement

CHAPTER 2
Motivating Forces

*I truly believe that we in this generation
must come to terms with nature, and I think
we're challenged, as mankind has never been
challenged before, to prove our maturity and our
mastery, not of nature but of ourselves.*
— RACHEL CARSON [1]

Wake-Up Calls

The adoption of DFE practices, beginning in the early 1990s, was the inevitable result of growing environmental awareness around the world. Attitudes in the U.S. have transformed from indifference into widespread recognition of environmental problems such as global warming and solutions such as renewable energy. This chapter traces the origins of the environmental movement in the United States against the larger backdrop of the emerging global consensus on the importance of sustainable development.

The urgent need to protect environmental resources was first recognized in the 1960s, yet scientists and policy-makers spent decades gathering data and debating the state of the environment. Environmental protection goals were typically viewed as secondary to the desire for economic growth. Finally, in the early 2000s two authoritative reports appeared, involving hundreds of scientists around the world, which left little doubt about the urgency of the situation. The International Panel on Climate Change [2] confirmed the rapid increase in global warming due to greenhouse gas emissions, and Al Gore wisely used the cinematic medium to sound a public alarm about the "inconvenient truth" of climate change.* Less well publicized, but equally significant was the Millennium Ecosystem Assessment, which confirmed the rapid degradation in ecosystems due to industrialization [3]. According to this report

---

*Al Gore became the first person in history to receive both a U.S. Motion Picture Academy Award and a Nobel Peace Prize.

> Over the past 50 years, humans have changed ecosystems more rapidly and extensively than in any comparable period of time in human history, largely to meet rapidly growing demands for food, fresh water, timber, fiber, and fuel.... Gains in human well-being and economic development... have been achieved at growing costs in the form of the degradation of many ecosystem services, increased risks of nonlinear changes, and the exacerbation of poverty for some groups of people. These problems, unless addressed, will substantially diminish the benefits that future generations obtain from ecosystems.

The many important ecosystem services that humans take for granted include

- Provisioning services that provide food, energy, water, and raw materials for industry
- Regulating services including climate regulation, waste decomposition, and nutrient cycling
- Supporting services including water purification, crop pollination, and pest control
- Cultural services including aesthetic inspiration, recreation, and learning
- Preserving services including genetic biodiversity and protection of future options

Today, the tide of opinion seems to have turned, and the global response to environmental concerns has begun in earnest. A remarkable part of this story is the increasing rate of voluntary participation on the part of the industrial community. Part 3 of this book demonstrates that environmental responsibility has been wholeheartedly embraced by companies that are recognized as leaders in their respective industries.

Contrary to the popular image of companies being hostile to environmental pressures, the logic of environmental innovation has been readily accepted in the United States with virtually no regulatory coercion. DFE exemplifies the willingness of companies to move "beyond compliance" in their environmental initiatives, provided that there are valid business drivers. Granted, pockets of resistance remain in the business world, and there are still advocacy groups intent on exposing laggard corporations. But generally speaking, the traditional hostility between environmentalists and the private sector has evolved into a collaborative engagement between industry, nonprofits, and other stakeholders to pursue their common interest in sustainable development. Examples of this growing collaboration are described in Chapter 3.

A Rediscovery of Ancient Values

Beginning in the mid-twentieth century, the relationship of humans to the environment emerged as a popular topic of concern. Some attribute the origins of this awareness to the 1962 publication of *Silent Spring* by Rachel Carson [1]. Many other writers and political figures have contributed to the rising tide of environmental consciousness. For example, E. F. Schumacher preached a new humanistic economics in *Small Is Beautiful* [4]. Theodore Roszak decried our alienation from nature in *Where the Wasteland Ends* [5]. Barry Commoner, in a series of books, argued persuasively for the development of new industrial technologies based on an understanding of ecological principles. Many of those who listened to these voices grew up to occupy positions of power and leadership. Advocacy groups such as Environmental Defense and Natural Resources Defense Council began to develop considerable influence and lobbying power. As environmental awareness blossomed, a sort of opposition movement sprang up as a "backlash" against environmentalist dogma [6].

While it sometimes has a fanatical ring, the passion of the environmental movement appears to be genuine, born from a profound realization of our intimate connections with the ecosystem that surrounds us. In fact, environmentalists have simply rediscovered an ancient mode of thought that can be traced back to the mythical beliefs of early civilizations. Ancient peoples revered the land and respected other creatures. Our "western" culture has taken a considerable detour in its cognitive development, thanks to philosophers such as Isaac Newton and René Descartes. They viewed the universe as an orderly mechanism that we could analyze logically and conquer through our intellectual powers. Descartes' famous utterance, "I think therefore I am," is perhaps the ultimate denial of our biological origins.

As twentieth-century scientists progressed beyond Newton's tidy theories and began to probe the mysteries of quantum physics, they acquired humility and in some cases became downright mystical. For example, Fritjof Capra, the physicist-turned-philosopher who authored the renowned book *The Tao of Physics,* wrote the following description of a "systems view" of life:

> The earth… is a living system; it functions not just *like* an organism but actually seems to *be* an organism—Gaia, a living planetary being. Her properties and activities cannot be predicted from the sum of her parts; every one of her tissues is linked to every other tissue and all of them are mutually interdependent; her many pathways of communication are highly complex and nonlinear; her form has evolved over billions of years and continues to evolve [7].

These observations were made within a scientific context, but they go far beyond science. They reflect a profound ecological awareness that is ultimately spiritual. Many laypeople have experienced similar revelations. Having been cut off from their organic roots, they feel a sense of reverence at rediscovering them.

In his influential first book, *Earth in the Balance,* Al Gore aptly described the crisis of the spirit that insulated our civilization from environmental awareness. He quotes the words of the American Indian Chief Seattle upon ceding his tribal lands to the government [8]:

> If we sell you our land, remember that the air is precious to us, that the air shares its spirit with all the life that it supports... if we sell you our land, you must keep it apart and sacred, a place where man can go to taste the wind that is sweetened by the meadow flowers... This we know: the earth does not belong to man, man belongs to the earth. Man did not weave the web of life, he is merely a strand in it.

As history can attest, this message was largely ignored. Those few who foresaw the unfortunate side-effects of industrial development—pollution, wilderness destruction, soil erosion—often adopted a Malthusian outlook, predicting catastrophic consequences if we did not mend our ways [9]. To this day, there are skeptics who believe that the only path to sustainability is a retreat from globalism to a simpler, locally based lifestyle.

In fact, Malthus' predictions of worldwide famine were incorrect because he did not foresee the incredible impact of technological change on human productivity. Innovations, such as the internal combustion engine, the telephone, the transistor, genetic engineering, and, of course, the Internet, revolutionized our ability to generate economic value. However, we cannot take false comfort in our ability to pursue unfettered economic growth. In 1992 a team of global modelers updated the famous Club of Rome study, *The Limits to Growth,* which was released 20 years earlier. They concluded that [10]

- Pollution and resource usage have already surpassed sustainable levels.
- We need drastic decreases in population growth and material consumption, and increased efficiency of material and energy usage.
- A sustainable society is still technically and economically possible.

These assertions have an eerily familiar ring today—we have made some progress in limiting environmental pollution, but consumption of energy and other resources has climbed briskly, and we are continuing to release greenhouse gases at an ever-increasing rate (see Chapter 3). Many economists have now acknowledged the need to seek a more sustainable path, although there is a continuing ideological debate over the need for growth [11]. Yet other priorities, such as politics and entertainment, have distracted us for decades from confronting the creeping threats of environmental degradation. We can no longer afford to procrastinate.

Is a sustainable society still possible? Just as previous technological revolutions helped to disprove the doomsayers of the past, there is still hope that Design for Environment will be a key technology enabling our civilization to achieve sustainable development and preserve the best aspects of our present way of life for future generations. From the DFE perspective, the utopia of the future is not a world of spartan lifestyles and isolated, tribal communities. It is an eco-efficient global village, where anthropogenic waste materials from each industrial process are ingeniously consumed as inputs to other processes, and we maintain a sophisticated, carefully designed balance with the natural resources that surround us.

The Global Sustainability Agenda

To better understand the motivations for DFE, we need to examine the global changes in environmental consciousness that swept through the international community in recent decades. The Earth Summit of 1992, held in Rio de Janeiro (officially known as the United Nations Conference on Environment and Development) was a landmark event that represented the culmination of many years of discussion and debate in various nations. The dimensions of the debate transcended national and industrial boundaries, touching upon issues such as export of pollution to developing countries, international equity of environmental regulations, and sustainability of population and industrial growth in the face of limited planetary resources [12]. A number of agreements about international cooperation were produced at the Rio Summit, along with voluminous documentation. For purposes of this book, there are a few fundamental principles worth noting among the 27 principles of the Rio Declaration (the numbering is this author's).

1. Development today must not undermine the development and environment needs of present and future generations.

2. Nations shall use the precautionary approach to protect the environment. Where there are threats of serious or irreversible damage, scientific uncertainty shall not be used to postpone cost-effective measures to prevent environmental degradation.

3. In order to achieve sustainable development, environmental protection shall constitute an integral part of the development process and cannot be considered in isolation from it.

4. The polluter should, in principle, bear the cost of pollution.

Principle 1 summarizes the basic tenet of sustainable development, that we should not compromise the quality of life for our descendants, which provides the foundation for most of the subsequent principles.

Principle 2 suggests a conservative posture toward uncertain environmental impacts, which has already been manifested in regulatory responses to issues such as carcinogens in food products. This principle is closely tied to traditional "command and control" notions of environmental protection and has often been resisted by the business community for justifiable reasons. There are numerous examples of over-regulation which erred on the side of caution.

Principle 3 articulates the seeds of a new approach, wherein environmental concerns are *integrated* into the development process. This book argues that the *only* way to meet the other principles in an economically viable fashion is to accomplish such integration at the earliest stages of product and process development.

Principle 4 represents the "stick" which has forced many reluctant business people to confront environmental issues. The threat of government decision makers assigning pollution taxes to selected industrial sectors is certainly a forceful "wake-up call." When the Clinton Administration floated a proposal for a "carbon tax" in 1993, it resulted in vociferous lobbying on the part of industries that felt they would be unfairly penalized. Although that proposal was eventually quashed, similar approaches were adopted in other nations, including extension of the notion of producer responsibility to waste recovery and "product take-back" regulations.

Senator Max Baucus, Chairman of the Committee on Environment and Public Works, gave a glimpse of future legislative trends when he stated in 1993:

> The cornerstone of my strategy rests on the principle that ... anyone who sells a product should also be responsible for the product when it becomes waste. Thus, the costs associated with collecting, sorting, transporting, reprocessing, recycling and returning materials back into commerce can be internalized and reflected in the price of the product.

Indeed, another document from the Rio Summit, Agenda 21, laid out a blueprint for how the "polluter pays" principle should be implemented. To reduce the rate of solid and hazardous waste generated, it recommended setting targets for specific waste reductions in industrial processes and promotion of "cleaner" production methods as well as recycling of wastes. It also recommended government support for research and development into environmentally sound technologies in order to stimulate businesses to practice sustainable development. The European Union (E.U.), Japan, and other nations have taken these prescriptions to heart. As discussed in Chapter 3, the E.U. has issued a series of directives aimed at waste reduction.

Recognizing that current patterns of resource consumption are not sustainable, the European Commission launched a comprehensive 25-year strategy in 2003 to develop an integrated policy for sustainable management of natural resources [13].

Finally, the Rio Summit produced an international environmental treaty called the United Nations Framework Convention on Climate Change, aimed at stabilizing greenhouse gas concentrations in the atmosphere at a level that would prevent dangerous anthropogenic interference with the climate system. The treaty led to the adoption of the Kyoto Protocol in 1997, which established commitments for industrialized nations to reduce their greenhouse gases, and was ratified by the great majority of nations, with the notable exception of the United States (see Chapter 3).

Although the United States has resisted pursuing new environmental regulations, many worthwhile initiatives have been realized in that country with minimal regulatory intervention. Federal government programs administered by the Department of Energy (DOE), the Environmental Protection Agency (EPA), and even the Defense Advanced Research Projects Agency (DARPA), have provided funding for industry-university cooperative research that has yielded significant advances in environmentally benign manufacturing technologies and alternative energy sources. EPA has also developed a number of voluntary programs to encourage industry adoption of environmentally sustainable practices (see Chapter 3). However, the level of funding and the pace of innovation are still inadequate, given the enormous scale of the environmental challenges that we face.

The Response of Industry

The sweeping goals of Agenda 21 were accompanied by equally ambitious estimates of multibillion dollar investments required by developed nations to support the agenda. Such investments did not materialize in subsequent years. However, apart from any government subsidies and incentives, there has been a quiet revolution in industry attitudes toward the environmental issues raised in Rio. A significant factor in this revolution was the formation of the World Business Council on Sustainable Development (WBCSD), an international group of business leaders established in 1990 to develop a global perspective on sustainable development (see Chapter 3). Their ground-breaking book, *Changing Course*, was an important manifesto describing both the challenges and the opportunities for profitability associated with sustainable development [14]. Today, WBCSD continues to be an influential champion for sustainable business practices, with participation from chief executives of over 150 global companies.

> ENVIRONMENTAL MANAGEMENT HAS MOVED FROM A PERIPHERAL ACTIVITY INTO THE MAINSTREAM.

As shown in Chapter 4, sustainable development is a logical outgrowth of industry practices, such as resource conservation and environmental stewardship, which have been ongoing for years. Leading manufacturing firms have broadened their focus from pollution prevention and waste minimization to a more integrated "life cycle" approach, replacing "end-of-pipe" pollution control methods with more cost-effective process design changes, such as reducing the quantities of toxic materials used or produced as by-products. The most striking change in recent years has been the elevation of environmental issues to an unprecedented level of strategic importance, with visibility at the Board of Directors level. Thanks in large part to rising energy costs and global warming concerns, environmental management has moved from a peripheral, tactical activity into the mainstream of business thinking.

The "early adopter" firms that leaped onto the environmental bandwagon during the 1990s were captained by visionaries who believed in a sustainable and profitable future. In North America, these enlightened individuals have included a number of influential chief executive officers; for example, Samuel Johnson of SC Johnson, Frank Popoff of Dow Chemical, Maurice Strong of Ontario Hydro, Chad Holliday of DuPont, and Ray Anderson of Interface. But visionaries are always a minority, and only when their vision proves credible do the more pragmatic members of the community begin to pay attention. Around the turn of the twenty-first century, corporate sustainability initiatives began to grow phenomenally, as more and more companies recognized that sustainability was an essential factor in their continued competitiveness.

There are good reasons for this—companies are receiving strong signals from all segments of their stakeholders that corporate responsibility, including environmental stewardship, is essential to long-term business success. The specific driving forces that influence corporate responsibility, and hence the adoption of DFE, are illustrated in Figure 2.1 and are the subject of the next two chapters. Chapter 3 describes the external drivers associated with changing societal norms and expectations, while Chapter 4 describes the internal business drivers for DFE, including contributions to competitiveness and shareholder value. Perhaps the most important factor in changing industry attitudes has been the realization that paying attention to environmental responsibility can actually increase profitability. Reducing pollution at the source and designing products and processes in ways that enhance environmental performance generally result in higher productivity and reduced operating costs, while improving brand image and reputation.

It must be emphasized that DFE is only one facet of corporate environmental responsibility, focusing on the innovation process. There are many other initiatives that companies must pursue to ensure that they are meeting their environmental commitments, ranging from

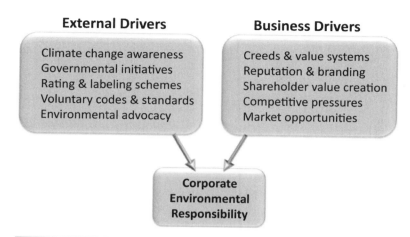

External Drivers

Climate change awareness
Governmental initiatives
Rating & labeling schemes
Voluntary codes & standards
Environmental advocacy

Business Drivers

Creeds & value systems
Reputation & branding
Shareholder value creation
Competitive pressures
Market opportunities

Corporate
Environmental
Responsibility

Figure **2.1**   Driving forces influencing the adoption of DFE.

everyday EH&S operations to broad community involvement. However, DFE-based innovation is perhaps the only real opportunity to produce quantum improvements in a company's environmental footprint while enabling continued growth and prosperity.

References

1. R. Carson, *Silent Spring* (New York: Houghton Mifflin, 1962).
2. Intergovernmental Panel on Climate Change (IPCC). *Third Assessment Report: Climate Change* (Cambridge, UK: Cambridge University Press, 2001).
3. Millennium Ecosystem Assessment, Synthesis Report, *Ecosystems and Human Well-Being* (Washington, DC: Island Press, 2003).
4. E. F. Schumacher, *Small Is Beautiful* (New York: Harper and Row, 1973).
5. T. Roszak, *Where the Wasteland Ends* (New York: Doubleday and Company, 1972).
6. Dixie Lee Ray with Lou Guzzo, *Environmental Overkill; Whatever Happened to Common Sense?* (New York: Harper Collins, 1993).
7. F. Capra, *The Turning Point* (New York: Simon and Schuster, 1982), p. 285.
8. A. Gore, *Earth in the Balance: Ecology and the Human Spirit* (New York: Houghton Mifflin, 1993).
9. Thomas R. Malthus was a British political economist who identified the potential for populations to increase faster than the available food supply. His famed "Essay on the Principle of Population" was first published in 1798.
10. D. H. Meadows, D. L. Meadows, J. Randers, *Beyond the Limits* (Post Mills, Vt.: Chelsea Green Publishing Company, 1992).
11. W. D. Sunderlin, *Ideology, Social Theory, and the Environment* (Lanham, Md.: Rowman & Littlefield, 2002).
12. M. Keating, *Agenda for Change*, (Geneva: Centre for Our Common Future, 1993).
13. Commission of the European Communities, "Towards a Thematic Strategy on the Sustainable Use of Natural Resources" (Brussels: Communication to the European Parliament, Oct. 2003). See www.europa.eu.int/comm/environment/natres/index.htm.
14. Stephan Schmidheiny with the Business Council on Sustainable Development, *Changing Course: A Global Business Perspective on Development and the Environment* (Boston: MIT Press, 1992).

CHAPTER 3

External Drivers: The Voice of Society

Design ... is the only deliberate way out of the
unsustainable dominating and addictive patterns
of individual and social behaviors that have
become the norms in the United States and in
other affluent consumerist societies.

JOHN EHRENFELD [1]

Green Expectations

Society has come to expect a great deal of corporations, beyond delivering good quality products at affordable costs. Chapter 2 described the groundswell of environmental awareness, beginning with concerns over environmental pollution and the risks posed to humans and wildlife, and culminating in the energy and climate change concerns that are central to contemporary policy making. This chapter examines in greater detail the external forces that are shaping today's business landscape. Chapter 4 describes the evolution in business thinking that has led to the acceptance of corporate environmental and social responsibility as staples of corporate governance, and Chapter 5 briefly discusses the emergence of "green" markets. Sustainability has become everyone's concern, and stakeholders are counting on the business community to be part of the solution.

Designing products with reduced environmental impacts is mainly an engineering challenge. However, as companies roll out the results of their DFE efforts in response to societal expectations, this poses communication challenges that are far more daunting. Conflicting political views, debates on methodology, scientific ambiguities, and stakeholder perceptions can easily cloud any discussion of environmental performance. Consumers lose confidence in manufacturers' claims when they are contradicted by competitors or public interest groups. Consequently, companies have sought to build confidence and trust by aligning themselves with third parties, such as government agencies, nonprofits, or environmental advocacy organizations.

As companies strive for greater transparency and external engagement, they soon recognize that there are many different categories of stakeholders whose needs and expectations they must consider. Table 3.1 lists eight categories of stakeholders, each of which has specific environmental interests and concerns. The table also lists corresponding performance indicators that would resonate most with these different audiences; Chapter 7 discusses performance

Stakeholder Group	Typical Expectations	Examples of Relevant Metrics
Investors, lenders, financial analysts	• Rate of return • Shareholder value • Acceptable risk profile	• Earnings per share • Price/earnings ratio • Volatility
Suppliers, value chain partners, customers	• Integrity • Reliability • Product stewardship	• Corporate governance rating • Customer satisfaction index • Eco-labeling & other certifications
Local community residents	• Corporate citizenship • Support for economic development • Public health & safety protection	• Number of community advisory panels • Volume of purchasing or engagement with local businesses • Philanthropic contributions
Governmental officials, regulatory agencies	• Compliance • Model for voluntary initiatives	• Notices of violation • Percent reduction in listed emissions • Awards & recognition
Employees, labor unions	• Commitment to employee rights and well-being • Health and safety practices • Wages & benefits	• Percent of employees participating in self-improvement programs • Percent reduction in incident rate • Employee satisfaction

TABLE 3.1 Stakeholders' Environmental Interests and Concerns

Stakeholder Group	Typical Expectations	Examples of Relevant Metrics
Religious groups, NGOs, advocacy groups, public interest groups	• Sensitivity to human rights & global environmental issues • Demonstrated improvements in performance • Transparency of decisions	• Commitment to codes of conduct (e.g., UN Global Compact) • Percent reduction in emissions and resource consumption • Partnerships with NGOs
Youth, women, indigenous peoples, minority groups	• Commitment to equal opportunity • Access to education and training	• Indigenous employment • Percent of workforce holding diplomas • Number and extent of diversity programs
Academic and research organizations	• Degree of innovation • Willingness to collaborate	• R&D investment as percent of sales • Environmental research partnerships

TABLE 3.1   Stakeholders' Environmental Interests and Concerns *(continued)*

measurement in greater detail. Through sincere commitment and authentic dialogue with stakeholders, companies are able to integrate their concerns more effectively into business decision making processes.

This chapter describes efforts by organizations around the world that have contributed to the establishment of DFE principles, practices, and performance goals, often working hand-in-hand with industry. It turns out that defining "green" products or processes is not an easy exercise. Among the many stakeholder communities of the world, there is a wide diversity of views on what it means to be environmentally responsible, sustainable, or eco-efficient.

How Clean Is Green?

In the early days of environmental regulation, government and business officials frequently debated the question of "how clean is clean?" This question arose in making decisions about pollution limits or environmental remediation for substances that might be present at levels as low as one part per billion. Assuming that the objective is to avoid human health risk or ecological risk, then the above question translates to "what level of residual risk can we accept?" Since zero risk is virtually impossible, some cutoff point was needed to avoid spending absurd amounts for pollution control and cleanup. For example, the U.S. Environmental Protection Agency (EPA) has often used the level of one in one million lifetime risk of cancer as a regulatory threshold.

To further confuse the issue, laboratory detection limits have improved greatly, so that we can now detect contaminants in air, water, or food at levels of 1 part per trillion or less. Some environmental and health advocates have invoked the precautionary principle (see Chapter 2) to support an extreme position that no amount of contamination is tolerable. One of the most frustrating aspects of dealing with the "how clean is clean?" question has been our inability to measure low levels of risk empirically. Instead, risk analysis practitioners are compelled to use dubious mathematical extrapolations (see Chapter 9), so that measuring "cleanness" can become a somewhat theoretical exercise.

Today, a similar question arises in a different context; namely, "how green is green?" The appellation "green" is used widely and gratuitously, but when applied to a product or a company it usually connotes environmental responsibility. However, "greenness" is even more difficult to measure than risk; in fact it suggests a multidimensional set of characteristics that include emissions, energy use, water use, and many other attributes. The expanding field of eco-labeling is discussed later in this chapter, and Chapter 7 describes how companies have dealt with the challenges of environmental performance measurement. Establishing meaningful standards of comparison for competing products is even more daunting than comparing hypothetical risks.

Finally, a new riddle can be posed; namely, "how clean is green?" In other words, when we speak of "green" design or "clean" manufacturing, what level of pollution or waste reduction should we strive for? There are definitely two camps. Some believe that zero waste is an attainable goal, as demonstrated in the practice of industrial ecology where one firm's wastes become another firm's feedstocks (see Chapter 8). Others argue that by the laws of thermodynamics we can only prolong, but not prevent, the inevitable progress of entropy, which causes matter to decay into waste. In either case, companies are increasingly forced to confront this question as they deliberate over what voluntary goals to set for future reductions in emissions and waste. As shown in Chapter 4, even if zero waste is achievable it may not be desirable because of diminishing economic returns.

Confronting Climate Change

Today, the term "green" is often equated with being "energy-efficient" or "climate-friendly." Concerns over the "greenhouse effect" were voiced as far back as the 1970s, but not until the early 2000s were these concerns fully accepted. Scientific doubts and economic concerns led the United States to decline participation in the Kyoto Protocol, the international agreement mentioned in Chapter 2 that set targets for industrialized countries to reduce their greenhouse gas (GHG) emissions* by 2012. There is now overwhelming evidence that the Earth's climate is changing and that anthropogenic emissions of GHGs are a major contributor to global warming [2]. Al Gore has called climate change a "global emergency," but it remains difficult for people to change their behavior when the consequences lie in the distant future.

There is still considerable uncertainty about how rapidly temperatures will rise and the extent of the resulting impacts, but the need for action is apparent. Figure 3.1, based on the Stern report issued by the United Kingdom government, illustrates the potential catastrophic consequences of rising global temperatures [3]. As it turns out, the targets set by the Kyoto Protocol were very modest, and it is generally agreed that a new international accord will be needed with more stringent reductions if we hope to stabilize

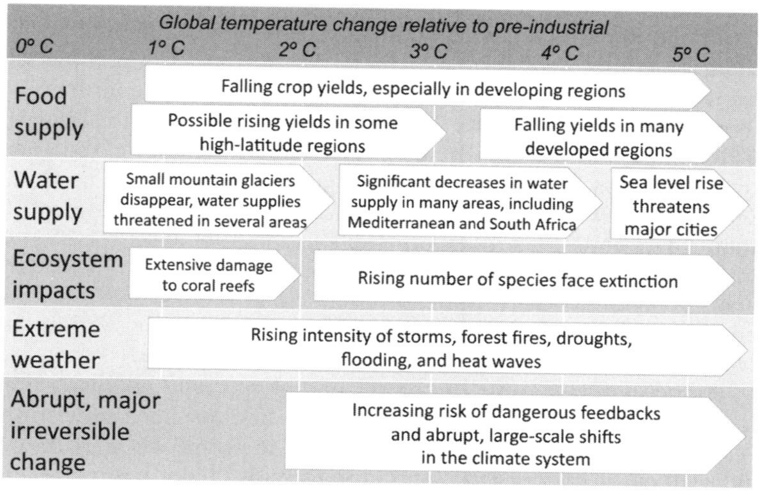

Global temperature change relative to pre-industrial					
0° C	1° C	2° C	3° C	4° C	5° C
Food supply	Falling crop yields, especially in developing regions				
	Possible rising yields in some high-latitude regions		Falling yields in many developed regions		
Water supply	Small mountain glaciers disappear, water supplies threatened in several areas	Significant decreases in water supply in many areas, including Mediterranean and South Africa		Sea level rise threatens major cities	
Ecosystem impacts	Extensive damage to coral reefs	Rising number of species face extinction			
Extreme weather	Rising intensity of storms, forest fires, droughts, flooding, and heat waves				
Abrupt, major irreversible change	Increasing risk of dangerous feedbacks and abrupt, large-scale shifts in the climate system				

FIGURE 3.1   Projected impacts of climate change [3].

---

*The six most common greenhouse gases are: carbon dioxide (CO_2), methane (CH_4), nitrous oxide (N_2O), sulfur hexafluoride (SF_6), perfluorocarbons (PFCs), and hydrofluorocarbons (HFCs). The *global warming potential* (GWP) of each gas can be expressed in terms of equivalent CO_2 emissions, for example methane and nitrous oxide have a 100-year GWP of 25 and 298, respectively, and thus are much more potent than CO_2.

GHG levels in the atmosphere (see Chapter 20). Moreover, the Kyoto Protocol placed the principal responsibility for GHG reduction on developed nations with a history of industrial activity, but it has now become essential to include participation by rapidly developing nations such as China and India.

There are actually a number of mechanisms whereby companies can strive to meet GHG reduction targets without undue economic hardship. These include emissions trading, purchase of offsets, and investments in GHG reduction projects such as wind farms. Many U.S. groups have advocated a mandatory national "cap and trade" system for GHGs, and voluntary regional carbon trading systems have already been established in New England and California. Already, many third parties are offering elaborate procedures for verifying the authenticity of carbon trading schemes. However, such mechanisms remain controversial because they may provide a means for delaying progress and preserving the *status quo*. There is a continuing debate over "additionality" of carbon offsets; in other words, whether the same GHG reductions would have been realized without the offset purchase.

In response to growing public expectations, even in the absence of regulatory drivers, many companies have initiated programs to assess their "carbon footprint," to set reduction goals, and to include GHG emissions as a key component of their environmental performance measurement systems. The World Business Council for Sustainable Development has teamed with the World Resources Institute to publish a detailed **GHG Protocol** for performing an inventory of GHG emissions at a corporate or facility level [4]. The "carbon-neutral" label has become increasingly popular to describe products, processes, services, or events which attempt to eliminate or offset their GHG emissions. For example, Terra-Pass enables travelers to purchase carbon credits that offset the emissions associated with the energy used by their transportation. Some companies have set aggressive goals for their operations to become carbon-neutral and waste-free, as described in Part 3 of this book.

In point of fact, most carbon offset schemes only address a company's *direct* use of energy in the form of fuel or electricity, so the "carbon-neutral" label may be misleading. If a company were to consider all of the energy expended in the supply chain to provide purchased goods and services, its overall carbon footprint could be as much as 10 to 20 times larger [5]. As pointed out in Chapter 1, the root cause of our enormous carbon footprint is not direct energy use but *material throughput*, which drives the consumption of energy throughout our economy. (Chapter 9 provides a more detailed discussion of life-cycle environmental footprint assessment.) The GHG Protocol recently launched a new, multi-stakeholder initiative to draft two new international standards for product life-cycle accounting and corporate value chain accounting, which were released in 2010. The focus of this book is on direct GHG mitigation through DFE, argu-ably the most effective means for a company to reduce its life-cycle carbon footprint while generating shareholder value.

Governmental Initiatives: Stick and Carrot

An important driver of increased environmental responsibility has been the influence of governmental initiatives in motivating improved environmental performance. These include broad regulatory direc-tives in the European Union aimed at curbing toxics use and waste, as well as government award programs in the U.S. that recognize environmental leadership. This book does not attempt to catalogue the complex field of global environmental regulation. While the regulatory "stick" may be effective at changing corporate behavior, environmental laws can also stifle innovation, especially when they prescribe specific technologies to be implemented. DFE is most valu-able in the competitive arena beyond compliance, where companies can differentiate their products and gain a technological advantage. In this arena, the most effective motivational "carrots" that govern-ments can provide are market-based incentives, such as tax breaks or public recognition of the successful innovators.

International Directives

In 2001 the European Union (E.U.) adopted a broad sustainable devel-opment strategy aimed at developing integrated policies to address climate change and energy; transport, production, and consumption; natural resource protection; and human health and well being. To promote sustainable production and consumption, the strategy calls for green procurement, environmental innovation, product labeling, and recycling and reuse. Accordingly, the European Union has issued a series of directives to its member states requiring the implemen-tation of this policy. Many leading companies were tracking these initiatives for years in advance and were prepared to demonstrate compliance as soon as the directives became law. The most signifi-cant directives influencing DFE practices include the following.

Packaging and Packaging Waste Directive (1994). This directive requires manufacturers to recover and dispose or recycle packaging associated with their products. Companies typically comply by pay-ing a license fee to join a nonprofit program, such as Green Dot, which has become a standard take-back program in most European countries. Member nations are required to implement systems to attain the following targets by the end of 2008: At least 60% by weight of packaging waste to be recovered or incinerated for energy recov-ery; minimum recycling targets for different packaging waste mate-rials as follows: 60% for glass, 60% for paper and board, 50% for metals, 22.5% for plastics, and 15% for wood.

End-of-Life Vehicles (2000). The purpose of the ELV directive is to reduce the quantity of solid or hazardous waste generated by motor vehicles when they are discarded. It restricts the use of heavy metals and sets a number of objectives for new vehicle design, including avoidance of hazardous substances, ease of disassembly and recycling, and use of recycled materials. Manufacturers are required to facilitate recycling and recovery by issuing detailed instructions that include material identification based on coding standards. Recovery facilities must be set up to safely disassemble vehicles at no cost to the owner. Vehicles must be stripped of recyclable parts or materials and drained of fluids, and polluting components such as batteries and tires must be properly disposed of. As discussed in Chapter 12, this directive has stimulated DFE practices throughout the automotive industry supply chain.

Waste Electrical and Electronic Equipment (2003). Known as the WEEE Directive, this requires each nation to set collection, recycling and recovery targets for electronic products, including televisions, computers, and cellular phones, and imposes the responsibility for take-back and disposal of the equipment on the original manufacturers. Companies are expected to establish an infrastructure for product take-back without charge to the consumer and to recycle or dispose of the collected waste in an ecologically-friendly manner. As discussed in Chapter 11, this has induced many companies to design their products in a way that facilitates take-back.

Eco-design of Energy-using Products (2005). Known as the EuP directive, this provides coherent EU-wide rules for eco-design to ensure that disparities among national regulations do not become obstacles to intra-EU trade. Rather than setting requirements for specific products, it defines conditions for setting requirements for environmentally relevant product characteristics.

Restriction on Hazardous Substances (2006). The RoHS Directive seeks to phase out the use of lead, mercury, cadmium, hexavalent chromium, and brominated flame retardants that are used in certain plastics, especially in the electronics industry. These substances are believed to have adverse environmental impacts, especially at product end-of-life. The directive applies to a wide range of products including consumer electronics, household appliances, computer and telecommunications equipment, light bulbs, toys, and sporting goods; medical devices and monitoring and control instruments received a temporary exemption. European laws often influence those in other nations; China, Korea, Taiwan, and some U.S. states such as California have already implemented legislation similar to RoHS.

Registration, Evaluation, and Authorisation and Restriction of Chemicals (2006). The REACH directive requires chemical suppliers, manufacturers, or importers of more than one metric ton of a chemical substance annually to register the substance with the European Chemical Agency central database. It is based on the presumption that low levels of toxic chemicals may bioaccumulate in the food chain and be harmful to humans. All substances (medicinal products are exempted) had to be pre-registered by December 1, 2008, to take advantage of the staggered implementation program. The regulation will expand significantly the number of substances that require authorization for use and has raised controversy because of the cost burdens imposed on industry, as well as the increased requirements for animal testing.

Japanese and Chinese Initiatives. Similar to the European Union, the government of Japan has adopted a comprehensive regulatory approach toward reducing material throughput [6]. Although Japan is already far ahead of the United States and European Union in resource efficiency, it established a goal of 50% reduction in final waste disposal by 2010. The Japanese legislative framework, adopted in 2000, includes laws governing waste management, resource recycling and green purchasing, with specific regulations targeting packaging, home appliances, construction materials, food recycling, and end-of-life vehicle recycling. The Chinese government has introduced similar legislation under a new policy framework designed to promote a "circular economy" and is pursuing a variety of new "green" initiatives [7].

United States Initiatives

Although the United States has seen substantial improvements in air and water quality, the traditional "command and control" approach of U.S. regulatory agencies is widely regarded as burdensome and ineffective. On the other hand, programs based on voluntary initiatives and market influence have proven extremely successful. The following are examples of programs relevant to DFE.

Toxic Release Inventory (1986). One motivation for the rise in U.S. voluntary programs may have been the surprising response of industry to the EPA's Toxic Release Inventory (TRI) reporting requirements. This program was originally established by the Emergency Planning and Community Right-to-Know Act under Title III of the Superfund Amendments and Reauthorization Act (SARA) of 1986. By merely requiring companies to report their hazardous material releases to the public, EPA gave industrial firms an incentive to reduce them. Before long, companies had established TRI reduction as a standard corporate objective, and overall releases decreased by roughly 40% from the baseline year of 1988 through 1992. In 2006, EPA reported that the original releases had decreased by 59% since 1988. Since 1995, EPA has widened the scope of the TRI program, including doubling the number of substances tracked to 650 and extending the requirements to nonmanufacturing sectors and Federal facilities. The "right-to-know" concept clearly demonstrated the power of disclosing environmental performance information.

Pollution Prevention Act (1990). This legislation directed the EPA to develop a national pollution prevention strategy, administered by the Office of Pollution Prevention and Toxics (OPPT). It set forth the well-known "pollution prevention hierarchy," which recommends first trying source reduction, then recycling, and finally waste treatment as alternatives to disposal. It also encouraged a multi-media, life-cycle approach that would cut across EPA's traditional regulatory programs. Since then, the EPA and other agencies have launched a host of innovative programs that rely mainly on industry coopera-tion rather than regulatory coercion.

Environmentally Preferable Products. The United States gov-ernment is one of the world's largest consumers, spending an estimated $350 billion for goods and services each year. This pur-chasing clout enables the government to provide a strong incentive for manufacturers to "green" their products. In 1993, President Bill Clinton issued an Executive Order directing Federal agencies to pur-chase "environmentally preferable" products or services, i.e. those with a lesser effect on human health and the environment compared to competing products or services that serve the same purpose. EPA subsequently issued procurement guidelines for a variety of prod-ucts including engine coolants, construction materials, paperboard, plastic pipe, geotextiles, cement and concrete, carpet, floor tiles, traf-fic cones, running tracks, mulch, and office supplies. This policy was later strengthened in 2007, recommending a life-cycle approach to sustainable procurement and establishing environmental steward-ship scorecards to measure progress.

ENERGY STAR is a joint program of the U.S. EPA and Depart-ment of Energy, first introduced by EPA in 1992 as a voluntary label-ing program, and later extended to cover new homes and commercial and industrial buildings. The objective of the program is to protect the environment through energy efficient products and practices. By establishing partnerships with more than 12,000 private and public sector organizations, ENERGY STAR has enabled both energy use and cost reduction. In 2007 alone the program was estimated to have saved businesses, organizations, and consumers 180 billion kilowatt-hours and $16 billion. As one of the most recognized labels, it has helped to achieve widespread adoption of innovations, such as effi-cient fluorescent lighting, power management systems for office equipment, and low-power standby mode for electrical devices.

EPA Design for Environment Program. For many years The U.S. EPA has worked in partnership with a broad range of stakeholders to encourage the use of safer chemical alternatives and to recognize safer chemical products. Its DFE program focuses on industries that combine the potential for chemical risk reduction and improve-ments in energy efficiency with a strong motivation to make lasting, positive changes. The program convenes partners, including industry representatives, nongovernmental organizations, and academia, to develop goals and guide the work of the partnership. Partnership projects evaluate the human health and environmental considerations of traditional and alternative technologies, materials, and processes. The DFE program encourages safer products through "informed substitution," a science-based transition from a chemical of particular concern to a safer chemical or nonchemical alternative. Since its establishment in 1994, the program has reached more than 200,000 business facilities and approximately 2 million workers. In 2007, DFE partnerships prevented the use of 114 million pounds of chemicals of environmental and human health concern. Examples of program accomplishments include assistance in phase-out of lead solder (see Chapter 11), identifying alternative flame retardants for furniture foam, and training auto refinishing personnel to reduce toxic exposures.

Green Suppliers Network is a collaborative venture among industry, the U.S. EPA, and the U.S. Department of Commerce's National Institute of Standards and Technology's Manufacturing Extension Partnership. The program works with large manufacturers to engage their small- and medium-sized suppliers in low-cost technical reviews that focus on process improvement and waste minimization. Teaching suppliers about "Lean and Clean" manufacturing techniques has helped them to increase energy efficiency, identify cost-saving opportunities, and optimize resources to eliminate waste.

Green Chemistry. Since the early 1990s, EPA has promoted sustainable chemical products and practices, primarily through its Green Chemistry Program. The Program promotes technologies that reduce or eliminate the use or generation of hazardous substances during the design, manufacture, and use of chemical products and processes. It supports fundamental research in environmentally benign chemistry as well as a variety of educational activities, conferences, and tool development through voluntary partnerships with academia, industry, government agencies, and other organizations. The annual Presidential Green Chemistry Challenge Awards are prized by industry leaders such as Dow Chemical (see Chapter 13).

SmartWay Transport is a voluntary alliance between various freight industry sectors and EPA, with the goal of reducing greenhouse gases and other air emissions as well as operating costs. By using EPA's freight logistics analysis tool and adopting fuel-efficient technologies, carriers can achieve performance improvements and be certified to display the SmartWay mark.

WasteWise, launched by EPA in 1994, encourages solid waste reduction and recycling through information, education, and recognition of member successes. As of 2008 the program had about 2000 partner members including large, multinational corporations, small- and medium-sized businesses, local governments, and nonprofit organizations.

Environmental Management System Standards

Along with governmental initiatives such as those described above, a major motivation for DFE has been the emergence of ISO 14000 and similar international standards that prescribe how companies should manage their environmental performance. Although compliance with such standards is voluntary, companies generally cannot afford to ignore them. The forerunner of environmental standards was the ISO 9000 series of quality management standards, which have achieved worldwide acceptance and became a *sine qua non* for international trade in industrial goods. Industrial customers feel secure in knowing that their suppliers' quality practices are ISO 9000-certified, and therefore feel less obliged to audit them carefully. While certification involves some costs, the end result is greater productivity for all concerned.

A similar rationale holds for the adoption of environmental standards. To the extent that companies are concerned about their own environmental performance, they will also want assurance that their suppliers are practicing environmental stewardship and pollution prevention. Certification, in theory, provides an efficient means for verifying a company's commitment to environmental management. However, it is important to note that, as in ISO 9000, certification cannot guarantee the actual *quality* of performance, but only the existence of a management system designed to enhance that quality. Also, the scope of environmental standards is far broader and more complex than that of quality standards. While the latter relate only to the quality of deliverable products, the former span all of the interactions between a company, its physical environment, and its stakeholders, including customers, employees, shareholders, regulators, suppliers, communities, and interest groups.

The following international standards activities have particularly influenced the design of environmentally-conscious products and services.

International Organization for Standardization (ISO). ISO is the world's largest standards-setting organization, with a membership consisting of national standards bodies from 143 countries. ISO 14000 is a series of international standards on environmental management that provide a framework for implementing corporate environmental management systems (EMS). It originated in the early 1990s as a way to harmonize the national standards that were emerging, led by the United Kingdom's BSI 7750 standard, and followed by a host of other countries, including Canada, Ireland, France, the Netherlands, and South Africa. The overarching standard is 14001, which defines the EMS principles against which an organization can be certified by a third party, called a "registrar." As shown in Figure 3.2, the EMS framework is based on the "plan-do-check-act"

FIGURE 3.2 Elements of an environmental management system.

model for continuous improvement. The series of 14000 standards and guidelines covers specific elements of the framework, such as environmental auditing, product labeling, life-cycle assessment, and environmental performance measurement; however, there is little guidance on the actual practice of DFE. Additional guidelines and standards under development by ISO include ISO 26000 on social responsibility, ISO 21930 on environmental declaration of building products, and ISO 50001 on energy management.

Eco-Management and Audit Scheme (EMAS). The EMAS certificate can only be awarded in the European Union and represents an alternative to ISO 14001 certification. The first version was published in 1995, and it was revised in March 2001 (EMAS II) to extend to all economic sectors, including public and private services. It is similar to ISO 14001 in embracing continuous improvement principles for environmental management, but the requirements are more stringent and involve greater transparency. To receive EMAS certification a company must publish an environmental management policy statement and undergo comprehensive eco-audits by external parties. The certificate is valid for a period of three years, after which the entire certification process must be repeated.

Sustainability Rating Schemes

One of the strongest external drivers influencing the adoption of corporate sustainability has been the growing interest on the part of the financial investment community. Portfolio analysts are increasingly recognizing the importance of sustainability as a leading indicator of future value. "Socially responsible" investing (SRI) has grown rapidly, and now accounts for an estimated 10 to 15 percent of the total assets under management in the U.S alone. There are a variety of SRI funds that screen investments based on sustainability criteria such as environmental risks and labor practices. In addition, many companies have been presented with shareholder resolutions demanding greater attention to sustainability issues such as climate change.

> **PORTFOLIO ANALYSTS RECOGNIZE SUSTAINABILITY AS A LEADING INDICATOR OF FUTURE VALUE.**

The United Nations Environment Program has established a Finance Initiative (UNEP FI) that has become a powerful force in promoting sustainability investing. This public-private partnership between UNEP and over 180 global financial companies aims to understand the impacts of environmental, social, and governance considerations on financial performance and sustainable development, and to promote the adoption of best practices. UNEP FI helped to develop the Principles for Responsible Investment, which have been adopted by over 400 institutional investors, representing more than $15 trillion in assets under management.

In addition, to serve the needs of sustainability investors, a number of organizations have been developing scoring and ranking tools that rate companies according to environmental, social, and economic criteria. For example, the FTSE4Good Index analyzes environmental and social responsibility activities, with the stated intent of promoting a stronger business commitment. Other well-known indexing and rating services include Ethibel and Innovest Strategic Value Advisors. The companies that are rated highly by these indexes have generally performed in line with or have outperformed the broader market averages.

One of the most widely recognized rating schemes is the Dow Jones Sustainability Indexes (DJSI) managed by Sustainable Asset Management (SAM) Research in Switzerland. This consists of a collection of indexes specifically designed to track the financial performance of the leading sustainability-driven companies worldwide. The intent is to provide asset managers with reliable and objective benchmarks to manage sustainability portfolios. Each eligible company is rated according to a comprehensive set of criteria that cover the "triple bottom line"—economic, environmental, and social performance. A major source of information for DJSI is the SAM questionnaire that is completed by companies participating in the annual review [8]. Further sources include company and third-party documents as well as personal contacts between the analysts and companies. The resulting information is analyzed by SAM using their proprietary sustainability assessment protocol, which provides a numerical ranking. The top ten percent of companies in each of about 60 industry groups are then selected for listing on the DJSI. Every year, these rankings are revised based on updated information. Most of the major companies described in Part 3 are listed on the DJSI.

Eco-Labeling Programs

Rising awareness of environmental issues has led to the practice of "green marketing," which involves positioning a product or service to communicate its environmental benefits (see Chapter 5). A natural consequence has been the proliferation of eco-labeling programs, which designate products or services that meet specified environmental performance criteria. In 2008 there were over 300 national and international eco-labels in use around the world, spanning a wide range of industries. ISO identifies three types of eco-labels:

- Type 1: Based on satisfaction of multiple environmental criteria, verified through third party certification, and normally requiring periodic recertification.
- Type 2: Self-declared environmental claims (e.g., manufacturer logos).
- Type 3: Quantified declarations of environmental parameters based on life-cycle assessment, primarily for business-to-business communication.

Unlike Type 2 claims made by manufacturers and service providers, the leading eco-labels are mainly Type 1, awarded by independent third-party organizations. For example, the ENERGY STAR label awarded by the U.S. government was described earlier in this chapter. The following are some of the most widely recognized labels around the world (see Figure 3.3).

- **Blue Angel.** This German label, introduced in 1978, is the oldest in the world. It is awarded to products and services that are environmentally beneficial over their life cycle and also meet high standards of occupational health and safety and fitness for use. It applies to 80 product categories, and about 3700 labels have been awarded.
- **EcoLogo.** This Canadian program, launched in 1988 as Environmental Choice, has certified over 7000 North American products and services based on compliance with life-cycle environmental criteria.
- **Green Seal.** This U.S. eco-label, launched in 1989, provides science-based environmental certification standards covering

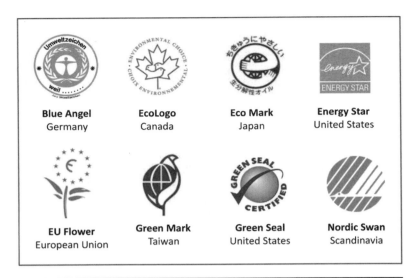

Blue Angel
Germany

EcoLogo
Canada

Eco Mark
Japan

Energy Star
United States

EU Flower
European Union

Green Mark
Taiwan

Green Seal
United States

Nordic Swan
Scandinavia

FIGURE 3.3   Examples of prominent eco-labels around the world.

a wide variety of products. It is favored by large institutional purchasers, including government agencies and universities.

- **Nordic Swan.** This eco-label, launched in 1989, is jointly supported by Norway, Sweden, Denmark, Finland and Iceland. It applies to 60 product categories, and about 1200 labels have been awarded.
- **Eco Mark.** The Eco Mark symbol was launched in 1991 by the Japan Environment Association and uses environmental criteria to determine environmental impacts throughout the product life cycle. It applies to 64 product categories, and about 5200 labels have been awarded.
- **EU Flower.** This is a European Union eco-label launched by the European Commission in 1992, based on evaluation of product life-cycle environmental impacts. Criteria have been published for 23 product groups and the label has been awarded to 350 products.
- **Green Mark.** This Taiwanese mark was launched in 1992 by Taiwan's Environmental Protection Administration to guide consumers in purchasing "green" products and to encourage manufacturers to design and produce them. It applies to 41 product categories, and about 450 labels have been awarded.

In addition, Part 3 mentions a number of established labeling programs that are associated with particular industry sectors, such as EPEAT for electronic products (Chapter 11), Marine Stewardship Council for seafood products (Chapter 15), and Forest Stewardship Council for paper products (Chapter 16). The Carbon Trust has launched a carbon footprint and labeling initiative to help companies measure, reduce, and communicate the life-cycle GHG emissions of products and services. They are collaborating with the British Standards Institute, which has developed PAS 2050, a specification for a common approach to GHG life-cycle assessment. Already, some retail products are beginning to display carbon labels.

While customer expectations may be one motivation for DFE, it is a mistake to view DFE as synonymous with green marketing. In particular, many DFE practices relate to the efficiency of resource usage in manufacturing and distribution, in which case the benefits will accrue mainly to the producer, not the consumer. Thus, DFE is valuable as an integral part of product and process design, whether or not marketing managers decide that they want to convey an environmental "message" about its benefits to their customers.

Voluntary Codes and Principles

As businesses have moved toward adoption of environmental and social responsibility principles, many have chosen to endorse existing codes of conduct or corporate citizenship principles developed by nongovernmental organizations. (See the sidebar below, which lists the Coalition for Environmentally Responsible Economies {Ceres} principles as an example.) Most of these codes include language about a commitment to design products and processes with attention to their life-cycle environmental impacts. The following is a partial list of well-known codes.

- Natural Step System Conditions
- Global Sullivan Principles
- Coalition for Environmentally Responsible Economies (Ceres) Principles
- UN Global Compact
- IISD Bellagio Principles
- Hannover Principles for Sustainable Design
- UN Universal Declaration of Human Rights
- Interfaith Center on Corporate Responsibility: Global Principles
- Social Venture Network's Standards of Corporate Social Responsibility
- Keidanren Charter for Good Corporate Behavior
- Caux Round Table Principles for Business

While the above codes tend to be quite abstract, the **Global Reporting Initiative** (GRI) Guidelines are more relevant to DFE because they provide a standardized framework for sustainability measurement. GRI was convened in 1997 by Ceres, a nonprofit coalition of over 50 investor, environmental, religious, labor and social justice groups. The Guidelines include a set of indicators intended to be applicable to all businesses, as well as sector-specific indicators and a uniform format for sustainability reporting. GRI has been lauded for its success in achieving collaboration among businesses and stakeholders across the world. A major revision of the guidelines, called G3, was introduced in 2006, with a number of enhancements for greater clarity, ease of use, and materiality (i.e., relevance). In subsequent iterations, GRI continues to evolve based on stakeholder feedback.

Ceres Principles

Protection of the Biosphere

- Eliminate harmful releases
- Safeguard habitats and biodiversity

Sustainable Use of Natural Resources

- Sustain renewables (water, soil, forest)
- Conserve nonrenewables

Reduction and Disposal of Wastes

- Source reduction and recycling
- Safe handling and disposal

Energy Conservation

- Conservation
- Energy efficiency
- Environmentally safe energy sources

Risk Reduction

- Minimize employee and community risks
- Safe technologies, facilities, procedures
- Preparation for emergencies

Safe Products and Services

- Reduce or eliminate environmental damage and safety hazards
- Inform customers about environmental impacts and safe use

Ceres Principles *(continued)*

Environmental Restoration

- Correct existing hazardous conditions
- Redress injuries or environmental damage

Informing the Public

- Inform those affected by hazardous conditions
- Maintain dialogue with communities
- Do not penalize employees for reporting problems

Management Commitment

- Ensure top management is informed and responsible
- Consider environmental commitment in selecting directors

Audits and Reports

- Conduct annual self-evaluation
- Create environmental audit procedures
- Complete Ceres report and make available to public

Environmental Advocacy Groups

Perhaps the strongest signals of change in industry-society dynamics are the budding relationships between the business community and environmental advocacy groups, mainly nongovernmental organizations (NGOs). Some environmental groups, such as Greenpeace, pursue confrontational tactics to expose the alleged shortcomings of major companies, often using electronic media with great skill. Today, while these conflicts have not vanished, there is far more philosophic alignment between NGOs and business in terms of broad environmental and social objectives, and a number of companies have worked with NGOs on environmental innovation (see Chapter 5). These engagements have helped to promote creative dialogue about potential solutions and have enabled companies to leverage the competencies and credibility of external groups with diverse perspectives. Some NGOs, such as Business for Social Responsibility, have established themselves as change agents that promote collaboration among companies and other segments of society. The following are selected examples of high-profile NGO partnerships.

- **Environmental Defense Fund** (EDF) formed an initiative called the Alliance for Environmental Innovation, which collaborates with companies such as FedEx and SC Johnson to help design environmentally benign products and supply chain processes.

- **Natural Resources Defense Council** (NRDC) created a Center for Market Innovation that works with companies such as Intel and Wal-Mart to promote positive, environmental change and economically sustainable, profitable growth.

- **Greenpeace,** traditionally an adversarial organization, is collaborating with Scottish wave power companies to promote progress in harnessing wave power.

- **Sierra Club** has partnered with Clorox, a household products company, on a new line of natural, nonpetroleum-based cleaning products called Greenworks.

- **World Wildlife Fund** (WWF) has partnered with a number of global companies, including Coca-Cola and HP, to promote progress on issues such as water conservation, climate protection, and sustainable agriculture.

- **United States Climate Action Partnership** (USCAP) was a group of businesses and leading environmental NGOs that joined together in 2007 to call on the U.S. Federal government to enact strong national legislation aimed at reducing greenhouse gas emissions. The founding members included Alcoa, BP America, Caterpillar, Duke Energy, DuPont, FPL Group, General Electric, PG&E, PNM Resources, and four NGOs—Environmental Defense Fund, Natural Resources Defense Council, Pew Center on Global Climate Change, and World Resources Institute. New members that have joined include Chrysler, ConocoPhillips, Deere, Dow Chemical, Exelon, Ford Motor, General Motors, Johnson & Johnson, PepsiCo, Rio Tinto, Shell, Siemens and Xerox.

- **Business for Innovative Climate and Energy Policy** (BICEP) is a coalition formed in 2008, including Ceres and five leading U.S. corporations, calling for strong U.S. climate and energy legislation in order to spur the clean energy economy. The group's key principles include stimulating renewable energy; promoting energy efficiency and green jobs; requiring 100% auction of carbon allowances; and limiting new coal-fired power plants to those that capture and store carbon emissions. The founding members are Levi Strauss & Co., Nike, Starbucks, Sun Microsystems and Timberland.

Many of the above USCAP and BICEP companies appear in Part 3 of this book, which describes the sustainability programs, NGO partnerships, and DFE initiatives of leading companies in a variety of industries. In an age of environmental awareness, businesses are paying attention to the voice of society because it increasingly represents the voices of their customers and shareholders. It simply makes good business sense. The alignment between environmental responsibility and shareholder value is explored further in Chapter 4.

References

1. J. Ehrenfeld, *Sustainability by Design* (New Haven: Yale University Press, 2008).
2. *Intergovernmental Panel on Climate Change* (IPCC) *Third Assessment Report: Climate Change.* Intergovernmental Panel on Climate Change (Cambridge, UK: Cambridge University Press, 2001).
3. Nicholas Stern, "The Economics of Climate Change: The Stern Review" (London, UK: Cabinet Office—HM Treasury, 2007).
4. WBCSD/WRI. (2004). *The Greenhouse Gas Protocol: A Corporate Accounting and Reporting Standard.* World Business Council for Sustainable Development/World Resources Institute.
5. The *Wall Street Journal* reported on December 30, 2008 that Dell Inc.'s claim of "carbon neutrality" was based on a carbon footprint that the company admits only covers 5% of the greenhouse gases resulting from the life cycle of manufacture and use of its products.
6. H. Minamikawa, "Japan's Plan for Establishing a Sound Material-Cycle Society," Presented at International Expert Meeting on Material Flow Accounts and Resource Productivity, Tokyo, 2003.
7. Z. Guomei, "Promoting Circular Economy Development in China: Strategies and MFA and Resource Productivity Issues," Presented at International Expert Meeting on Material Flow Accounts and Resource Productivity, Tokyo, 2003.
8. More detailed information about the Dow Jones Sustainability Indexes can be found at www.sustainability-indexes.com.

CHAPTER 4

Business Value Drivers

The multinational corporation, with its efficiency, assets and capacity for innovation, is the only institution with the resources necessary to produce a sustainable global economy.

STUART HART [1]

Sustainability Goes Mainstream

In the early days of the environmental movement, capitalism was perceived to be in opposition to environmental protection. The caricature of "big business" was a smokestack belching out pollution, and those involved in manufacturing industries jokingly referred to chemical odors as "the smell of money." Thanks to environmental legislation and pollution control technologies, most of the odors are long gone, but the image persists. Every so often, an unfortunate industrial accident has occurred to reinforce that image—notably, the Union Carbide isocyanate release in Bhopal, India and the Exxon-Valdez oil spill in Alaska. The Enron scandal of 2001, as well as WorldCom and other cases of corporate fraud, further contributed to the notion that businesses are avaricious and amoral. In response, the Sarbanes-Oxley Act of 2002 imposed new requirements for corporate governance, accounting, and financial reporting [2].

Although crises and scandals attract media attention, the vast majority of business managers are scrupulous and civic-minded citizens who are genuinely concerned about environmental and social well being. Virtually all major companies have adopted the concept of *corporate citizenship*, which suggests that businesses have an ethical responsibility to society in addition to their statutory obligations. The notion of citizenship is entirely consistent with the creeds and value systems that are central to the corporate culture of most multinational corporations. These creeds often hark back to the founding

49

of the company and express the fundamental principles and values by which employees are expected to conduct themselves, such as integrity, respect, and teamwork, which become part of the corporate identity [3]. Examples include the HP Way and the Johnson & Johnson Credo (see Chapter 14).

While values are important, the oft-repeated phrase "doing the right thing" is merely a platitude, not a business strategy. The motivation for adopting practices such as DFE goes beyond ethics and good citizenship—it is ultimately a strategic business decision. In simple terms, *business value is the most important driver of DFE*. In fact, corporations all over the world have recognized that sustainability makes good business sense and is essential for their survival and growth [4]. Many CEOs have asserted a belief that sustainable business practices will improve both enterprise resource productivity and stakeholder confidence. At the same time, corporations are beginning to consider the interests of a broader range of stakeholders, including not only customers and shareholders, but also employees, local communities, regulators, lenders, suppliers, business partners, and advocacy groups (see Table 3.1). All of these stakeholders have the power to help or hinder the success of the business.

It is clear that U.S. corporations are eager to communicate their sustainability commitments to their stakeholders. According to a report by the Sustainable Investment Research Analysts Network covering the period from mid-2005 through the end of 2007, over 50% of America's 100 largest publicly traded companies (the S&P 100) report on their sustainability efforts. Over a third of those reports integrate elements of the GRI sustainability reporting guidelines mentioned in Chapter 3. Moreover, 86 of the S&P companies have corporate sustainability websites, a 48% increase since 2005. In recent years, the wave of sustainability adoption has accelerated, as some of America's most influential companies have joined the parade, including Procter & Gamble, General Electric, and even Wal-Mart (see Chapter 19). Some of the factors that explain this phenomenal growth are described below.

Evolution of Environmental Strategy

In the course of about fifty years, environmental sustainability has migrated from an obscure fringe concept to a mainstream concern at the highest levels of corporate governance. The emerging public awareness of environmental sustainability challenges, beginning in the 1960s, was the first wave that heralded a transformation of industry attitudes toward environmental and social responsibility. The second wave, beginning in the late 1980s, was the codification of principles of conduct and best practices—a vital prerequisite to broad adoption of corporate sustainability goals. The third wave, which boosted both environmental awareness and codification of practices, was the sudden growth in public concern over climate change, which began in the early 2000s.

Figure 4.1 traces the evolution of thinking in the business community though the decades, beginning in the 1960s when the predominant mindset was compliance with the law. Despite the emergence of environmental advocacy groups, environmental issues were addressed in a reactive fashion and only caught the public's attention when crises occurred, such as the Cuyahoga River in Cleveland catching fire in 1969. The 1970s marked an era of change, with the formation of the U.S. Environmental Protection Agency and the enactment of a series of laws aimed at cleaning up the environment. Companies began to think systematically about environmental risk management to prevent unplanned incidents, such as the loss of radioactive coolant at Three Mile Island in 1979.

In the 1980s, many businesses began to see a connection between cleaner production and operational efficiency. This gave rise to the voluntary practice of *pollution prevention* (known as P2), i.e., modifying production processes and technologies so that they generate less pollution and waste. Proactive P2 practices included better housekeeping to assure efficient use of resources, elimination of toxic or hazardous substances, process simplification, source reduction, and recycling of process wastes. These techniques formed the basis for many of the DFE guidelines described in Chapter 8.

As the total quality movement took hold during the 1990s, the next logical step was extension of P2 and DFE concepts to the full product life cycle. Companies in the chemical and other industries began to recognize that a defensive posture toward environmental, health, and safety issues no longer made sense. Rather, they decided that it was important to affirm their values and articulate a constructive approach

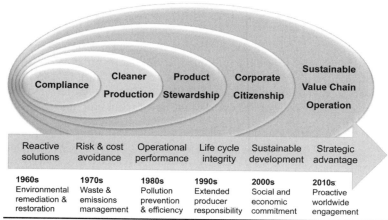

Figure 4.1 The scope of environmental responsibility has extended beyond compliance and beyond the enterprise boundaries.

toward stakeholder concerns. Environmental management system standards and codes of conduct were developed to promote best practices. The concept of *product stewardship* emerged as an ethical commitment by companies to integrity and care in the management of their assets and products, extending to all phases of manufacturing and distribution. Table 4.1 is an example of a product stewardship checklist from this era, which DuPont distributed freely.

Around the turn of the twenty-first century, the progressively expanding scope of producer responsibility led to broad adoption of corporate citizenship and sustainable development principles, including stewardship of the "commons"—protection of natural resources and ecosystem services. By this time, most corporations had embraced a broader commitment to social and economic well-being and had initiated stakeholder outreach and dialogue. During this period, the financial community began to recognize the important linkages between sustainability and shareholder value, discussed later in this chapter; and socially responsible investing began to accelerate.

Today, in search of strategic advantage, companies are expanding the scope of their sustainability initiatives to their full value chains. As discussed in Chapter 10, proactive management of the full product life cycle implies that companies not only need to implement green purchasing and operations policies but also must ensure that their suppliers adopt sustainable business practices. These efforts have already begun, as illustrated by the Electronic Industry Code of Conduct described in Chapter 11. One of the key factors reinforcing this trend is the expansion of multinational companies into developing nations, where they must confront poverty reduction and quality of life issues. At this point, environmental responsibility becomes inseparable from social responsibility. At the leading edge of corporate sustainability, companies are exploring how they can assure safe and ethical labor practices in developing nations and how they can partner with communities at the "base of the pyramid" to create viable new businesses [5].

In the course of this evolutionary journey, the role of environmental issues in decision making has shifted from a tactical risk management approach that emphasizes cost avoidance to a strategic life-cycle management approach that emphasizes value creation. Figure 4.2 uses marginal economic analysis to illustrate this shift:

- Under the traditional risk management paradigm, companies are largely concerned with identifying and mitigating sources of risk that may result in financial liability. They will invest in risk-control expenditures to the extent that the next marginal dollar of expenditure does not exceed the corresponding reduction in potential liability, although the latter is difficult to assess due to the presence of large uncertainties.

- What are the principal safety, health, and environmental (SHE) hazards?
- What competitive products may be substituted for ours? What are their SHE hazards? What advantages do we have?
- Does the customer have our MSDS and product information?

Product and packaging use

- How does the customer use the product? For what purpose? Are there unique or new users?
- Do any uses or handling raise potential SHE concerns?
- Do the customer's employees have access to product information?
- Are the customer's employees using recommended personal protective equipment?
- Are recommended protective systems, including local ventilation, in place?
- Has the customer done workplace monitoring? Should this be done?
- Is the product stored properly? Storage tanks labeled? Spill containment facilities, e.g., dikes?
- Does the customer have emergency response procedures in place?
- What happens to product packaging? Is it reduced, reused, recycled?

Product issues

- Have there been any incidents involving our product? What was learned?
- Have allegations been made regarding health effects of the product? Environmental effects?
- Any other issues associated with the product?

Product emissions and disposal

- Does any part of the product become waste? Regulated hazardous waste? How is it disposed of?
- Is any part of the product discharged to a wastewater treatment system? What is its fate?
- Are there any air emissions from the product's use or disposal? What is their fate?
- How do the discharges/emissions affect customer permits, compliance?
- How can we continuously reduce all emissions and waste?

Distributor questions

- Does the distributor open and repackage or blend our product?
- If so, answer relevant questions above for the distributor.
- Does the distributor provide product information to all customers? How do we know this?
- Does the distributor visit customers to confirm proper use and disposal?

Other

- Can we help in SHE? (Waste management and minimization; SHE training).
- Does the potential exist for exposure to downstream users of our product? If so, answer relevant questions above for downstream users.

TABLE 4.1   Product Stewardship Checklist Developed by DuPont

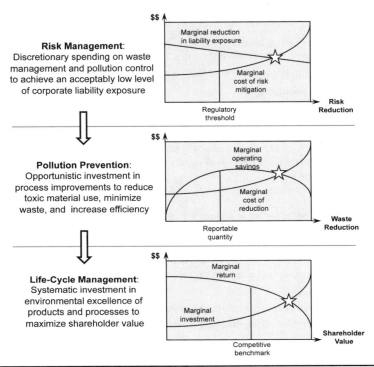

Risk Management:
Discretionary spending on waste management and pollution control to achieve an acceptably low level of corporate liability exposure

Pollution Prevention:
Opportunistic investment in process improvements to reduce toxic material use, minimize waste, and increase efficiency

Life-Cycle Management:
Systematic investment in environmental excellence of products and processes to maximize shareholder value

FIGURE 4.2   The paradigm shift in environmental management decision making.

This approach may lead companies to move "beyond compliance"; for example, if there are significant residual risks associated with emissions that are exempt or below the regulatory threshold.

- Under the pollution prevention paradigm, companies are largely concerned with identifying opportunities for improving efficiency while reducing waste and emissions. On a case-by-case basis, they will invest in pollution prevention opportunities to the extent that the next marginal dollar of expenditure does not exceed the corresponding savings in operating costs. Eventually, they will reach a point of diminishing returns where it is not cost-effective to continue reducing waste using existing technologies. However, new product and process technologies may change the economics to the point where zero waste or closed-loop recycling is attainable.

- Under the life-cycle management paradigm (see Chapter 10), companies are largely concerned with assuring environmental excellence and stakeholder satisfaction over the full life cycle of their products and facilities. This leads them to move from an opportunistic approach to a systematic approach, with environmental considerations factored into virtually all decisions. In fact, environmental decision making is no longer a separate exercise—it becomes an integral part of business decision making. These companies will invest in environmental technology and performance improvements, to the extent that they earn an adequate return on investment. If they are able to leverage their skills and technologies to profitably achieve superior performance, environmental or otherwise, this will constitute a competitive advantage.

The life-cycle management approach leads naturally to the practice of DFE in the context of marketing innovation and new product development, while also incorporating the traditional risk management concerns associated with environmental health and safety. Companies that create value by leveraging their creativity and environmental insights have found the experience rewarding in many ways—intellectually, emotionally, and financially.

Pathways to Value Creation

Corporate sustainability commitments should not be based on citizenship alone. Every company that has invested substantial resources in sustainability improvement has done so because of a persuasive business case. The drivers of business value relevant to DFE include productivity, profitability, enhanced reputation, and competitive advantage. Short-term economic drivers are often the catalyst, and there is no question that rising energy prices have helped to motivate companies to take a harder look at their resource efficiency. But besides contributing to the bottom line, sustainable business practices create shareholder value by strengthening intangible factors such as brand equity, reputation, and human capital.

A significant driving force for corporate sustainability has been the World Business Council for Sustainable Development (WBCSD), mentioned in Chapter 2, a Geneva-based consortium of over 150 leading companies [6]. WBCSD has published a series of studies that demonstrate the business value of sustainable practices and present agendas for change in industries such as pulp and paper, mining, cement, transportation, and electric power. WBCSD has also been instrumental in developing tools and best practices for eco-efficiency and environmental footprint assessment. Like many other organizations, WBCSD conceives of the goals of economic, social, and ecological well-being as a "triple bottom line" that expands upon the financial bottom line [7]. The three elements can be defined as

- Economic prosperity and continuity for the business and its stakeholders,
- Social well-being and equity for both employees and affected communities,
- Environmental protection and resource conservation, both local and global.

This metaphor has been used as the basis for many sustainability assessment tools including the GRI guidelines and the Dow Jones Sustainability Indexes.

In the U.S., corporate environmental, health and safety excellence has been the principal goal of the Global Environmental Management Initiative (GEMI), a consortium of over 40 multinational companies that collaborate closely on codification of best practices [8]. Since 1990, GEMI has developed and published a series of studies that provide guidance to corporations on environmental management practices, including climate and water strategies, performance measurement, and supply chain management. The SD Planner™ tool developed by GEMI provides a concise summary of the key elements of sustainability, shown in Table 4.2, which synthesizes the multitude of principles and codes of conduct discussed in Chapter 3.

Category	Element	Definition
Social	1. Employee Well-Being	Protecting and preserving the fundamental rights of employees, promoting positive employee treatment, and contributing to employee health, safety, dignity, and satisfaction.
	2. Quality of Life	Working with public and private institutions to improve educational, cultural, and socio-economic well being in the communities in which the company operates and in society at large.
	3. Business Ethics	Supporting the protection of human rights within the company's sphere of influence and promoting honesty, integrity, and fairness in all aspects of doing business.

TABLE **4.2** Elements of Sustainable Development [8]

Category	Element	Definition
Economic	4. Shareholder Value Creation	Securing a competitive return on investment, protecting the company's assets, and enhancing the company's reputation and brand image through integration of sustainable development thinking into business practices.
	5. Economic Development	Building capacity for economic development in the communities, regions, and countries in which the company operates or would like to operate.
Environmental	6. Environmental Impact Reduction	Minimizing and striving to eliminate the adverse environmental impacts associated with operations, products, and services.
	7. Natural Resource Protection	Promoting the sustainable use of renewable natural resources and conservation and sustainable use of nonrenewable natural resources, including ecosystem services.

TABLE 4.2 Elements of Sustainable Development [8] *(continued)*

One of GEMI's major contributions has been to establish a business case for sustainability, including an understanding of how *intangible value drivers,* such as leadership, brand equity, and human capital, create shareholder value. Intangibles include people, relationships, skills, and ideas that add value but are not traditionally accounted for on the balance sheet. It is estimated that between 50% and 90% of a company's market value can be explained by intangibles rather than traditional measures, such as earnings and tangible assets [9]. As shown in Figure 4.3, there are three major pathways whereby sustainability contributes to shareholder value.

1. Sustainable business practices can contribute directly to **tangible financial value** by enabling top-line growth, reducing operating costs, conserving capital, and decreasing risks. For example, DuPont plans to expand into new markets by developing products that focus on human safety and personal protection (see Chapter 13). Many companies are implementing GHG emission reduction programs that will also reduce energy costs.

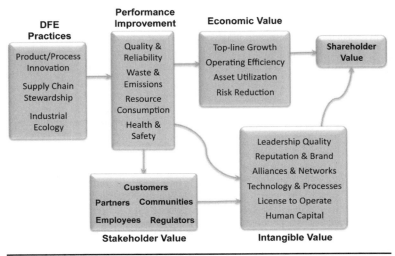

FIGURE 4.3   Pathways to shareholder value through environmental performance Improvement.

2. Sustainability can directly improve **intangible assets** such as reputation, brand equity, strategic alliances, human capital, and innovation. For example, Xerox's pioneering efforts to design products for reverse logistics and asset recovery have both improved its manufacturing technologies and strengthened its customer relationships (see Chapter 11).

3. Sustainability can provide strategic advantage by creating **value for stakeholders**. Each company can choose to focus on the stakeholder issues that are best aligned with its own interests and core competencies. For example, Procter & Gamble has focused its sustainability efforts on creating innovative products that address worldwide needs for water, health, and hygiene (see Chapter 16).

Shareholder value is driven by a combination of Economic Value, corresponding to financial performance, and Intangible Value (see sidebar). Past efforts at building a business case for sustainability have focused mainly on estimation of financial returns associated with EH&S initiatives, such as converting wastes into by-products. However, these contributions tend to be incremental in nature, and are generally seen as tactical rather than strategic. The more strategic contributions of sustainability tend to be associated with nonfinancial value drivers, such as relationships and reputation, which provide a prospective, rather than retrospective, view of shareholder value. To portfolio managers and investment analysts, these intangible strengths are often the hidden clues that differentiate companies with comparable financial statements. In other words, improvements in environmental and social performance can strengthen a company's intangible assets in ways that lead to sustained long-term shareholder value.

The third pathway, creating value for stakeholders, is especially important to consider. As shown in Chapter 3, the demand for greater transparency and improved corporate governance has led to an increase in voluntary disclosure, as well as greater scrutiny from major investors. In addition to customers, shareholders, and employees, there is a broader collection of external stakeholders that can influence the success of a business and are interested in environmental performance. These include suppliers and business partners; regulators and government officials at the local, state, and federal levels; neighboring communities; religious groups, advocacy groups and other NGOs; academic and research organizations; and, of course, the media (see Table 3.1). By responding to the diverse interests of these stakeholders, companies can strengthen their key relationships, reputation, and license to operate.

Figure 4.3 can serve as a template for companies to identify high-priority pathways whereby DFE efforts can deliver shareholder value. For example, one of the most important intangible assets is the ability to attract talent. A Stanford University survey of about 800 graduating MBAs at 11 top business schools has shown that these future business leaders rank corporate social responsibility high on their list of personal values and are willing to sacrifice a significant part of their salaries to find a similarly minded employer. Another important intangible asset is brand equity, and it is no accident that the top companies on Interbrand's list of the most valuable brands—Coca-Cola, IBM, Microsoft, GE, Nokia, Toyota, Intel, McDonalds, Disney and Google—have all invested considerable effort in developing sustainability programs.

Components of Shareholder Value

The components of Shareholder Value shown in Figure 4.3 merit closer examination. Economic Value can be measured on an annual basis through the following commonly used formula:*

Economic Value Added (EVA) =
After-tax operating profit − Capital charge

The basic concept is that additional value is created either by increasing cash flow or by reducing the capital required to

---

*Some chief financial officers prefer to replace after-tax operating profit with related measures such as earnings before interest, taxes, depreciation and amortization (EBITDA) or return on net assets (RONA).

Components of Shareholder Value *(continued)*

generate cash flow. Conversely, value is destroyed by a decrease in cash flow or an increase in capital requirements. Note that the "bottom line," corresponding to the profit and loss statement, is just one part of the EVA equation. The other part is concerned with capital and corporate assets, reflected in the balance sheet.

Figure 4.4 breaks down these two terms, indicating how environmental excellence can contribute to value creation. After-tax operating profit can be improved either by increasing revenues or by reducing costs. Capital charge can be reduced either by managing assets more effectively or by decreasing the weighted average cost of capital (WACC) through favorable financing rates. Accordingly, DFE practices can deliver four major types of tangible value creation:

1. **Top-Line Growth**: DFE contributes to increased revenues through environmentally preferred product differentiation, improved license to operate in existing markets, or penetration of new markets. In addition, the practice of concurrent engineering helps to reduce cycle time for new product development, which leads to earlier introduction and increased market share.

2. **Operating Efficiency**: DFE applied to manufacturing and supply chain technologies leads to leaner and cleaner processes, resulting in greater productivity and lower operating costs. Resource conservation and recovery help to reduce material and energy costs, as well as

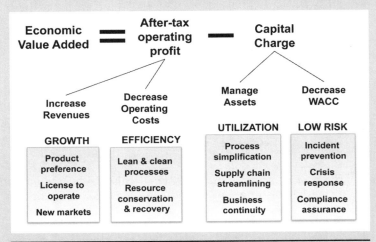

FIGURE 4.4 Economic value creation opportunities.

Components of Shareholder Value *(continued)*

waste disposal costs. Other indirect benefits of DFE include reduced insurance premiums and training costs (see Life-Cycle Accounting in Chapter 9).

3. **Asset Utilization**. DFE can reduce the complexity and fixed costs of assets through process simplification and streamlining of the supply chain. In addition, DFE can improve the utilization of assets through increased business continuity and equipment reliability, thus reducing the total asset base required to support the business.

4. **Risk Reduction**. DFE can help to reduce overall enterprise risk, which is the main driver of WACC. Prevention of incidents that may lead to business interruption, improved responsiveness to crises, and assurance of regulatory compliance all contribute to lowering both the actual and perceived risks associated with business operations.

Intangible value is less amenable to precise quantification, but a number of studies have actually measured the strength of intangibles across a variety of industries. Based on considerable research, the following characteristics depicted in Figure 4.3 have been identified as among the most important intangible assets [10].

- **Leadership Quality**: Management capabilities, experience, vision for the future, transparency, accountability, and trust.

- **Reputation**: How the company is viewed globally in terms of stakeholder concerns, inclusion in "most admired company" lists, and sustainability performance.

- **Brand Equity**: Strength of market position, ability to expand the market, perception of product/service quality, and investor confidence.

- **Alliances and Networks**: Customer and supply chain relationships, strategic alliances and partnerships.

- **Technology and Processes**: Strategy execution, information technology, inventory management, flexibility, quality, and internal transparency.

- **License to Operate:** Regulatory positioning, relationships with local communities, ability to expand operations.

- **Human Capital:** Talent acquisition, workforce retention, employee relations, compensation, and perception as a "great place to work."

The Sustainability Landscape

While the logic of shareholder value creation is compelling, it is not necessarily a sufficient motivation to break old habits. Many forces of change in the business environment have converged during the late twentieth and early twenty-first centuries, resulting in a sort of "tipping point" for adoption of corporate sustainability. Most of these changes have been discussed in previous chapters, and are summarized here:

- **Climate change anxiety.** Once climate change was finally acknowledged as a reality, governments, NGOs, and corporations began to seriously explore policies and technological solutions for mitigation of greenhouse gas emissions; and carbon offset schemes flourished.

- **Energy security.** Concerns over depletion of fossil fuels and dependence on imported petroleum, coupled with the problem of carbon emissions, drove investments in alternative fuels; this trend was further intensified by a sudden rise in oil prices in 2007.

- **Customer awareness.** Both retail and industrial customers became increasingly concerned about the environmental performance of products that they purchased. Major corporations began to systematically review the environmental performance of their suppliers, and many governments introduced environmentally preferred procurement policies.

- **Legislative requirements.** A series of government directives in the European Union forced global multinationals to change their practices with regard to product design and life-cycle management; similar measures were adopted in many other countries.

- **Voluntary codes and standards.** Voluntary codes of conduct such as the Ceres principles, as well as environmental management system standards such as ISO 14001, were widely adopted by the business community as a way to demonstrate environmental responsibility.

- **Eco-labeling programs.** A number of eco-labeling initiatives have gained acceptance by consumers around the world, and companies in the electronics, consumer products, food and beverage, and other industries are now compelled to qualify in order to remain competitive.

- **Sustainability-driven investing.** The financial investment community has begun to recognize sustainability as an indicator of overall superior management, as exemplified by the increasing interest in the Dow Jones Sustainability Indexes and other rating systems.

- **Globalization**. Rapid economic growth in emerging economies, such as Brazil, Russia, India, and China (known as BRIC), as well as globalization of supply chains, have forced multinational companies to grapple with the challenges of energy, environmental protection, human rights, poverty, and social responsibility.

- **Transparency**. Public expectations for information disclosure, as well as the explosive growth of electronic communication, have made it essential for global companies to increase their level of accountability, transparency, and stakeholder engagement.

Given all these changes, it is clear that corporate sustainability is not just altruism. Rather, it is an enlightened response to emerging market forces. Instead of merely listening to the voice of the individual customer, companies are beginning to listen to the collective voice of the larger Customer, namely, human society. Because of the great impact that multinational companies can have on society and the environment, stakeholders expect them to do business in a socially and environmentally responsible manner. This landscape creates opportunities for companies to respond to stakeholder expectations with new technologies, products, and services. The potential role of DFE is evident—enabling companies to simultaneously increase shareholder value and meet the needs of their stakeholders, thereby gaining competitive advantage.

Those companies that recognized these trends and became early adopters have established highly visible and successful sustainability programs. Typically, there are several levels of sustainable business practices. The most basic level involves corporate initiatives, such as philanthropic programs aimed at solving community social problems. The next level often involves reducing the "ecological footprint" associated with the product life cycle, including manufacturing, use, and end-of-life disposition. The most challenging level involves enhancing the *inherent social value* created by the firm's operations, products, and services, which may range from assuring human health and nutrition to stimulating consumer education and growth of new businesses. At this level, challenging trade-offs may arise—for example, balancing job creation and economic development against community concerns about industrial pollution and environmental justice, i.e., equitable distribution of risks and benefits.

> **CORPORATE SUSTAINABILITY IS AN ENLIGHTENED RESPONSE TO EMERGING MARKET FORCES.**

Rather than following a "cookie-cutter" approach, the early adopters have explored how they can integrate sustainability into their own business innovation strategy. Electronics companies have used information technology to bridge the digital divide between haves and have-nots,

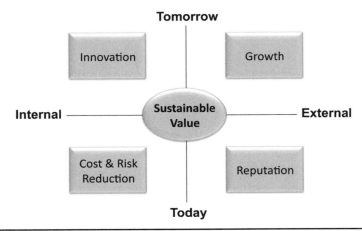

FIGURE 4.5   Framework for corporate sustainability strategy development [11].

consumer products companies have dramatically reduced packaging to minimize solid wastes, chemical companies have invented new processes to eliminate the use of toxic substances, and so forth. Part 3 of this book elaborates on these stories and many more. The common purpose of all of these programs has been to shift enterprise operations from a traditional, resource-intensive, and volume-maximizing business model to a more eco-efficient, socially responsible, and value-maximizing model. This shift aligns nicely with the goal of increasing shareholder value by raising profits while reducing the use of capital and resources—i.e., "doing more with less."

Looking to the future, Stuart Hart and Mark Milstein have developed a useful framework, shown in Figure 4.5, to characterize the strategic landscape that companies face in their search for sustainable value creation [11]. The different quadrants of this framework echo many of the themes discussed in this chapter. The lower left quadrant represents a traditional short-term focus on internal cost and risk reduction, which is a necessary part of doing business. The lower right quadrant is also concerned with short-term sustainability but is outwardly directed toward stakeholder engagement and protection of the firm's reputation. The upper left quadrant represents a focus on innovation to assure that the firm is positioned to meet future demands for sustainable technologies, products, and services—this is the main thrust of DFE. Finally, the upper right quadrant involves external engagement to understand growth opportunities and market needs, which provides critical feedback to the innovation process. Achievement of global sustainability is dependent on the ability of companies in every industry to balance these strategic thrusts, creating shareholder value for themselves while assuring the continued well-being and prosperity of human societies.

References

1. S. L. Hart, *Capitalism at the Crossroads: The Unlimited Business Opportunities in Solving the World's Most Difficult Problems* (Philadelphia: Wharton School Publishing, 2005).
2. Sarbanes-Oxley Act of 2002: "Public Company Accounting Reform and Investor Protection Act of 2002," Public Law 107–204, 107th Congress of the United States of America.
3. R. Van Lee, L. Fabish, and N. McGaw, "The Value of Corporate Values," *strategy+business*, Summer 2005.
4. A. Hoffman, *Heresy to Dogma: An Institutional History of Corporate Environmentalism* (Stanford, Calif: Stanford University Press, 2001).
5. E. Simanis and S. Hart, *The Base of the Pyramid Protocol: Toward Next Generation BoP Strategy*, 2nd ed. (BoP Network, Cornell University, 2008).
6. The World Business Council for Sustainable Development (WBCSD) portal at wbcsd.org provides access to a large portfolio of reports and case studies.
7. J. Elkington, *Cannibals with Forks: The Triple Bottom Line of 21st Century Business* (Oxford, U.K.: Capstone Publishing, Ltd., 1997).
8. The GEMI portal at www.gemi.org offers free access to informative publications and interactive tools.
9. J. Fiksel, J. Low, and J. Thomas, "Linking Sustainability to Shareholder Value," *Environmental Management*, June 2004.
10. J. Low and P. Kalafut, *Invisible Advantage: How Intangibles Are Driving Business Performance* (New York: Basic Books, 2002).
11. S. L. Hart and M. B. Milstein, "Creating Sustainable Value," *Academy of Management Executive*, Vol. 17, No. 2, 2003.

PART 2

Charting the Course: The Art and Science of Design for Environment

CHAPTER 5

Managing Environmental Innovation

Disruptive technologies, though they initially can only be used in small markets remote from the mainstream, are disruptive because they subsequently can become fully performance-competitive within the mainstream market against established products.

CLAYTON CHRISTENSEN [1]

The Rise of Green Markets

There was a time, not so long ago, when environmentalists were derided as "tree-huggers" and products with environmental benefits appealed only to a fringe market of true believers. The generally accepted wisdom was that mainstream consumers were primarily cost-driven and would not tolerate the higher cost of environmentally beneficial products. A classic example of an environmental product was the pioneering "green PC" introduced in 1993 by IBM. Despite its thoughtful, energy-efficient design, the premium price tag prevented it from capturing any appreciable share of the market.

Fifteen years later, "green" has become a ubiquitous marketing buzzword, not just in personal computers but in virtually every product category. (Sadly, green language is often overused to the point of being gratuitous or even misleading.) As discussed in Part 1, this tidal shift is a consequence of many factors from rising energy prices to global warming fears. Environmental sensitivity has moved from being a luxury to a necessity. And greener products need not necessarily cost more.

69

There have been many efforts to characterize the different market segments that comprise "green" consumers [2]. One segment that has generated a great deal of attention in the green marketing community is LOHAS, an acronym for Lifestyles of Health and Sustainability, which includes products and services related to health and fitness, the environment, personal development, sustainable living, and social justice. According to the Natural Marketing Institute, the LOHAS market in the United States exceeds $200 billion. Consumers in this segment are known as "Cultural Creatives" and tend to be progressive, thoughtful individuals who are concerned with corporate social responsibility. (While understanding consumer attitudes is important for setting DFE requirements, the intricacies of green marketing are beyond the scope of this book.)

As customer expectations and economic conditions change, new technologies can disrupt existing markets. Once-obscure techniques for renewable energy generation, such as biologically based fuels derived from plant residues or algae, have attracted the hordes of venture capitalists who previously were captivated by the digital revolution. Large pick-up trucks and sport-utility vehicles are being abandoned, as automakers hastily retool their plants to build smaller cars with fuel-efficient or hybrid engines. Manufacturers of clothing, cosmetics, foods, beverages, office products, and many other consumer staples have rushed to embrace natural, organic, and eco-friendly materials while reducing and recycling their packaging. Video-conferencing solutions have exploded as business travel becomes more expensive and inconvenient. Many of these trends are examined in Part 3 of this book.

Sustained innovation is always a challenge for established companies who are market leaders, and even more so in times of rapid change. Those that excel at innovation, companies, such as 3M and Apple, have spent considerable effort at establishing disciplined yet flexible internal processes that encourage creativity while systematically weeding out and refining the truly worthy concepts. In contrast, companies that react impulsively to external changes may be unable to achieve genuine innovation and find that they cannot compete effectively. In particular, adding environmental features to products cannot be approached as a quick fix or a simple overlay. To be truly successful at DFE, it is essential that companies embed an understanding of environmental performance into their core product development processes.

Integrated Product Development

Manufacturing firms in the United States have almost universally recognized the need to establish clearly defined business processes for new product development and introduction. With increasing competitive pressures and rapidly changing markets, reduction in product development cycle time has become an essential goal. Equally important is the goal of continuous improvement in product quality and responsiveness to the "voice of the customer." Many firms have introduced new management methods, such as "stage-gate" processes designed to accelerate time to market, standardize product characterization, and improve decision making. A classic publication by the National Research Council summarizes the prevailing wisdom of modern product development [3]. It cites four requirements for using design as a source of competitive advantage:

1. Commit to continuous improvement both of products and of design and production processes.

2. Establish a corporate product realization process (PRP) supported by top management.

3. Develop and/or adopt and integrate advanced design practices into the PRP.

4. Create a supportive design environment.

Moreover, it defines an effective PRP as incorporating the following steps:

- Define customer needs and product performance requirements.
- Plan for product evolution beyond the current design.
- Plan concurrently for design and manufacturing.
- Design the product and its manufacturing processes with full consideration of the entire product life cycle, including distribution, support, maintenance, recycling, and disposal.
- Produce the product and monitor product and processes.

In this spirit, the term "Integrated Product Development" (IPD) describes a process that has been adopted by most progressive manufacturing firms, even though they may have different names for it [4]. IPD is commonly understood to mean development of new products using cross-functional teams from inception to commercialization. More specifically, we define IPD as follows: *a process whereby all functional groups (engineering, manufacturing, marketing, etc.) that are involved in the product life cycle participate as a team in the early understanding and resolution of key issues that will influence the success of the product.*

> **BREAKTHROUGH PRODUCTS MAY LIKELY BE BASED ON TRANSFORMATIVE ENVIRONMENTAL TECHNOLOGIES.**

Integrated product development does not tolerate the traditional "sequential engineering" approach, in which a design is developed and refined in successive stages by different engineering groups (design, layout, test, etc.), with each group "throwing results over the wall" to the next group. Instead, companies that have adopted IPD use the approach of "concurrent engineering" or simultaneous engineering, in which the different engineering *disciplines* work in a parallel, coordinated fashion to address life-cycle requirements. Examples of these disciplines include quality, manufacturability, reliability, maintainability, safety, and the newest member—environmental sustainability.

The motivation for IPD is extremely strong when one considers the economics of product development. For example, in the automotive and electronics industries, up to 80% of product life-cycle costs are committed during the concept and preliminary design stages, and the cost of design changes increases steeply as a product proceeds into full scale development and prototyping. The implication is that product developers should use concurrent engineering to examine a design from multiple perspectives and to anticipate potential problems or opportunities. By getting it right the first time, they will avoid costly changes and delays due to design iterations. Another advantage of IPD is that it helps to speed time to market, i.e., the time interval between the launch of a new product development effort and the market introduction of the product. It is well known that early market entry tends to increase ultimate market share.

One of the most challenging aspects of IPD is the "fuzzy front end" of the process, where a wide variety of ideas are screened, initial concepts are formulated and evaluated, and decisions are made as to whether to move these concepts into a more structured development process [5]. This phase often involves major commitments of time and money, and determines the essential features of the ultimate product. Therefore, it is imperative that professionals with an understanding of environmental engineering and design issues be involved in the fuzzy front end. Too often companies fall into the trap of treating environmental issues as an afterthought to be addressed once the product concept is established. In an age of environmental awareness, breakthrough products that are "game-changers" may likely be based on transformative environmental technologies. That is why 3M, among others, has adopted a disciplined life-cycle management approach that includes DFE as a mandatory part of new product development (see Chapter 10).

Organizing for Environmental Excellence

Enterprise integration, as described in Chapter 1, encourages effective teamwork and coordination between marketing, engineering, and manufacturing groups. In this context, IPD can be seen as a strategy for agile product development, enabling companies to release higher-quality products while reducing time to market. With the growing emphasis on environmental performance as an important aspect of total quality management, DFE fits naturally into the IPD process. DFE is essentially an application of the above IPD approach to environmental performance. As stated above, the basic imperative of DFE is undeniably valid: "Get it right the first time." In other words, anticipate environmental performance issues during design, thus avoiding costly changes in the future. It sounds simple, but to put this principle into systematic practice typically requires a substantial organizational commitment.

Despite the many examples of successful DFE efforts described in Part 3 of this book, the current state of practice can be characterized as mainly opportunistic. Well-motivated and well-informed teams may be able to identify product improvements that are environmentally beneficial or that reduce life-cycle costs. However, many of the early successes in environmental performance improvement were the result of companies finding the "low-hanging fruit"—projects for which benefits were evident and barriers were few. In order for DFE to mature and be integrated into company practices, two types of permanent change are needed:

1. Organizational norms must be established that encourage superior environmental performance. This requires evolving from a broad corporate "mission statement" to setting achievable environmental improvement goals and to making employees accountable for meeting or exceeding these goals.

2. Business processes must be modified to accommodate environmental performance criteria. This requires integrating environmental quality metrics and assessment tools into standard engineering practices, as well as developing managerial accounting systems that recognize environmental costs and benefits.

Even though many companies have announced high-level commitments to sustainability and environmental excellence, they often encounter passive resistance at middle management levels. Implementation of environmental innovation processes can be hampered by the same types of barriers that afflict any change initiative:

- Resources are limited for starting new projects.
- Organizational and cultural inertia tends to favor "business as usual."
- Environmental issues are poorly understood among both managers and employees.
- Existing accounting systems are inadequate for reflecting environmental value.
- Product teams have a fear of compromising product quality or production efficiency.

In addition, design teams find it challenging to perform DFE consistently and effectively for several reasons:

- Environmental expertise is not widely available among product development engineers.
- The complex and open-ended nature of life-cycle environmental impacts makes them difficult to analyze.
- The economic systems in which products are produced, used, and recycled are much more difficult to understand and manage than the products themselves.

Overcoming these challenges requires effective communication, credible tools, and a clear implementation strategy. To establish a successful DFE program, there are a number of key steps that a company should follow:

Program Definition

- Establish top management commitment to legitimize the program.
- Invite inputs from external stakeholders, including regulators, customers, and communities.
- Fit the program to the existing organizational structure and cross-functional teams.
- Ensure adequate staff training, incentives, and empowerment.

DFE Implementation

- Establish appropriate DFE metrics, guidelines, and supporting tools (see Chapters 7 to 9).
- Assess the baseline environmental performance of existing products and processes.
- Identify alternative technologies and set development priorities, using systematic criteria.
- Implement high-priority changes and measure the improvements relative to baseline.

Ongoing DFE Practice

- Integrate new technologies and lessons learned into other products and processes.
- Institutionalize the metrics, guidelines and other tools, using computer support as appropriate.
- Communicate the beneficial results of DFE actions to stakeholders.
- Recognize and reward both team and individual accomplishments.

Once DFE becomes an integral part of the product development process, it will reveal new opportunities for companies to reduce life-cycle costs and improve overall product quality and profitability, while helping to assure that they are meeting their sustainability commitments.

The organizational strategy for pursuing the above steps need not involve establishing a whole new bureaucracy to enforce the practice of DFE. On the contrary, many leading companies with effective DFE programs, such as Intel and DuPont, utilize a small corporate staff group of specialists who provide support to cross-functional teams. The driving force for adoption of DFE goals and practices typically flows down from line management, while the core sustainability teams act as change agents throughout the business units, meeting with leaders of business and functional groups to set strategy and coordinate implementation.

Practicing Concurrent Engineering

As described above, concurrent engineering is at the heart of IPD, and involves simultaneous consideration of manufacturability, reliability, and many other "ilities." (The corresponding design disciplines are sometimes described as "Design for X," or DFX.) There are three essential elements of any concurrent engineering discipline:

1. **Indicators and metrics,** which reflect fundamental customer needs or corporate priorities and provide the basis for establishing product objectives and measuring performance.

2. **Design rules and guidelines,** which are based on an in-depth understanding of relevant technologies and supported by accepted engineering practices.

3. **Analysis methods,** which enable engineers to assess proposed designs with respect to the above metrics and to analyze cost and quality trade-offs.

For example, consider the design process for printed circuit boards, which has become highly systematized in order to enable rapid time to market. Among the many "ilities" addressed in this field is Design for Reliability (DFR). Table 5.1 illustrates the above three elements for both DFR and DFE. Chapters 6 through 9 provide a thorough discussion of these three main elements in the context of DFE—environmental performance indicators and metrics, DFE guidelines, and environmental analysis methods.

Figure 5.1 illustrates how the elements of concurrent engineering are deployed to support achievement of rapid cycle time by an integrated product development team. These are usually deployed as part of an iterative stage-gate process that begins with the analysis of customer needs and the establishment of product requirements

	Performance Indicator	Practice Guideline	Analysis Tool
Design for Reliability	Mean time between failures (MTBF)	Allow adequate spacing for components with high power consumption	Thermal simulation program for specified board layout
Design for Environment	Average annual energy consumption	Utilize low-power components and enable sleep mode	Life-cycle energy use assessment based on typical user patterns

TABLE 5.1   Application of DFR and DFE to a Printed Circuit Board

and ends with the release of the final design. In a true concurrent engineering approach, requirements address not only product performance but also manufacturing, service, and other downstream issues including environmental performance. Like any other objectives, environmental goals and requirements should be established using measurable and verifiable indicators.

In general, a *requirement* can be defined as a description of a set of testable conditions applicable to products or processes. A requirement

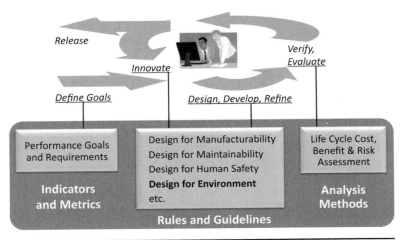

FIGURE 5.1   Incorporating DFE into the integrated product development process.

is said to be *satisfied* by a product (or process) if a test or observation reveals that the described conditions are met by that product (or process). The conditions may be represented in various forms, including

- behavioral   e.g., "system shall shut down when left idle"
- qualitative   e.g., "shall be *stable* in extreme heat"
- quantitative   e.g., "*width* ≤ (*inner radius*) × 1.9 cm"
- pictorial   e.g., "shall resemble this zigzag pattern"
- logical   e.g., "*process type* shall only be *acid* or *thermal*"

The requirements management process consists of three main functions that are performed repeatedly in an iterative fashion. These are requirements *analysis*, requirements *tracking*, and requirements *verification* [6]. Each of these functions is described below:

1. **Requirements analysis** is the process of interpreting customer needs and deriving explicit requirements that can be understood and interpreted by people and/or computer programs. It is usually carried out by a select group of program managers, chief engineers, and senior project engineers. Complex designs such as automotive systems can have thousands of requirements, which are often represented in a hierarchical fashion.

2. **Requirements tracking** involves continuous interchange and negotiation within a project team regarding conflicting and changing objectives. Design decisions must be weighed in terms of a variety of factors, including project risk, schedule, cost constraints, and performance goals. An organizing scheme such as a *traceability* hierarchy is helpful in locating specific types of requirements, navigating through large volumes of requirements, and adding new ones at appropriate points.

3. **Requirements verification** is the process of evaluating whether a product design complies with a designated set of requirements. This can be accomplished through actual testing of a prototype or, at an earlier stage, through predictive methods. Concurrent engineering teaches that the earlier in the design process verification can be performed, the more likely it is that design flaws will be detected before a large prototyping investment has been made.

After requirements have been defined, the product development cycle begins in earnest. As the team works on developing a detailed design, they employ **design rules and guidelines** drawn from each of the relevant design disciplines, e.g., manufacturability, maintainability, etc. While automated aids such as expert systems can facilitate the practice of DFX, the innovative capabilities of human engineers are the key to success at this stage. Once an initial design has been formulated, it is possible to begin design **verification**; the earlier this takes place, the sooner the team will recognize design shortcomings and take steps to overcome them. As mentioned above, early and systematic design verification is critical to reducing both the number of design iterations and the time to market.

The use of automated tools at the verification stage is helpful to assure coverage of all product requirements and to identify design flaws or omissions. In particular, DFE requires a verification system because of the complexity of environmental issues associated even with simple products. Many of the first-generation DFE practices were limited to the use of simple material selection checklists or single-dimensional metrics (e.g., total waste), which provide little insight into the true *value* of product improvements. Instead, design teams need the ability to "close the loop" by receiving feedback regarding the anticipated benefits of proposed design changes.

One of the remaining barriers to concurrent engineering, especially in high technology industries, is the reliance of engineers on design and manufacturing automation tools that do not allow cross-functional integration. Specialized tools include CAE/CAD systems for systems engineering, design characterization, modeling, simulation, and requirements verification. Although these tools are effective at supporting the detailed work of individual engineers, there is still a need for tools that help to evaluate trade-offs, provide design advice, and capture the accumulated experience of development teams. New software technologies, such as "intelligent assistant" design tools, have the potential to facilitate a company's transformation from traditional ways of doing business to a more concurrent approach, and can provide ongoing support for IPD. For example, Motorola has developed a "Green Design Advisor" tool for applying DFE to mobile phone products (see Chapter 11).

Finally, in a truly integrated approach, DFE must be balanced against other cost and quality factors that influence design trade-off decisions. The mark of a successful team is the ability to innovate under pressure, rather than compromising product quality. A "win-win" outcome is the introduction of environmentally beneficial innovations that also improve the cost and performance of the product when viewed as part of an overall system. Ideally, a single design innovation may contribute to achieving several, different types of goals. For example, reducing the mass of a product can result in (1) energy and material use reduction, which contributes to resource conservation, and (2) pollutant emission reduction, which contributes to health and safety.

Making trade-off decisions is the most challenging part of the process because of the need to simultaneously consider so many different criteria. While not mandatory, it is useful to have a well-structured *decision framework* that supports the exploration of uncertainties and unintended consequences as well as satisfaction of the original product requirements. Integrated decision making involving DFX trade-offs is usually addressed through conceptual tools such as "house of quality" or Quality Function Deployment (QFD) matrices, which are often grouped under the general category of Design for Six Sigma tools [7]. Chapter 10 describes a framework that enables consideration of multi-stakeholder perspectives on sustainability issues for purposes of business decision making.

Understanding Product Life Cycles

The term "life cycle" has become so much in vogue that it bears closer examination. It turns out that one person's definition of life cycle may be quite different from another's. Consider the following widely differing interpretations:

1. **Business life cycle:** the product life cycle is a sequence of activity phases, including the creation of a product concept, its development, launch, production, maintenance, maturity, reevaluation, and renewal in the form of a next-generation product.

   Similarly, the process life cycle is a sequence of activity phases, including the development of facility and process designs, architecture and construction, operation and maintenance, and eventual upgrading or retirement.

2. **Physical life cycle:** the product life cycle is a sequence of transformations in materials and energy that includes extraction and processing of materials, product manufacture and assembly, distribution, use, and recovery or recycling of product materials.

   Similarly, the process life cycle is a sequence of transformations in materials and energy that includes extraction and processing of materials used for process equipment and supplies, process operation and control, equipment cleaning and maintenance, and waste disposal or recovery.

Note that a single process may be involved in producing a variety of different products, while producing a single product may involve a variety of different processes. Most of this book is oriented toward the *product* life-cycle perspective.

As depicted in Figure 5.2, the dual product life cycles—business and physical—are closely interwoven, and therefore easily confused. They intersect at the point of product launch and deployment, where product and process designs are released for ramp-up to full-scale production. Yet, they are very different in several respects.

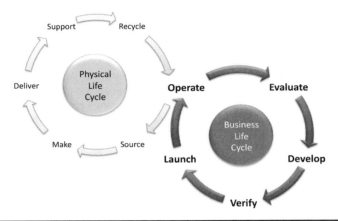

Figure **5.2** Dual life cycles associated with a product.

- The *time scale* of the business life cycle is determined by the rate of technological change and other market conditions that determine product obsolescence; the time scale of the physical life cycle is determined by the length of service of a typical product unit, which may range from days or weeks for consumable products to years or decades for durable goods.

- The *responsibility* for the business life cycle and the business impacts (profits or losses incurred) is borne fully by the producing company; the responsibility for the physical life cycle and the associated impacts is distributed among many companies or individuals involved at different stages of transformation; and, in some cases, the responsibility for adverse impacts may be ambiguous (e.g., liability for waste impacts).

- The *sustainability* of the business cycle depends upon the ability of a company to innovate by developing and delivering product extensions or upgrades that serve the needs of its markets; the sustainability of the physical cycle depends upon the ability of various interested parties, including the manufacturer, to cost-effectively acquire the necessary energy and material inputs, and to recover, recycle or refurbish discarded products.

From a conventional product development perspective, the "business life cycle" interpretation is the more meaningful one, since it provides a framework for making business decisions regarding desirable product features and cost or effort trade-offs. The concurrent engineering approach, described above, traditionally has been based on this model of the business life cycle, with "ilities" associated with different phases. On the other hand, the notion of the "physical life cycle" has been the primary focus of recent efforts at ecological footprint assessment, environmental labeling, standard-setting, and performance evaluation.

It is no wonder then, that product teams are often puzzled by the increasingly widespread use of life-cycle assessment (LCA), which seeks to quantify the "cradle to cradle" impacts of the physical life cycle. (Chapter 9 describes various approaches to LCA.) While relevant to environmental sustainability, this type of life-cycle assessment may be quite unrelated to the business decisions addressed by product development groups, in which life-cycle analysis focuses on cost and performance trade-offs. For example, the increased cost of making a product more durable may be offset by reduced warranty costs. On the other hand, the durability of a product may decrease its recyclability; for example, high-strength, multilayer composite materials are difficult to reprocess for secondary uses. Ultimately, the design decision framework must have a "return on investment" perspective, with environmental costs and benefits explicitly included. Chapter 10 describes how companies can practice an integrated life-cycle management (LCM) approach, based on the *business* life cycle depicted in Figure 5.2.

Collaborating with Stakeholders

The broad life-cycle scope of DFE includes many activities that are beyond the direct control of a product manufacturer, including upstream supplier operations and downstream customer behaviors. Therefore, it is only natural that companies who are developing DFE strategies should seek the advice and assistance of external stakeholders, including business partners, customers, communities, environmental groups, and even regulatory agencies (see Chapter 3). The trend toward sustainability reporting and increased transparency of environmental strategies has paved the way for in-depth stakeholder involvement and collaboration. The following are only a few examples of such practices:

- Intel has worked closely with its customers to develop eco-efficient, reusable packaging for transporting microchips through the different stages of manufacturing and assembly (see Chapter 11).
- Coca-Cola has teamed with other food and beverage companies to encourage more rapid development and adoption of sustainable refrigeration technologies (see Chapter 15).
- General Motors works with the U.S. EPA's Green Suppliers Network to encourage environmentally beneficial practices on the part of automotive suppliers.
- FedEx Express collaborated with Environmental Defense to introduce hybrid delivery trucks in North America, helping to stimulate the market for hybrid technology (see Chapter 19).

- Dow Chemical has established community advisory panels to improve communication between plant managers and local neighboring communities.

The availability of modern information technology is a key enabler for the types of collaborations described above. Sustainability reporting has gone digital, and many companies are issuing abbreviated versions of hard copy reports, with much more detailed information published on their websites. Likewise, information about DFE goals and product designs can be shared across time and space through a variety of mechanisms, ranging from electronic meetings to sophisticated product data management systems. Global companies are learning to integrate their global design teams through electronic means and are becoming increasingly comfortable at sharing information with their supply chain partners and consultants in order to analyze existing operations and design improved solutions.

One critical stakeholder group that is often overlooked is company employees. In their haste to communicate externally, some companies may neglect internal communication and alignment. As a result, employees may be skeptical about the business relevance and authenticity of a company's sustainability commitments. For a DFE program to be credible and successful, the engagement and enthusiasm of employees—both managers and the workforce—are essential [8]. This means involving a variety of functional groups in understanding customer expectations, exploring alternative technologies, driving environmental innovations, and taking credit for the resulting competitive advantages. By promoting internal collaboration, engaging employees and recognizing their DFE-related achievements, companies can create a "multiplier effect"—DFE stories are transmitted via informal communication networks, thereby enhancing the company's reputation and helping to attract and retain talent [9].

References

1. C. Christensen, *The Innovator's Dilemma: When New Technologies Cause Great Firms to Fail* (Cambridge: Harvard Business Press, 1997).
2. J. Makower, *Strategies for the Green Economy: Opportunities and Challenges in the New World of Business* (New York: McGraw-Hill, New York, 2009).
3. National Research Council, *Improving Engineering Design: Designing for Competitive Advantage* (Washington, D.C.: National Academy Press, 1991).
4. D. Rainey, *Product Innovation: Leading Change through Integrated Product Development* (Cambridge, U.K.: Cambridge University Press, 2005).
5. A. Khurana and S. R. Rosenthal, (1997), "Integrating the fuzzy front end of new product development," *Sloan Management Review*, Vol. 38, No. 2, 1997, pp. 103–20.
6. J. Fiksel and F. Hayes-Roth, "Computer-Aided Requirements Management," *Concurrent Engineering Research and Applications*, Vol. 1, No. 2, June 1993, pp. 83–92.
7. K. Yang, *Design for Six Sigma: A Roadmap for Product Development* (New York: McGraw-Hill, 2003).
8. J. Fiksel, R. Axelrod, and S. Russell, "Inside Out: Sustainability Communication Begins in the Workplace," *green@work*, Summer 2005.

CHAPTER 6

Principles of Design for Environment

The following definition was given in Chapter 1.

> **Design for Environment** is the systematic consideration of design performance with respect to environmental, health, safety, and sustainability objectives over the full product and process life cycle.

This chapter describes seven basic principles for companies that wish to integrate DFE into their innovation practices. These are:

1. Embed life-cycle thinking into the product development process.
2. Evaluate the resource efficiency and effectiveness of the overall system.
3. Select appropriate metrics to represent product life-cycle performance.
4. Maintain and apply a portfolio of systematic design strategies.
5. Use analysis methods to evaluate design performance and trade-offs.
6. Provide software capabilities to facilitate the application of DFE practices.
7. Seek inspiration from nature for the design of products and systems.

Chapters 7 through 10 elaborate further on these principles.

Life-Cycle Thinking

DFE Principle 1. Embed life-cycle thinking into the product development process.

When considering the environmental implications of product and process design, think beyond the cost, technology, and functional performance of the design and consider the broader consequences at each stage of the value chain. This fundamental principle, life-cycle thinking, has motivated the development of life-cycle assessment and the related analysis methods described in Chapter 9. The physical product life cycle, depicted in Figure 6.1, involves five major stages: Source, Make, Deliver, Support, Recycle. These stages are consistent with logistics models broadly used in the supply chain management community [1]. The five stages encompass all of the business activities necessary to serve the market responsibly; note that some of these activities may fall outside the boundary of the enterprise:

1. **Source**—acquire the raw materials, components, energy, and services required to manufacture the product.
2. **Make**—manufacture and/or assemble the product, inspect, package, and stockpile or prepare the product for delivery.
3. **Deliver**—transport the product via distribution channels to warehouses, wholesalers, and/or retail customers.
4. **Support**—provide services to customers or users of the product, including supplies, repair, replacement, maintenance, or upgrading.
5. **Recycle**—recover used, obsolete, or defective products and extract residual value through reuse, refurbishment, or recycling.

Figure 6.1 Each life-cycle stage consumes natural resources, generates environmental impacts, and creates costs or benefits for stakeholders.

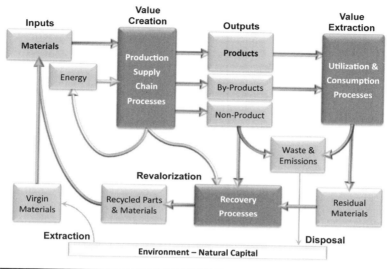

FIGURE 6.2   Overview of the life cycle for industrial materials.

In each of these stages, as described in Chapter 7, there may be positive or negative consequences in terms of financial performance, human health and safety, and environmental impacts, such as greenhouse gas emissions, biodiversity, and natural resource depletion.

To understand the environmental consequences, it is often helpful to develop a life-cycle material flow diagram, illustrated in Figure 6.2, for industrial materials such as plastics. Raw materials are transformed into products though value-adding processes, and then value is extracted when these products are utilized in industrial applications. There are several points along the life cycle where waste and emissions are generated, representing opportunities for recovery and "revalorization" of discarded materials (see Principle 4 below). For example, many companies are working to minimize their "non-product output" by recycling post-industrial scrap.

When considering environmental consequences, a firm should consider not only potential impacts on its own assets, but also the economic, environmental, health, and safety concerns of affected stakeholders, including employees, customers, suppliers, contractors, and local communities. Stakeholder expectations were described in Chapter 3, and examples of potential life-cycle considerations are shown in Table 6.1. Chapters 7 to 9 discuss approaches for measuring and analyzing both economic and environmental considerations, but social issues such as community quality of life are beyond the scope of this book. Chapter 10 provides a framework for integrating these broad considerations into business decision making.

Economic life-cycle considerations	• Total life-cycle costs incurred by the enterprise, including capital and operating costs, and impact on long-term profitability • Economic efficiency in terms of resource productivity and net energy production over the product life cycle • Total cost of ownership to the customer, including purchase or leasing of equipment and ongoing supplies or services • New jobs created both directly and indirectly via the multiplier effect • Regional economic benefits of production due to sourcing of materials, supplies, and services • Economic growth or entrepreneurship opportunities enabled by the product introduction (e.g., related professional services)
Ecological life-cycle considerations	• Energy consumption and energy efficiency of supply chain operations • Depletion of nonrenewable resources, including materials and fuels, involved in transportation, production, and distribution • Impacts upon local and regional ecosystems, including habitat integrity (e.g., wetlands), biodiversity, and disruption of natural cycles • Potential loss or degradation of agricultural lands, forests, water bodies, fisheries, or other natural resources critical to human subsistence • Airborne emissions, including hazardous air pollutants, particulates, smog-forming chemicals, and greenhouse gases • Solid or liquid waste streams associated with supply chain operations, maintenance, disposable supplies (e.g., batteries), as well as facility construction and demolition • Potential risks associated with accidental spills, leakage, fire, explosion, or other incidents that could threaten human safety or ecosystem integrity
Social life-cycle considerations	• Benefits of product or service availability upon community quality of life, including improvements in health, nutrition, education, access to resources, sanitation, mobility, and recreation • Impacts upon employees and families, including skill development, education, and personal health and safety • Potential adverse effects of new business operations and facilities upon existing cultural and community activities (e.g., traffic disruptions) • Potential impacts upon esthetics, including landscape changes, noise, odor, or other effects of industrial activities

TABLE 6.1   Potential Life-Cycle Consequences of Product or Process Design Decisions

System Perspective

DFE Principle 2. Evaluate the resource efficiency and effectiveness of the overall system.

Every product, process, or service is part of a larger economic system, so try to understand how different design choices can influence the overall environmental performance of that system. No product can be deemed "sustainable" without reference to its broader context. As suggested in Figure 6.1, the enterprise and its stakeholders all depend on natural resources for materials, energy, and ecosystem services (such as fresh water), and they all deposit wastes into the environment. The challenge of DFE is to consider the entire system and determine how the needs and expectations of these stakeholders can be met in the most resource efficient, effective, and environmentally benign manner. Therefore, the system boundaries associated with DFE are broader than those in the customary definition of a product "system." Rather than merely considering how the product is handled and used, one must consider the entire value-added chain—the "upstream" processes involved in producing the components, raw materials and energy to fabricate the product, as well as the "downstream" processes involved in its distribution, use, and disposal. One must also consider how by-products or releases from these processes may transmute, migrate, and affect humans or the environment.

Eco-efficiency is a concept that many companies have adopted as a way to capture the environmental performance of their overall production systems. The World Business Council on Sustainable Development (WBCSD), a global coalition of companies, first defined the term in its 1992 manifesto, *Changing Course* [1]. WBCSD defines eco-efficiency as "delivery of competitively priced goods and services that satisfy human needs and bring quality of life while progressively reducing environmental impacts of goods and resource intensity throughout the entire life cycle." In more concise terms, eco-efficiency means generating more value with less adverse ecological impacts. This concept is useful for several reasons:

- It combines economic value creation with environmental resource protection
- It is essentially a resource productivity measure and, so, is correlated with profitability
- It can be scaled to any system boundary, from a specific process to the full life cycle

As stated in Chapter 4, eco-efficiency makes business sense. Through eliminating waste and using resources more wisely, eco-efficient companies are able to reduce costs and become more competitive. Moreover, as environmental performance standards become commonplace, eco-efficient companies are gaining advantages in penetrating new markets and increasing their share of existing markets. Chapter 7 provides examples of eco-efficiency performance indicators that companies have used to establish design objectives and criteria. A common approach toward measuring eco-efficiency is to take the ratio of value produced, including products and by-products, to resources consumed; for example, sales per BTU of energy.

While the logic of eco-efficiency is compelling, a *caveat* is in order. If the concept is applied too narrowly, companies can simply improve eco-efficiency by gradually decreasing their waste generation or resource use, and may become complacent about incremental achievements that do not substantially benefit the overall systems in which they operate. As outlined in Chapter 1, the challenge of reversing current environmental trends is enormous. Rather than just managing resource intensity with conventional products, companies should keep in mind the "value" side of the equation, and try to design innovative, game-changing products that provide significant improvements in quality of life and/or quantum reductions in their environmental footprint.

Even more radical is the notion that designers should try to enlarge the "positive" footprint by seeking restorative and beneficial impacts upon the environment. For example, Bill McDonough and Michael Braungart have hypothesized a "nutrivehicle" that cleanses the air by trapping pollutants and captures its own emissions for conversion into useful by-products [2]. They prefer to design for *eco-effectiveness*—i.e., doing more good rather than less bad and working in harmony with natural systems. In their "cradle to cradle" model, waste becomes "food," either entering the environment as "biological nutrients" or recirculating in industrial systems as "technical nutrients." This type of transformative innovation is actually beginning to take shape in some of the *industrial ecology* examples discussed later in this chapter (see DFE Principle 7).

Indicators and Metrics

DFE Principle 3. Select appropriate metrics to represent product life-cycle performance.

To guide product development decisions, identify key environmental performance indicators and metrics that are aligned with evolving customer needs and corporate sustainability goals. As described in Chapter 5, performance measurement is a critical element of new product development because it assures that the product will meet a variety of customer requirements as well as corporate priorities and regulatory constraints. The choice of high-level environmental indicators is extremely important, in that it determines what types of signals are sent to engineering and manufacturing staff responsible for meeting operational goals. In addition, it determines the available options for communicating company performance to outside audiences. Once environmental performance goals have been expressed in terms of specific metrics (e.g. 50% solid waste reduction), the next step in product development is to decompose these metrics into quantitative parameters that can be estimated and tracked for a particular product design. This process is discussed further in Chapter 7.

A variety of environmental performance measures, including eco-efficiency metrics, are used in various industries around the world, reflecting regional and industry-specific environmental issues. Examples of different types of metrics include toxic use measures (e.g., total kg of solvents purchased per unit of production), resource utilization measures (e.g., total energy consumed during the product life cycle), atmospheric emission measures (e.g., greenhouse gases and ozone-depleting substances released per unit of production), and waste minimization measures (e.g., percent of product materials recovered at end-of-life), to name a few. A more exhaustive list is provided in Chapter 7.

Design Strategies

DFE Principle 4. Maintain and apply a portfolio of systematic design strategies.

Build upon past experiences to assemble a portfolio of design strategies that can be codified, communicated through training, and systematically applied by your design teams. This will encourage a repeatable and consistent innovation process rather than anecdotal successes based on individual ingenuity. There has been a great deal of knowledge developed worldwide about DFE strategies, including many useful design rules and guidelines, suitable for various industries and product categories. For example, Anastas and Zimmerman published a set of general principles for green engineering, shown in Table 6.2, which can be applied from the micro to the macro scale [3]. Drawing upon their work and many other sources, Chapter 8 provides a comprehensive catalogue of DFE guidelines, organized according to the four principal DFE strategies depicted in Figure 6.3. These are:

1. **Design for Dematerialization**—Minimize material throughput as well as the associated energy and resource consumption at every stage of the life cycle. This can be achieved through a variety of techniques, such as product life extension, source reduction (i.e., downsizing), process simplification, remanufacturing, use of recycled inputs, or substitution of services for products. Dematerialization represents the best opportunity for decoupling economic growth from resource consumption.

1. Designers need to strive to ensure that all material and energy inputs and outputs are as inherently nonhazardous as possible.
2. It is better to prevent waste than to treat or clean up waste after it is formed.
3. Separation and purification operations should be designed to minimize energy consumption and materials use.
4. Products, processes, and systems should be designed to maximize mass, energy, space, and time efficiency.
5. Products, processes, and systems should be "output pulled" rather than "input pushed" through the use of energy and materials.
6. Embedded entropy and complexity must be viewed as an investment when making design choices on recycle, reuse, or beneficial disposition.
7. Targeted durability, not immortality, should be a design goal.
8. Design for unnecessary capacity or capability (e.g., "one size fits all") solutions should be considered a design flaw.
9. Material diversity in multicomponent products should be minimized to promote disassembly and value retention.
10. Design of products, processes, and systems must include integration and interconnectivity with available energy and materials flows.
11. Products, processes, and systems should be designed for performance in a commercial "afterlife."
12. Material and energy inputs should be renewable rather than depleting.

TABLE 6.2   Twelve Principles of Green Engineering [3]

2. **Design for Detoxification**—Minimize the potential for adverse human or ecological effects at every stage of the life cycle. This can be achieved through replacement of toxic or hazardous materials with benign ones; introduction of cleaner technologies that reduce harmful wastes and emissions, including greenhouse gases; or waste modification using chemical, energetic, or biological treatment. Note that, while detoxification can reduce environmental impacts, it may not substantially reduce resource consumption.

3. **Design for Revalorization**—Recover residual value from materials and resources that have already been utilized in the economy, thus reducing the need for extraction of virgin resources. This can be achieved by finding secondary uses for

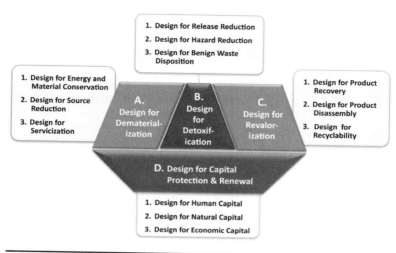

1. Design for Release Reduction
2. Design for Hazard Reduction
3. Design for Benign Waste Disposition

1. Design for Energy and Material Conservation
2. Design for Source Reduction
3. Design for Servicization

A. Design for Demateriali-zation

B. Design for Detoxif-ication

C. Design for Revalor-ization

1. Design for Product Recovery
2. Design for Product Disassembly
3. Design for Recyclability

D. Design for Capital Protection & Renewal

1. Design for Human Capital
2. Design for Natural Capital
3. Design for Economic Capital

FIGURE **6.3** Design for Environment: four principal strategies.

discarded products, refurbishing or remanufacturing products and components at the end of their useful life, facilitating disassembly and material separation for durable products, and finding economical ways to recycle and reuse waste streams. Industrial ecology approaches fit within this strategy and are discussed separately below. Revalorization goes hand in glove with dematerialization, since repeatedly cycling materials and resources within the economy reduces the need to extract them from the environment.

4. **Design for Capital Protection and Renewal**—Assure the availability and integrity of the various types of productive capital that are the basis of future human prosperity. Here "capital" is used in the broadest sense. *Human capital* refers to the health, safety, security, and well being of employees, customers, suppliers, and other enterprise stakeholders [4]. *Natural capital* refers to the natural resources and ecosystem services that make possible all economic activity, indeed, all life. *Economic capital* refers to tangible enterprise assets including facilities and equipment, as well as intellectual property, reputation, and other intangible assets that represent economic value (see Chapter 4). Capital protection involves maintaining continuity and productivity for existing capital, while renewal involves restoring, reinvesting, or generating new capital to replace that which has been depleted. Thus renewal may include attracting new talent, revitalizing ecosystems, and building new factories.

Analysis Methods

DFE Principle 5. Use analysis methods to evaluate design performance and trade-offs.

Develop and apply rigorous quantitative tools to analyze the environmental, economic, and other consequences of design decisions and to weigh the trade-offs of alternative choices. Every product requirement must have a corresponding verification method, whether through physical tests or numerical calculations. Design teams will systematically apply these methods to assess the acceptability of a design and the degree of improvement expected with respect to product cost, quality, environmental performance, and other metrics of interest. During the product development cycle described in Chapter 5, analysis methods are needed for a number of different purposes:

- **Screening** methods are used to narrow design choices among a set of alternatives. These can range from material checklists to life-cycle environmental footprint indicators.

- **Performance assessment** methods are used to estimate the expected performance of designs with respect to particular indicators. These include risk assessment, life-cycle assessment, and life-cycle cost accounting.

- **Trade-off analysis** methods are used to compare the expected cost and performance of several alternative design approaches. These can range from scoring matrices to quantitative cost-benefit analysis.

As described in Chapter 9, there are a variety of techniques and tools available to support the above needs, including tangible evaluation, qualitative assessment, quantitative environmental analysis, risk analysis, and financial analysis. Some of these techniques are very simple, while others may require the use of sophisticated computer-based models.

Information Technology

DFE Principle 6. Provide software capabilities to facilitate the application of DFE practices.

Enable the systematic practice of DFE by integrating the relevant metrics, guidelines, and analysis methods into the computing environment that is routinely used by design teams. As mentioned in Chapter 5, concurrent engineering teams can benefit from an integrated software environment. The information technology infrastructure for product development has advanced rapidly, and sophisticated commercial tools are available for key tasks, such as requirements management, product data management, computer-aided engineering and design (CAE/

CAD), and performance simulation. While some environmentally-oriented tools are available, they are nowhere nearly as well developed, and many companies have resorted to building their own. Several types of information technology can be helpful for DFE:

- **Online advisory tools** such as context-sensitive "expert systems" can help designers and engineers to benefit from accumulated wisdom about design guidelines, useful technologies, or lessons learned from previous successes and failures.

- **Computer-based modeling** can be useful for assessing DFE consequences because of the complexity of the life-cycle systems associated even with simple products. Examples include physical modeling of environmental emissions, assessment of carbon footprints, and estimation of end-of-life recovery costs (see Chapter 9).

- **Linkage to the CAE/CAD framework** can encourage regular use of DFE tools by enabling convenient and "seamless" application as part of the normal workflow. To avoid the "islands of automation" syndrome, DFE tools ideally should be compatible with the host environment and facilitate data sharing and interoperability with other CAE/CAD tools.

Although this is an important and rapidly-developing topic, a full discussion of information technology and software tools for DFE is beyond the scope of this book.

Learning from Nature

DFE Principle 7. Seek inspiration from nature for the design of products and systems.

Try to emulate the sustainability and resilience of natural systems in designing product and process technologies as well as collaborative industrial networks. Many product innovations have been based on observation of natural systems, where elegant solutions have evolved over millions of years. Janine Benyus coined the term *biomimicry* to describe the practice of adapting nature's designs to solve human problems [5]. For example, Velcro® was inspired by the structure of insects' feet, while new designs for turbine blades are mimicking the flippers of whales. Natural systems manage to accomplish physical tasks and chemical transformations without the excessive baggage of industrial supply chains. One company that applied biomimicry successfully is Interface, which designed a new brand of carpet tiles to mimic the random patterns of natural ground cover, thus facilitating installation and replacement. Biotechnology companies such as Dow Agro-Sciences often turn to natural organisms for sources of innovation (see Chapter 13).

Beyond discovering innovative technologies, another valuable lesson that humans can learn from nature is how to organize a complex web of material and energy transformations in a way that generates absolutely no waste. In ecological systems, food webs have evolved so that every bit of biomass is consumed by some organism occupying a particular niche. Even the *detritus* that falls on the forest floor provides nutrients for microorganisms, which in turn enrich the soil. The concept of *industrial ecology** suggests that industrial systems can operate in a similar fashion—converting waste materials from production or consumption activities into "food" for industrial processes. As shown in Figure 6.4, every company is part of a broader network of material and energy flows that provide pathways from natural resources to consumers of goods and services. Rather than discarding their wastes, it is likely that companies, as well as consumers, could discover alternative uses in their own supply chain or in other industry sectors, and thus convert them into by-products. To achieve this type of system-level innovation will require new forms of collaboration among different industries, as well as support from state and local regulatory authorities and other stakeholders. However, today the environmental and economic motivations for seeking industrial ecology opportunities are stronger than ever. Around the world, some promising examples have already emerged, as discussed in Section C.3, Design for Recyclability, in Chapter 8.

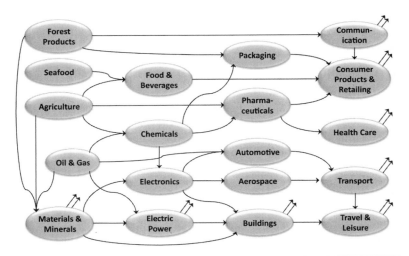

Figure 6.4   Industrial ecosystems convert natural resources into economic goods and services.

---

*Other metaphorical terms related to this concept include industrial *metabolism* and industrial *symbiosis*.

From Principles to Practices

The above principles are elaborated in more detail in the next several chapters, which include many examples of progressive industry practices:

- Chapter 7 describes the use of DFE performance indicators and metrics (Principles 2 and 3).

- Chapter 8 provides a catalogue of DFE design strategies and guidelines (Principles 4 and 7).

- Chapter 9 discusses a variety of DFE analysis methods and tools (Principle 5).

- Chapter 10 describes the practice of integrated life-cycle management (Principle 1).

Part 3 of this book shows how these DFE principles have been put into practice by a wide range of companies in different industries, including complex assembly industries such as electronics (Chapter 11) and transportation systems (Chapter 12); process-intensive industries such as chemicals (Chapter 13), materials (Chapter 17), and electric power (Chapter 18); high-volume discrete manufacturing industries such as pharmaceuticals (Chapter 14), food and beverages (Chapter 15), and consumer products (Chapter 16); and, finally, service industries such as retailing, package delivery, and tourism (Chapter 19).

References

1. See the supply chain operations reference model (SCOR) developed by the Supply Chain Council, www.supply-chain.org.
2. W. McDonough and M. Braungart, *Cradle To Cradle: Remaking the Way We Make Things* (New York: North Point Press, 2002).
3. P. T. Anastas and J. B. Zimmerman, "Design through the 12 Principles of Green Engineering," *Environmental Science & Technology,* March 1, 2003.
4. Also important is the preservation of *social capital*; namely, the institutions, relationships, and norms that underpin human society, including bonds of mutual trust. See the World Bank discussion: go.worldbank.org/C0QTRW4QF0.
5. J. Benyus, *Biomimicry: Innovation Inspired by Nature* (New York: William Morrow, 1997).

CHAPTER 7

Performance Indicators and Metrics

DFE Principle 3. *To guide product development decisions, identify key environmental performance indicators and metrics that are aligned with evolving customer needs and corporate sustainability goals.*

Measure Up!

Companies that commit to corporate responsibility and environmental sustainability goals need to demonstrate measurable progress. Quantitative indicators are important, not only for stakeholder communication, but also for motivating internal process improvement. Therefore, they should reflect creation of customer and shareholder value as well as value to society. As described in Chapter 5, environmental performance indicators and metrics are necessary to support goal-setting, monitoring, and continuous improvement in the design of products and processes.

Performance measurement has become a *mantra* of modern management, embraced by nearly every major school of thought including Six Sigma, total quality management, business process reengineering, high-performance teams, and value-based management. "You can't manage what you can't measure" is an inescapable fact. Moreover, assigning accountability to individuals and teams based on specific performance targets generally leads to significant improvements without the need for close oversight.

As discussed in Chapter 3, with increasing commitments to corporate responsibility, the business community has adopted global environmental management system standards such as ISO 14001, which specify performance measurement as a required element. The emergence of the Global Reporting Initiative (GRI) and other sustainability reporting schemes has placed renewed emphasis on the selection, monitoring, and verification of environmental performance indicators and metrics. As time passes, the only way for companies to demonstrate continuing progress in environmental performance is through innovation—quantum improvements in the design and delivery of products, processes, or services. Hence DFE has become a necessary competency for environmental performance improvement.

As illustrated throughout this book, environmental performance is just one aspect of enterprise value creation. Instead of viewing environmental management as an unavoidable cost of doing business and treating it as an overhead expense, leading companies are beginning to view environmental management as an important business function, and to allocate funds to this function based on its contribution to corporate value. This new perspective implies two major changes in the business processes of an enterprise:

1. Environmental performance metrics and assessment methods need to be integrated into engineering practices, and

2. Accounting systems need to be developed that explicitly recognize and track environmental costs and benefits.

This chapter describes environmental performance measurement frameworks, indicators and metrics, particularly as they relate to the practice of DFE. One of the key challenges in this field is to incorporate a life-cycle view of environmental performance into metrics that can be easily computed and tracked. The methodologies of life-cycle assessment and life-cycle cost accounting, discussed in Chapter 9, are essential for this purpose.

Why Measure Environmental Performance?

Corporations, especially those that are publicly held, are accountable for financial performance, as measured by earnings per share, return on net assets, or other indicators. Shareholders and analysts have become acutely aware of small changes in performance and may often overreact to unexpected slippages. As a consequence, managing the productivity of existing resources—human, technological, and physical—has become an important priority. During mergers or economic downturns, many companies go through a period of "re-engineering" or "rightsizing" that involves consolidation of resources in order to improve performance.

However, as shown in Chapter 4, there are nonfinancial or intangible considerations that may be just as important as the financial indicators of performance. Businesses large and small are motivated to move beyond compliance and embrace corporate responsibility, not just because of external pressures, but also because it is in the interests of their owners and shareholders. In particular, environmental sustainability has become a key factor in long-term competitiveness. And for those companies that commit to sustainability principles and wish to integrate them into their business practices, it is essential to have an effective and credible performance measurement system. As illustrated in Figure 7.1, there are three important facets of performance measurement in the context of product life-cycle management:

1. Measuring continuous improvement in performance over time
2. Communicating with customers and stakeholders about product advantages
3. Benchmarking of performance among comparable products and processes

Since measurement is an act of communication, companies should strive for clarity in both the purpose and meaning of this communication. In selecting performance metrics for product development, they should ask the following types of questions:

- **Clarity of purpose**—Who is the intended audience, and why are we communicating with them? Are we providing information to customers that will sway their purchase decision? Are we reporting to external stakeholders in order to demonstrate environmental leadership? Are we developing internal insights to guide material, technology, or process selection? Or, are we seeking to benchmark our performance with respect to competing products or similar processes in other industries?

- **Clarity of meaning**—What do the selected performance indicators and results signify to the audience? What is the intended message? Are the results relevant to the interests of the audience? Are the results expressed in transparent and

FIGURE 7.1 Facets of performance measurement.

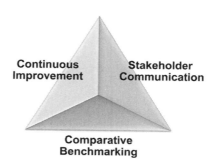

meaningful terms, or are they obscured by numerical manipulation and scientific jargon? Will they withstand the test of changing market conditions over an extended time period?

Because of these complexities, selection of an appropriate measurement framework and specific performance indicators has become an important strategic decision. Product development metrics need to be consistent with overall enterprise sustainability metrics. Once these choices are made and performance goals are announced, a company must live with them for years to come.

Enterprise Performance Frameworks

Most U.S. companies have moved beyond strictly financial measures and adopted more diverse performance metrics. For example, the **Balanced Scorecard** framework is widely used for balancing financial indicators with nonfinancial or intangible value indicators [1]. This framework defines four major performance quadrants: Financial Performance, Operational Excellence, Customer/Stakeholder Relationships, and Learning and Growth. There are important environmental indicators associated with each of these quadrants; for example, a company might define the following indicators:

1. Financial—life-cycle cost per unit of product, including emission costs or credits
2. Operational—process eco-efficiency in terms of material and energy use
3. Relationships—community and stakeholder perceptions of responsibility
4. Learning and growth—alignment and training toward the sustainability vision

Less well-known is the **Eco-Compass** framework, which provides a visual method for overlaying and comparing the multidimensional performance of alternative products or other entities [2]. It uses a circular array of performance indicators corresponding to different compass points, and indicator values can be shown as polygons of different colors (see Figure 7.2). This approach, similar to a "radar chart," can be adapted to virtually any collection of sustainability or other types of indicators.

With this type of multidimensional framework, integrating the different aspects of performance becomes more challenging. Leading corporations have learned some key lessons about performance measurement and achieving continuous improvement:

- The important dimensions of performance, financial or otherwise, need to be understood and characterized in terms of quantifiable metrics.

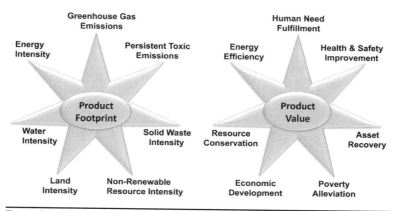

FIGURE 7.2 Examples of environmental "footprint" and "value" aspects.

- Each individual and team in the company must have a means of relating their own objectives to the achievement of these metrics, so that they receive correct signals regarding their contributions to shareholder value.

- A system must exist for tracking the achievement of performance targets and providing appropriate recognition to those responsible.

- As decisions are made at all levels of the company, a means must exist for understanding their impacts upon the affected performance metrics.

Thus, it is not sufficient to select and announce the metrics of interest—they must be made operational through deliberate intervention into the organizational culture, processes, and reward systems. Many companies define a small set of *key performance indicators* (KPIs), such as revenue growth, that evaluate progress toward their critical business goals. Having environmental indicators included in these KPIs can provide a powerful incentive for genuine progress. To accomplish this, companies are augmenting the conventional indicators favored by external stakeholder-driven organizations. As discussed below, leading companies such as 3M, DuPont, and General Motors have devised internal measurement tools that are *value-driven*, reflecting the positive impact of sustainability improvements upon shareholder value.

The principles of performance measurement are set forth in environmental management systems standards such as ISO 14001 and EMAS (see Chapter 3). They require that the organization evaluate its environmental "aspects," that performance goals be established by senior management, and that progress toward these goals be monitored using an environmental performance evaluation system.

It is important, however, to avoid the trap of treating environmental sustainability as a separate dimension of performance. The "triple bottom line" metaphor encourages an unfortunate separation of sustainability into environmental, economic, and social dimensions. When measuring environmental performance, the majority of manufacturing firms tend to focus on conventional environmental, health and safety indicators (emissions, incidents, etc.) associated with manufacturing operations, and do not address shareholder or stakeholder value creation. Yet, the three dimensions are closely intertwined; for example, waste elimination efforts may simultaneously result in environmental enhancement, operating cost reduction, and human health improvement.

> THE ENVIRONMENTAL, ECONOMIC, AND SOCIAL DIMENSIONS OF SUSTAINABILITY ARE CLOSELY INTERTWINED.

Ideally, sustainability goals should be woven into the existing managerial decision making, monitoring, and reward systems that drive continuous improvement. This is why the GRI Guidelines mentioned in Chapter 3, while helpful for public reporting of sustainability performance, are less useful for internal performance management. For cross-functional product development teams, careful selection of environmental performance metrics will help assure that the product design strategy is consistent with enterprise goals.

A powerful and practical indicator concept that combines environmental performance with economic value creation is *eco-efficiency*, first introduced by the World Business Council for Sustainable Development. Eco-efficiency can be defined as the ratio of product/service value delivered to the environmental burdens of product/service creation or consumption [3]. Eco-efficiency metrics are commonly used in product development; for example:

- General Motors has measured eco-efficiency in terms of "resource productivity," e.g., number of vehicles produced per kilowatt-hour of energy used, gallon of water used, or ton of greenhouse gas emitted.

- Sony calculates the economic value added over the lifetime of its batteries and divides it by the sum of the nonrecyclable material consumed and energy used in production.

- DuPont encourages its businesses to substitute intellectual capital for physical capital by measuring product performance in terms of shareholder value added per pound of product sold; this favors technology-based businesses over high-volume commodities.*

---

*Not surprisingly, DuPont divested its petrochemical business in 1999 and its nylon business in 2002.

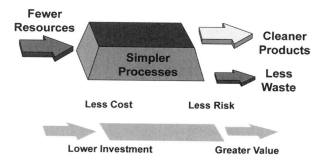

FIGURE 7.3 Eco-efficiency improvement increases the value delivered per unit of resource input.

Eco-efficiency indicators are attractive because they combine environmental and economic aspects into a single ratio that is positively correlated with profitability and shareholder value (see Figure 7.3). There are a variety of practical methods to improve eco-efficiency, including reducing material and energy intensity, reducing toxics dispersion, enhancing material recyclability, using renewable resources, extending the durability of products, and increasing service or knowledge intensity. In effect, eco-efficiency metrics encourage product designers to minimize their resource consumption while maximizing value creation; in other words, to "do more with less" [4].

Identifying Aspects and Indicators

The foundation of any environmental performance measurement system is identifying what aspects of performance the company wishes to emphasize. Since this is a voluntary practice, companies have great latitude in what aspects they identify and how they assign priorities. While stakeholder dialogue is an important input to this selection process, it needs to be coupled with an understanding of opportunities for value creation. Figure 7.2 illustrates some of the typical aspects that companies have used to characterize the environmental footprint of their products or services. Note that footprint reduction is only half the story. The other half relates to value creation—the delivery of human or environmental benefits as a consequence of DFE and other sustainable business practices. Examples of both footprint and value metrics corresponding to these indicators are provided in the next section.

As discussed in Chapter 4, companies that adopt a value-driven approach try to focus on those environmental aspects that are well matched to their core competencies and business strategy [5]. For example, Procter & Gamble (P&G) has built its sustainability efforts around two major themes:

1. Water—since 85% of P&G products involve household water use, the company has focused on how it can enhance both water conservation and water quality.
2. Health and hygiene—the company has sought to enhance the global contributions of its products to cleanliness, sanitation, health care, infant care, and education.

Once a company has identified the environmental aspects to be measured, the next step is to select measurable performance indicators. It is common to distinguish between *indicators* (which define *what* is to be measured) and *metrics* (which define *how* it will be measured). Thus, an indicator of environmental improvement might be "reduced energy consumption," and associated metrics might include:

- BTU/year or BTU/ton produced (for manufacturing)
- Kilowatt-hours consumed per operating year (for product design)
- Total net energy consumed per unit over the product life cycle

As discussed in Chapter 3, different stakeholder groups have a variety of differing concerns and will respond to different types of indicators. The choice of appropriate performance indicators will usually differ depending upon the intended communication purpose. For example, in measuring energy efficiency, a company might choose to focus on operational measures, such as *BTU of fuel per ton of product* for product benchmarking purposes, while reporting to stakeholders in terms of *percent annual decrease in net energy consumption.*

An important consideration in choosing performance indicators is selection of goals and targets. While it is not essential to announce targets for all sustainability indicators, stakeholders increasingly expect corporations to commit to specific, measurable goals. Some leading companies have established aggressive, long-term goals that "stretch" the organization (e.g., DuPont's well-known Goal of Zero for injuries, incidents, and waste), while setting realistic annual milestones to track incremental progress.[†]

The following is a list of *selection criteria* that can be used to choose sustainability performance indicators. The set of indicators should be

- Relevant to the interests of the intended audiences, including enhancement of social and environmental conditions for

---

[†]One popular term for "stretch" goals is Big Hairy Audacious Goals (BHAGs).

concerned stakeholders, and advancement of business interests for shareholders and management

- Meaningful to the intended audiences in terms of clarity of indicator definition, comprehensibility, and transparency

- Objective in terms of measurement techniques and verifiability, while allowing for regional, cultural, and socio-economic differences

- Effective for supporting benchmarking and monitoring over time, as well as deciding how to improve performance

- Comprehensive in providing an overall evaluation of the company's products and services, and recognizing issues that can influence supplier or customer sustainability performance

- Consistent across different sites or facilities, using appropriate normalization and other methods to account for the inherent diversity of businesses

- Practical in allowing cost-effective, non-burdensome implementation and building on existing data collection where possible

While every indicator need not satisfy all of these criteria, a credible portfolio of sustainability indicators should have the above characteristics. The most effective performance measurement programs are those that focus upon a small number of quantifiable KPIs covering the key aspects of sustainability. Based on generally accepted accounting principles, an overarching criterion in selection of indicators is "materiality"—the significance of performance results for purposes of decision making by management, shareholders, and other stakeholders.

Finally, companies should strive to use a combination of leading and lagging indicators. *Lagging* indicators, also referred to as *outcome* indicators, measure the actual results (e.g., emission reductions) attributable to operational improvements. While such indicators are more meaningful to external stakeholders, they represent a retrospective view of performance and do not provide managers with foresight about future performance expectations. It is important to couple lagging indicators with *leading* indicators, which measure internal business process improvements that are expected to drive future performance. One example is Motorola's success in reducing loading-dock occupational injuries (a lagging indicator) by discovering that the root cause was defective wooden pallets. Tracking and reducing pallet defects (a leading indicator) not only reduced injuries by 60%, but also enabled the company to save over $5 million annually in Workmen's Compensation, transportation, handling, and waste disposal costs [6].

Choosing Environmental Performance Metrics
Relationship of Metrics to Indicators

In the context of DFE, environmental performance metrics are quantitative parameters used to measure design improvement with respect to environmental goals. Because of their fundamental role in the development process, metrics are essential to the successful practice of DFE.

As mentioned above, any environmental indicator (such as energy usage) can correspond to a variety of metrics. Examples of environmental metrics that can be used to establish product or process design objectives include the following:

Energy usage metrics
- Total energy consumed during the product life cycle
- Renewable energy consumed during the life cycle
- Power used during operation (e.g., for electrical products)

Water usage metrics
- Total fresh water consumed during manufacturing
- Water consumption during product end-use (e.g., for laundry products)

Material burden metrics
- Toxic or hazardous materials used in production
- Total industrial waste generated during production
- Hazardous waste generated during production or use
- Air emissions and water effluents generated during production
- Greenhouse gases and ozone-depleting substances released over life cycle

Recovery and reuse metrics
- Product disassembly and recovery time
- Percent of recyclable materials available at end-of-life
- Percent of product recovered and reused
- Purity of recyclable materials recovered
- Percent of recycled materials used as input to product

Source volume metrics
- Product mass (for specified functionality)
- Useful operating life
- Percent of product disposed of or incinerated
- Fraction of packaging or containers recycled

Economic metrics
- Average life-cycle cost incurred by the manufacturer
- Purchase and operating cost incurred by the customer
- Cost savings associated with design improvements

Value creation metrics
- Utilization of renewable resources
- Avoidance of waste or pollution
- Reduction in energy requirements
- Human health and safety improvement
- Fulfillment of human needs (nutrition, mobility, etc.)
- Restoration or reuse of nonproductive assets
- Enhancement in environmental quality, land use, or biodiversity
- Enhancement in community quality of life
- Enhanced economic development and job creation
- Improvement in customer environmental performance

At the aggregate level, these metrics represent the overall performance of the manufacturing enterprise; they are sometimes called *primary* or *high-level* metrics. Typically they are driven either by a fundamental customer need or by important internal constraints (e.g., process capability). The relationship of primary metrics to product goals and objectives is illustrated in Table 7.1.

Primary metrics can be used to establish measurable overall objectives for a product development team. However, in order to be useful for quality improvement, primary metrics generally need to be decomposed into *operational metrics,* which represent observable and controllable measures associated with a product or process. These operational metrics can be used not only to monitor continuous improvement but also to develop incentives and reward systems for individual employees or teams.

Operational metrics can be defined through a business model that relates the performance of company assets and functional activities (e.g., fuel procurement) to the environmental performance of the company as a whole (e.g., hazardous air emissions). In particular, *product* metrics become operational when they are associated with a specific feature, module, or component of a design, and can, therefore, be estimated, tested and verified. Similarly, operational *process* metrics represent observable and verifiable measures associated with particular operations and business processes (e.g., number of defects per million units produced). Figure 7.4 illustrates how operational metrics can be derived from primary metrics related to recyclability of durable goods.

As mentioned above, there are several alternative uses for environmental performance metrics, and different types of metrics may be preferable for each use:

- *Performance tracking* metrics are used internally to increase productivity and to assure that both product and enterprise objectives are fulfilled.
- *Decision-making* metrics are used to evaluate competing options (e.g., product designs) when performance factors in addition to cost must be considered.
- *External reporting* metrics are used to communicate improvements in product or enterprise performance to customers and stakeholders.

Environmental performance measurement systems may address any or all of these uses. Because of the obvious data management challenges, modern information technology is an essential tool for

Goals	Examples of Metrics	Examples of Specific Objectives
Reduce or eliminate waste	• lb. of emissions over the life cycle • % of product weight disposed of in landfills	• Reduce life-cycle emissions by 30% annually • Reduce sol' * waste disposed c. to 1 lb. per product unit
Develop "green" recyclable products	• % of product weight recovered & recycled • toxic or hazardous constituents	• Achieve 95% recycling • Eliminate use of brominated compounds
Reduce life-cycle cost of product	• manufacturing cost • distribution and support cost • end-of-life cost	• Reduce total life-cycle cost to $7,500 per product unit • Reduce end-of-life cost (or increase value) by 20%
Cost of ownership for customers	• annualized purchase and operating cost ($)	• Must be less than $500 per year
Conserve energy consumption over the life cycle	• total energy (BTU) to produce one unit • average power use (for devices)	• Reduce to 1000 BTU • Reduce by 10% annually • Power less than 30 watts
Conserve natural resources by raising the recycled content	• % by weight of product materials from sources	• Achieve 20% or greater total recycled content • Achieve 30% recycled plastics

TABLE 7.1   Selected Environmental Goals and Corresponding Metrics

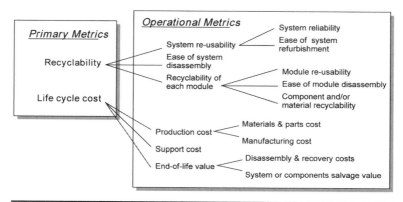

Figure 7.4 Decomposition of primary metrics into operational metrics related to product recycling.

implementing, computing, tracking, and converting these metrics to meet various needs.

Categories of Metrics

Environmental performance metrics can be classified according to the following three distinctions:

Qualitative vs. Quantitative

Qualitative metrics are those that rely upon semantic distinctions based on observation and judgment. While it is possible to assign numerical values (or scores) to qualitative metrics, such numbers have no intrinsic significance. An advantage of qualitative metrics is that they impose a relatively small data collection burden and are easy to implement. However, a disadvantage is that they implicitly incorporate subjective information and, therefore, are difficult to validate.

Quantitative metrics are those that rely upon empirical data and derive numerical results that characterize performance in physical, financial, or other meaningful terms. An example is the Toxic Release Inventory (TRI) system mandated by U.S. EPA. The advantage of quantitative metrics is that they are objective, meaningful, and verifiable. However, a potential disadvantage is that the required data may be burdensome to gather or simply unavailable. Moreover, there are some important environmental attributes, such as "brand perception" or "environmental commitment," which are inherently qualitative and cannot be precisely quantified.

Absolute vs. Relative

Absolute metrics are those that are defined with respect to a fixed measurement scale. An example is "total annual hazardous waste generated."

Relative metrics are those that are defined with respect to another metric or variable. A common approach is to use *time-based relative metrics*, i.e. those which compute the change in a particular quantitative metric over a given time period; for example "percent reduction from 2006 to 2007 in total hazardous waste generated per unit produced." Another common approach is to use "intensity" metrics, where flows of resources or emissions are *normalized* by total sales or production volume; for example, "total hazardous waste generated per unit produced."

Many stakeholder groups advocate the use of absolute metrics because intensity metrics reflect efficiency of production rather than actual environmental impacts. On the other hand, the use of absolute metrics may lead to inappropriate comparisons, whereas relative metrics are generally less biased by differences in the organization characteristics. For example, the largest companies in a given industry will typically be the largest emitters of airborne pollutants, even though their emission intensity may be significantly lower than others. Also, if sales happen to decline due to cyclical changes, absolute environmental performance appears falsely to improve. The examples in Part 3 show that companies use a mixture of absolute and relative metrics and, in some cases, report both types of metrics.

Source vs. Impact

With regard to environmental performance, *source* metrics are those that address the presumed root causes or origins of environmental consequences associated with an organization's activities. An example is the Toxic Release Inventory (TRI) mentioned in Chapter 3, which measures the quantity of toxic materials released at a given site. An advantage of source metrics is that they are both readily observable and controllable. A disadvantage is that they are an indirect indicator of potential impacts and generally ignore differences in fate, transport, exposure, and effect pathways among different organizations.

Impact metrics are those that address the actual environmental consequences which may result from an organization's activities. An example is the use of exposure assessment and dose-response assessment to calculate the "increased cancer risk in the exposed population." While impact metrics have the obvious advantage of directly addressing the impacts of concern, the development of environmental impact metrics is generally challenging due to the technical and statistical uncertainties involved in both assessing impacts and attributing them to specific sources (see Chapter 9).

In practice, the most efficient means of performance measurement is to select company-specific indicators (typically source-oriented) which are believed to be correlated with broad categories of environmental impacts. A common example is the measurement of airborne emissions of suspected toxic substances. Although the magnitude (or even existence) of the impacts may be speculative, the use of a source metric allows companies and regulatory agencies to establish clear targets for improvement.

Example: As part of its pioneering 3P initiative described in Chapter 10, Pollution Prevention Pays, 3M developed a measure of eco-efficiency called a *waste ratio* that is useful for application to manufacturing facilities [7]. It is calculated as follows:

$$\text{Waste ratio} = \frac{\text{Waste}}{\text{Product} + \text{By-product} + \text{Waste}}$$

This is a *quantitative relative source* metric, since waste volume is measured relative to the total material output of the plant. In order to minimize this ratio, engineers can either convert waste to useful by-products or reduce the waste generated.

Applicability and Scope of Metrics

Environmental metrics ideally should be assessed with respect to the life cycle of the product or process being developed. Table 7.2 illustrates how various types of metrics are typically related to life-cycle stages. Each row represents a class of primary environmental metrics; the arrows represent the direction of desired improvement (up or down). A star in a given cell indicates that the corresponding metric is *relevant* to the life-cycle stage.

In many cases, practical limitations of data resources or methodology may hinder the ability of a development team to evaluate all of the relevant cells. In other cases, companies may wish to exclude

	Materials	**Fabrication**	**Transport**	**End Use**	**Disposal**
Energy usage ↓	★	★	★	★	★
Water usage ↓	★	★		★	
Source volume ↓	★	★	★	★	★
Recycling & reuse ↑		★		★	★
Waste & emissions ↓	★	★	★	★	★
Recycled materials ↑	★	★			

TABLE 7.2   Relevance of Environmental Performance Metrics to Various Product Life-Cycle Stages

certain life-cycle stages from consideration because they are not relevant to business decision making (see Chapter 10). Therefore, the intended scope and rationale for metrics should always be clarified. For example, rather than speaking of "energy use reduction" we should specify "reduction in energy use during manufacturing and distribution" or "reduction in power consumption during product end use".

Another important consideration in the selection of metrics is recognizing the interactions among environmental and other product and process metrics. For DFE to be truly integrated with product development, engineers must understand the synergies and trade-offs among environmental performance attributes and other design attributes. Specifically, product development teams should analyze the pairwise interactions between environmental metrics and those that relate to cost, performance, and customer satisfaction. In the "house of quality" approach discussed in Chapter 5, this is equivalent to examining the "roof" of the house.

Table 7.3 illustrates interactions that might be identified for a hypothetical consumer product such as a cleaning agent. Again, each row represents a class of primary environmental metrics; the arrows represent the direction of desired improvement (up or down). In this case, interactions are shown as either favorable (star) or potentially unfavorable (question mark). Note that it is also possible to indicate the degree or strength of interaction in qualitative terms. When applied to specific products, this type of interaction matrix reveals some important insights. For example, one can readily see the synergies between life-

	Performance	Cost	Safety	Convenience	Aesthetics
Energy usage ↓		★			
Water usage ↓		★		★	
Source volume ↓	?	★	?	★	
Recycling & reuse ↑		★		?	
Waste & emissions ↓	★	★	★	★	
Recycled materials ↑		?			?

TABLE 7.3   Example of Interactions among Environmental Quality and Other Customer Benefits for a Consumer Product

cycle cost reduction and a number of environmental performance metrics. Likewise, one can identify environmental improvement options which may compromise other desirable features (e.g., recycled materials may have poorer aesthetic qualities than virgin materials).

Aggregation and Scoring Schemes

A common practice in environmental performance measurement is to use scoring or weighting techniques to aggregate together various specific performance measures. For example, a frequently-used approach to circumvent the challenges of environmental impact analysis is to rely upon source measures but to assign them priorities or weights based on an assessment of their relative importance, taking into consideration the available information about environmental impact pathways. Scoring schemes may be adopted to reflect a variety of different considerations, including

- Expectations of different stakeholder groups (e.g., customers vs. community)
- Relative importance of environmental impacts (e.g., human health vs. ecology)
- Internal business priorities (e.g., strategic advantage)

For example, the ISO 14040–43 guidelines for life-cycle assessment include an intricate scheme for quantifying the impacts of substance emissions: classification of substances according to their effects (e.g., carcinogens), characterization of their collective impacts based on environmental exposure and effect modeling, normalization of the effects relative to a benchmark, and weighting of effect scores based on relative importance (see Chapter 9).

While the aggregation of performance metrics may be desirable for purposes of simplifying decision making, there are a number of problematic aspects to the use of scoring schemes for environmental metrics:

- There are usually implicit policies and value judgments embedded into the weighting system that are not apparent, yet may skew the results in unintended ways.
- Performance metrics are much more meaningful when considered separately, whereas the significance of improvement in an aggregated score is unclear.
- Aggregated measures can invite comparisons among dissimilar products, facilities, or activities, while concealing important differences between them.

By applying good practices, it is possible to avoid some of the above abuses or pitfalls; for example, the traceability hierarchies used in requirements management provide transparency for purposes of product design modification (see Chapter 5). A measurement system that captures the sources and rationales for all aggregated scores will allow "drilling down" for purposes of examining contributions of individual components to overall performance. In general, no universal scoring scheme will suit the needs of diverse organizations, and each company should develop a scheme that suits its business characteristics and priorities.

Future Challenges and Opportunities

Environmental performance measurement remains a new and challenging arena, requiring significant commitment on the part of senior management. The first hurdle that a company faces is establishing a set of performance goals and indicators that are consistent with its business philosophy. If common indicators can be adopted within an industry, it will be easier for companies to benchmark their performance and to communicate improvements to stakeholders. Such industry-wide consensus is most easily reached on regulatory-driven indicators (e.g., airborne emissions). However, multinational companies must cope with international variations in regulations and standards; for example, the United States has stricter standards than Europe regarding incineration of solid wastes. Companies need to decide whether their sustainability policies and goals will be applied uniformly worldwide, or adjusted to reflect local constraints.

The breadth and complexity of environmental sustainability issues makes it difficult to reduce the number of indicators, as evidenced by the GRI Guidelines. It is tempting to try to aggregate various performance indicators into an environmental index, which could be used as a simple decision-making tool. However, as mentioned above, aggregation raises a number of practical challenges. Efforts to weight the indicators based on relative importance can potentially introduce controversial value judgments. Indeed, performance metrics are much more meaningful when considered separately, whereas the significance of improvement in an aggregated index is unclear. Moreover, an aggregated index may invite comparisons among dissimilar products or facilities, while concealing important differences.

Once corporate goals and indicators are selected, companies then face the challenge of implementing environmental performance measurement throughout their operations. In order to make genuine progress, a company will need to establish specific performance metrics that can be "flowed down" to their operating managers, along with targets for improvement, accountabilities, and programs of action. In addition, a methodology needs to be established for *normalization* (e.g., per unit of production) and "roll-up" of performance improvement results to measure overall company performance. Implementation of such schemes will often encounter barriers, and companies may require organizational alignment efforts to build understanding and commitment among the workforce. One of the most effective means of assuring progress toward sustainability goals is to incorporate sustainability KPIs into the executive compensation scheme.

Finally, it is helpful for companies to work collaboratively with NGOs and interested parties to establish national or even worldwide standards for environmental performance measurement and improvement. The collaboration between the World Business Council for Sustainable Development and the World Resource Institute on developing a greenhouse gas emissions inventory protocol is an excellent example of such an effort (see Chapter 9). In the United States, the Electric Power Research Institute (EPRI) has developed guidelines for environmental performance measurement and worked with member utility companies to help implement them [8]. Another noteworthy example is the chemical industry, discussed in Chapter 13, which has established a Responsible Care® program in several countries as a focal point for environmental, health and safety performance improvement and stakeholder communication [9].

In the long run, as sustainability becomes integrated into a company's operations, the distinction between environmental goals and business goals should vanish. Rather than utilizing separate KPIs that reflect environmental progress, companies can develop business value-driven KPIs such as eco-efficiency, that incorporate sustainability principles. With increasing awareness of sustainability among both the financial community and the general public, adopting clear indicators of economic and environmental value should eventually lead to a clear advantage in the marketplace.

References

1. R. S. Kaplan and D. P. Norton, *The Balanced Scorecard* (Cambridge: Harvard Business School Press, 1996).
2. C. Fussler with P. James, *Driving Eco-Innovation: A Breakthrough Discipline for Innovation and Sustainability* (London: Pitman, 1997).
3. H. A. Verfaillie and R. Bidwell. "Measuring Eco-Efficiency: A Guide to Reporting Company Performance," WBCSD, June 2000.
4. J. Fiksel, "Measuring Sustainability in Eco-Design," in M. Charter & U. Tischner, *Sustainable Solutions: Developing Products and Services for the Future* (Sheffield, U.K.: Greenleaf Publishing, September 2000).
5. J. Fiksel, J. McDaniel and D. Spitzley, "Measuring Product Sustainability," *Journal of Sustainable Product Design*, Issue No. 6, July 1998, pp. 7–18.
6. J. Fiksel, D. M. Lambert, L. B. Artman, J. A. Harris, and H. M. Share, "Environmental Excellence: The New Supply Chain Edge," *Supply Chain Management Review*, July/August 2004, pp. 50–57.
7. T. W. Zosel, "Pollution Prevention Pays: the 3M Approach," *Proc. 1st Intl. Congress on Environmentally-Conscious Design & Manufacturing*, Management Roundtable, Boston, 1992.
8. Electric Power Research Institute, *Environmental Performance Measurement: Design, Implementation, and Review Guidance for the Utility Industry*, EPRI Technical Report TR-111354, Palo Alto, Calif., 1998.
9. CEFIC, European Chemical Industry Council, "Responsible Care Status Report, Europe 2000," Brussels, June 2001.

CHAPTER 8

Design Rules and Guidelines

DFE Principle 4. *Build upon past experiences to assemble a portfolio of design strategies that can be codified, communicated through training, and systematically applied by your design teams.*

DFE Rules!

The basic premise of this book is that environmental sustainability is compatible with economic growth—that companies can redesign products and processes in a way that is both environmentally responsible and profitable. Accordingly, the goal of DFE is to enable design teams to create eco-efficient and eco-effective products while adhering to their cost, quality and schedule constraints. Chapter 5 argues that, for a company to be successful in this goal, DFE must be integrated seamlessly into the development process, from the analysis of customer needs and establishment of product requirements to the verification that these requirements have been fulfilled. The availability of *guidelines* for practicing DFE was identified as the second key element needed to support this process.

The integrated product development process defined in Chapter 5 begins with setting of objectives during the concept phase. Once the product objectives have been defined, the product development cycle begins in earnest. It is an exploratory process during which ideas are generated, considered from various perspectives, and either pursued or rejected. As manufacturing companies have refined their product development processes, they have increasingly recognized the need to adopt *guidelines*. These guidelines may be expressed in a variety of forms, ranging from verbal rules of thumb to multidimensional look-up tables to pictorial maps or diagrams. There at least two types of guidelines:

1. *Prescriptive* guidelines are definite statements about what designers should or should not do, and are sometimes called *design rules*. There are many such guidelines related to environmental health and safety—for example, lists of banned materials.

2. *Suggestive* guidelines represent accumulated knowledge, including best practices and lessons learned, but they are not strict rules. They merely point in useful directions, or conversely, indicate characteristics that should be avoided.

Virtually all of the DFE guidelines presented here are suggestive. There are several reasons for this—DFE is continually evolving, DFE is mainly a voluntary, nonregulated practice, and DFE involves many complex trade-offs so that general rules are difficult to find. Individual companies typically establish design rules that are appropriate for their products. There are a number of benefits from using guidelines, whether for DFE or any other DFX discipline:

- They encourage consistency among different development teams in areas where consistency is desirable or necessary, e.g., standard material labeling schemes.

- They promote continuity through the accumulation of knowledge ("lore") over successive design cycles and allow that knowledge to be preserved and passed down.

- They lead to a more systematic design process that is less dependent upon the idiosyncrasies or particular biases of individual designers.

- They expand the scope of issues considered during design, allowing the team to anticipate downstream pitfalls or constraints that they may have ignored.

Despite these benefits, in a fast-paced product development environment it is often difficult to ensure that product teams pay attention to guidelines, especially if they are only suggestive and not strict requirements. Various approaches are used to make developers aware of such guidelines, ranging from printed guidance manuals to online decision aids. To the extent that such tools can be built into a "stage-gate" development process, they can be extremely effective in influencing product designs. A classic example of successful introduction of DFX guidelines is the Design for Manufacture and Assembly (DFMA) methodology, originally developed by Boothroyd and Dewhurst at the University of Rhode Island. This methodology has been encoded into software tools that are routinely used by major companies to simplify designs and, hence, reduce their assembly costs. Design automation companies such as AutoDesk are currently developing decision support tools that can play an analogous role for DFE.

Catalogue of DFE Guidelines

This chapter provides a compilation of DFE guidelines that are commonly practiced by manufacturing firms in a variety of industries. The guidelines are divided into four principal strategies, which were introduced in Chapter 6 (see Figure 8.1). Each of these strategies is described at length in the numbered sections below:

A. **Design for dematerialization** seeks to reduce the required amount of material throughput, as well the corresponding energy requirements, for a product and its associated processes throughout their life cycle.

B. **Design for detoxification** seeks to reduce or eliminate the toxic, hazardous, or otherwise harmful characteristics of a product and its associated processes, including waste streams that may adversely affect humans or the environment.

C. **Design for revalorization** seeks to recover, recycle, or otherwise reuse the residual materials and energy that are generated at each stage of the product life cycle, thus eliminating waste and reducing virgin resource requirements.

D. **Design for capital protection and renewal** seeks to ensure the safety, integrity, vitality, productivity, and continuity of the human, natural, and economic resources that are needed to sustain the product life cycle.

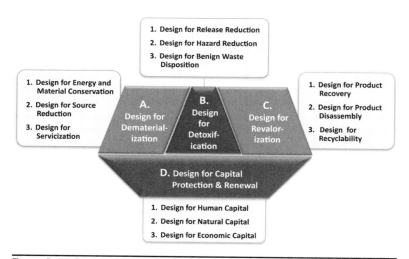

FIGURE 8.1 Four major strategies of Design for Environment.

There are obviously a broad range of guidelines and practices that can be considered within these DFE disciplines. However, to be truly useful to a particular company and product team, these types of guidelines should be converted from the general statements listed here to more specific approaches that are applicable to a particular industry, enterprise, and product category. There are several points to note about this catalogue of guidelines:

- The list is by no means exhaustive. Although it covers a majority of common industrial practices, new approaches are constantly being devised.

- The classification into four disciplines is not necessarily the only way to organize these guidelines. Any hierarchical scheme must recognize that many of the DFE practices are inter-related, as shown in Figure 8.2.

- Each DFE guideline may have beneficial impacts in one or more stages of the product life cycle, as illustrated in Table 8.1.

- There is considerable overlap with other DFX disciplines such as Design for Manufacture and Assembly. Indeed, one strength of DFE is its synergy with other design disciplines. For example, reducing design complexity leads to fewer parts, lower assembly costs, and easier disassembly, resulting in reduced energy and material use as well as increased recyclability.

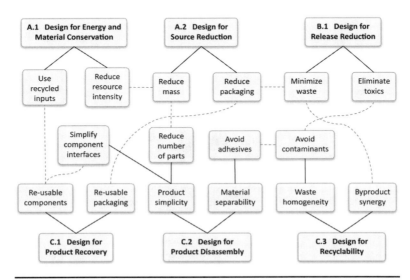

Figure 8.2   Interrelationships among DFE practices.

	Procurement	Assembly	Distribution	Use/Support	Recovery
Reduce number of distinct parts	•	•			•
Simplify component interfaces		•		•	•
Use similar or compatible materials	•				•
Use recyclable or recycled materials	•		•		•
Design for material separability		•			•
Reduce packaging mass	•	•	•		•
Reduce mass of components	•	•	•		
Reduce product size and mass	•	•	•	•	
Design upgradable components				•	•
Design re-usable components	•			•	•

TABLE 8.1 Benefits at Various Life-Cycle Stages for Electronic Assemblies

These DFE guidelines are consistent with the traditional Pollution Prevention (P2) hierarchy, a series of methods for dealing with waste streams in descending order of priority:

- Reduce the quantity of waste generated
- Reuse or reprocess the waste as an economically viable resource
- Recycle the waste for an environmentally beneficial purpose
- Incinerate the waste to recover energy
- Dispose of the waste safely

The main difference is that P2 methods assume that the product is already defined, and that process environmental improvements can only be implemented after the fact. DFE offers much greater latitude to designers, so that they can anticipate and prevent, reduce, or modify the implied resource burdens and waste streams. This enables product innovation to fundamentally change the environmental footprint and supports the highest and best use of residuals that are not incorporated into the product.

Design for Environment guidelines.

B.1   Design for Release Reduction

B.2   Design for Hazard Reduction

B.3   Design for Benign Waste Disposition

- Toxic and hazardous substance removal
- Process emission and waste reduction
- Life-cycle waste stream reduction

- Responsible treatment and disposal
- Waste sequestration
- Ecosystem adsorption
- Biodegradation

- Product reformulation
- Toxic and hazardous material use reduction
- Water-based technologies

C.1   Design for Product Recovery

C.2   Design for Product Disassembly

C.3   Design for Recyclability

- Secondary utilization
- Re-usable components and packaging
- Component or product refurbishment
- Remanufacturing

- End-of-life material recovery
- Closed-loop material recycling
- Waste composition and homogeneity
- By-product synergy & industrial ecology

- Product simplicity
- Disassembly sequencing
- Component accessibility
- Component and material separability

D.1   Design for Human Capital

D.2   Design for Natural Capital

D.3   Design for Economic Capital

- Workplace health, safety, & ergonomics
- Product safety, integrity, & efficacy
- Public health, safety, & security

- Process reliability, safety, & security
- Business continuity & supply chain resilience
- Asset utilization & resource productivity
- Reputation & brand protection

- Climate change mitigation
- Water resource protection
- Ecosystem integrity and biodiversity protection
- Land conservation and restoration

A. Design for Dematerialization

The rate of material throughput is perhaps the most important indicator of unsustainability in our global economy, as stated in Chapter 1. We are largely sheltered from the enormous flows of materials that are extracted from nature, converted into products, and finally released into the environment in order to support our lifestyles. Material flow is a fundamental driver of energy use, water use, greenhouse gas emissions, and most other environmental indicators—the more we use, the more we waste. Therefore, a principal strategy for improving sustainability is *dematerialization*, defined as the reduction of material throughput in an economic system. Dematerialization has been popularized in proposals such as Factor 4, which suggests a doubling of global economic wealth while halving material resource use [1]. Some argue that a Factor 10 transformation is necessary for industrialized nations to reach long-term sustainability [2].

Dematerialization includes a variety of techniques, such as increasing material efficiency in operations; designing products with reduced mass, packaging, or life-cycle energy requirements; replacement of virgin materials with post-industrial or post-consumer wastes; reducing transportation requirements in the supply chain, thus reducing fuel and vehicle utilization; substitution of electronic services for material-intensive services; and substitution of services

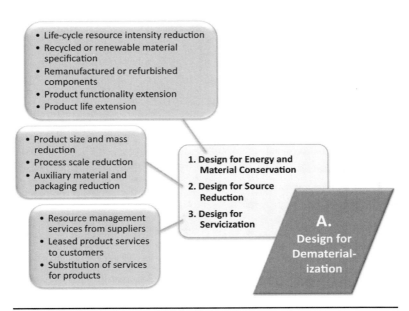

FIGURE 8.3 Design for dematerialization.

for products. These techniques are complemented by other DFE practices, such as recovering value from obsolete or discarded products (see Section C, Design for Revalorization).

A.1 Design for Energy and Material Conservation

Reducing energy and material consumption is the most direct way to improve eco-efficiency, i.e., utilizing fewer resources to deliver equivalent or greater value. As discussed in Chapter 4, decreasing resource intensity results in higher resource productivity, provides immediate reductions in operating costs, and, thus, is synergistic with business goals. In other words, the quantity and costs of purchased energy and materials are reduced by increasing operating efficiency (see Section D.3, Design for Economic Capital). Moreover, energy conservation reduces overall material consumption in the supply chain, since generating energy requires some type of fuel and/or equipment. Although energy management is often pursued as a separate program, energy and material resource conservation should, ideally, go hand-in-hand. Finally, energy conservation that reduces fossil fuel use will also reduce greenhouse gas emissions (see Section D.2, Design for Natural Capital).

Life-Cycle Resource Intensity Reduction

As stated in Chapter 6, DFE needs to consider the *full life cycle* of a product, including all of the processes involved in sourcing, production, distribution, use, and recovery of the product. Thus, the investigation of opportunities for energy and material conservation should consider both supplier and customer processes. Depending on geographic locations and type of facilities, certain companies in the supply chain may have much better opportunities than others for energy and material conservation. The following types of opportunities should be explored:

- **Reduce the procurement footprint**—Many companies have begun to examine the environmental practices of their suppliers and encourage greater energy and material efficiency. This can reduce the life-cycle footprint of their own products and potentially lower their costs. The most prominent example is Wal-Mart, which has developed sustainability scorecards for packaging and energy use and is requesting environmental performance improvements from all of its suppliers (see Chapter 19). This exercise goes beyond measuring direct energy and material consumption. See Chapter 9 for a discussion of the challenge of understanding a company's full life-cycle carbon footprint.

- **Reduce the operational resource footprint**—Companies have found a great deal of "low-hanging fruit" by tightening up energy management practices, e.g., heating, cooling, and lighting systems, and materials management practices, e.g., maintenance, inventory, and waste management. Newer facilities are being designed with recycled materials and advanced energy-saving features, as interest in "green building" has mushroomed. But the largest gains in resource conservation come from redesigning production processes to reduce throughput requirements and install more efficient equipment.

> **Example:** From 2005 to 2007, General Electric (GE) conducted a "Lean and Energy" initiative that identified over $100 million in potential energy savings through over 200 "energy treasure hunts" at GE facilities worldwide. This effort resulted in 5,000 related kaizen[*] projects, most of which are funded and in various stages of implementation. GE was able to reduce greenhouse gas emissions by 250,000 metric tons and realized $70 million in energy cost savings from implemented projects.

- **Reduce transportation requirements**—An often overlooked aspect of energy and material use is the distribution chain of products, including shipments from component vendors to manufacturers or assemblers, from manufacturers to distribution centers or retail stores, and from these intermediate points to customers. It is not uncommon for a product to go through a half dozen shipment stages by various modes before arriving at its ultimate destination. Each leg of such a journey may entail significant cost as well as packaging and energy consumption; the most extreme case is shipping a missing part by courier to meet a delivery deadline. Some products may have physical characteristics, such as thermal and vibration tolerance, that limit the available distribution options, but in most cases there are a number of options for increasing transportation efficiency:

    ○ Reduce the total transportation distance for a product or its components, e.g., by shipping outsourced modules directly from the supplier to the final customer.

    ○ Reduce transportation urgency by allowing greater lead times.

    ○ Reduce the shipping volume required by redesigning the product geometry, packaging volume, or stacking configuration so that less space is wasted.

    ○ Reduce temperature requirements or other energy-consuming constraints.

---

[*]In the practice of Lean, *kaizen* activities are intensive team exercises used to solve problems and eliminate waste, based on the Japanese philosophy of continuous improvement.

- **Reduce end-use power consumption**—Laptop computers and many other electrical or electronic devices have *power management* features that power down a unit when it has not been used for some length of time. In other types of products, such as refrigerators, energy is conserved through developing more efficient motors and reducing the energy load. The U.S. government has helped to promote energy efficiency of electronic devices and procurement of "environmentally preferable" products (see Chapter 3).

- **Reduce end-use material consumption**—Products requiring maintenance materials or replacement parts, such as ink cartridges and batteries, can consume significant resources over their lifetime. In many cases, the end-use environmental footprint of a durable product far outweighs the energy and materials required to produce it. Design efforts that reduce or eliminate these requirements will not only benefit the environment, but also will increase convenience and reduce "cost of ownership" for the customer.

  Example: Kyocera introduced a new printer design which eliminates the need for a disposable or recyclable toner cartridge. The printer uses a self-cleaning printing drum coated with superhard amorphous silicon. With this more elegant design, users need only add toner to maintain the device.

Recycled Material Specification

An important aspect of sustainable development is the conservation of nonrenewable resources. Manufacturing firms have long been in the habit of specifying virgin materials that are well-characterized because they were manufactured through a precise process with known feedstocks. Driven by growing environmental awareness, both government and industry organizations have begun to specify more "environmentally conscious" materials that have significant levels (25% to 100%) of recycled content. This is feasible to the extent that substitution of recycled materials with potential impurities is cost-effective and does not compromise the quality of the final product. For example, metals are easily recycled, because they can be purified in a molten state. Likewise, paper, glass, and many other materials can achieve substantial levels of recycled content.

However, the situation is different with engineering thermoplastics. The cost of separating the individual components is high, and the thermo-mechanical properties associated with recycled resins developed from mixed waste can be significantly compromised. One approach is to utilize virgin materials only for critical components, and recycled materials for less demanding applications such as base assemblies. Resin manufacturers have begun adding compatibilizing agents to strengthen both the physical and chemical bonds among mixed plastics. One of the key differences between virgin and recycled resins is color tolerances, which may be important if two parts are immediately adjacent. Producing all mating parts from the same lot of raw materials can minimize variability in physical and aesthetic attributes.

> **Example:** Mohawk Industries, a carpet manufacturer, converts over 3 billion bottles annually—25% of all the bottles collected in North America—into 160 million pounds of recycled carpet fiber. In addition, Mohawk recycles about 10 million pounds per year of crumb rubber from about 720,000 scrap tires into designer door mats. Finally, Mohawk has replaced cardboard carpet cores with those made from recycled carpet edge trim, plastic bottle tops, and stretch films. These new cores last longer, are less likely to damage the carpet, and can be recycled into new cores.

Renewable Material Specification

Instead of recycling nonrenewable materials, an alternative approach is substitution of renewable materials such as agricultural products. Materials are considered renewable if the rate at which they are replenished is sufficient to compensate for their depletion. For example, as discussed in Chapter 12, many automakers are beginning to use natural materials derived from crops such as flax or soybeans for car interiors (see Figure 8.4). Other examples of products that use renewable materials include soy inks and wooden furniture.

Remanufactured or Refurbished Components

Durable products can be manufactured with refurbished components, resulting in lower costs and reduced material consumption,

FIGURE 8.4 Many recent Ford models, such as the pictured Mustang, use soy-based foam for seatbacks and cushions.

often with the same level of quality as products manufactured with brand new components (see Section C.1, Design for Product Recovery). Ideally, companies can develop a reverse logistics system whereby spent materials and used components can be recovered, reprocessed, and recycled back into their supply chain. Establishing a *closed-loop* infrastructure provides greater assurance about uniformity, homogeneity, and reliability of the recycled assets. However, this type of closed-loop recovery is often not feasible, and a preferred alternative may be to find other companies that can utilize these material streams as process inputs (see Section C.3, Design for Recyclability). A leading practitioner of remanufacturing is Caterpillar, as described in Chapter 10.

Product Functionality Extension

Products that have multiple uses are by nature eco-efficient, in that the same amount of material achieves a higher level of functionality. The greater the proportion of time during a product life that the product is actually in use, the greater the ratio of value delivered to resources consumed. There are essentially two types of multiple functionality: *parallel* functions, in which the same product is designed to simultaneously serve several different purposes; and *sequential* functions, in which a product is retired from its primary use and then applied to a secondary and tertiary use (see "Resource Cascading" below).

Examples of multifunctional products include

- All-in-one copiers that double as printers, scanners, and fax machines
- Cell phones that also serve as portable music players and personal digital assistants
- Solvents that are used for metal parts cleaning and then reused for plant maintenance.

Product Life Extension

Increased product longevity or durability is another strategy to increase the amount of functionality delivered by a product over its useful life. This is one of the most direct ways to improve environmental performance because it decreases the average life-cycle resource consumption per product use. There is usually a trade-off between product cost and longevity, and customers may not be prepared to pay a higher purchase cost for a more durable product. Moreover, there are some products, notably cell phones and computers, whose life expectancy (measured in hours of operability or duty cycles) is much longer than their actual duration of primary use, due to technological obsolescence.

Ironically, even when the long-term economics are clearly favorable, as in the case of compact fluorescent light bulbs, some consumers may still behave irrationally and opt for a lower-priced, less cost-effective product. This has long been a challenge of "green marketing"—if consumers fail to accept the value proposition for a higher-performance product, then companies will continue to manufacture cheaper products that are quickly discarded.

Increased longevity implies a need for greater durability, which may require denser materials and stronger fastening methods. Interestingly, these objectives conflict with the design criteria used in designing for disassembly, separability, and waste reduction. You can't have it both ways, and designers need to grapple with the life-cycle cost and environmental trade-offs between durability and recoverability.

Apart from life extension for the product as a whole, another way to achieve longevity is to extend the life of product components. There are at several ways to accomplish this:

- **Design upgradable components**—for example, desktop computer graphic cards can easily be upgraded, as technological capabilities improve.

- **Design a reusable platform**—for example, a less wasteful alternative to disposable insulin pens for diabetics is a refillable pen which accepts cartridges, similar to a fountain pen.

  Example: In 2002 General Motors introduced a radical concept car called the AUTOnomy, built around a fuel cell system with drive-by-wire functionality, which allows steering, braking, and other vehicle systems to be controlled electronically instead of mechanically (see Figure 8.4). The vehicle body connects to the "skateboard" chassis by means of a "docking port," making the body lighter and freeing it from traditional design constraints. This can lead to the development of customized bodies that are easy to switch in and out of the basic platform. However, the concept was never commercialized.

- **Improve serviceability and repairability**—reducing the cost and difficulty of service and repair, including do-it-yourself repair, can extend the useful life of a product.

A.2 Design for Source Reduction

Source reduction is routinely practiced as the most desirable alternative in the pollution prevention hierarchy. Reducing the mass of a product is the surest and most direct way to achieve waste reduction and usually results in lower life-cycle costs as well. In many industries, notably electronics, the mass of products has been steadily decreasing due to consumer demand and technological advances. In other industries, such as food and beverages, the mass of the product itself is inflexible, and ingenious methods must be found to reduce the mass of containers and packaging. The following is a list of guidelines representing the most common source reduction practices:

- **Reduce the product's physical dimensions**—In the digital age, size reduction is a natural consequence of technological improvement. For example, Eastman Kodak's DFE efforts have helped its digital cameras grow lighter and more energy-efficient, while performance has dramatically improved.

- **Reduce the mass of key components**—For example, reduced power consumption in electronic devices has helped to reduce energy demand, while new battery technologies have enabled smaller and lighter battery packs.

- **Specify lighter-weight materials as substitutes**—For example, composite materials are attractive because they have superior strength-to-weight ratios, although this property may present recyclability challenges (see Section C.3, Design for Recyclability).

- **Design thinner enclosures with existing materials**—For example, U.S. EPA estimates that, since 1977, the weight of 2-liter plastic soft drink bottles has been reduced from 68 grams each to 51 grams, removing 250 million pounds of plastic per year from the waste stream.

- **Increase the concentration in liquid products**—For example, Procter & Gamble has been a pioneer in developing new product forms that reduce both water use and solid waste; ultra-concentrated detergents have rapidly grown in market share (see Chapter 16).

- **Use electronic documentation instead of paper**—This has become standard practice in the electronics industry. Availability of web-based technical support reduces the need for manufacturers to supply comprehensive hard copy documentation.

- **Reduce the weight or complexity of packaging**—The Sustainable Packaging Coalition has developed a comprehensive set of packaging design guidelines [3]. Like any product component, packaging can benefit from the full range of DFE strategies.

  Example: In 2008, in response to a challenge from Wal-Mart to reduce packaging, HP introduced the Pavilion dv6929 notebook PC in a recycled laptop bag with 97% less packaging than typical laptops. The carrying bag contains no foam, only some plastic bags for consumers to dispose of. The bag itself, save for the buckle, strap, and zipper, is made out of 100% recycled fabric. HP is able to fit three bags in a box for shipping the product to stores, thus reducing energy use and costs related to logistics.

- **Reduce consumption of shipping containers**—In many supply chains, disposal of used containers such as shipping pallets is a major source of solid waste. Significant cost savings and material efficiency can be achieved by designing containers that can be recovered and reused for the same application. In some cases, containers can be eliminated; for example, Duke Energy was able to redesign the cable storage and handling systems used for electric power lines so that wooden reels were no longer necessary, saving over $650,000 per year.

- **Reduce the scale of manufacturing processes**—By developing more eco-efficient processes with higher yields, lower temperature and pressure requirements, and smaller physical footprints, companies can reduce the amount of feedstocks and auxiliary materials necessary to operate these processes. This also results in greater productivity of assets (see Section D.3, Design for Economic Capital).

The ultimate in source reduction is represented by the science of nanotechnology, which seeks to develop products on a molecular scale. However, this new technology raises concerns about health effects of tiny particles, and nano-manufacturing may have a large life-cycle footprint.

A.3 Design for Servicization

The most radical approach to dematerialization is to eliminate products altogether and provide services instead. A simple example in the consumer realm is the substitution of a voice mail service for a physical device that records telephone messages. Some argue that "servicization" is an essential strategy for decoupling economic growth from environmental impacts. However, it must be remembered that service industries also have an environmental footprint, since they require physical facilities, equipment, energy, and labor (see Chapter 19). The potential benefits of servicization lie in the ability and motivation of service providers to maximize the efficiency of resource utilization, thereby increasing their own profitability.

One popular service-based business model is the "leased product" concept, in which the manufacturer retains ownership and responsibility for the physical product. For a fixed lease cost, customers receive the functionality and technical support of the product until they are ready to upgrade, and manufacturers can then recover the used product and maximize its residual value. Xerox has successfully used this model to maintain customer loyalty while saving millions of dollars in life-cycle costs (see Chapter 11). Another example is the emergence of short-term transportation leasing services which offer instant availability of automobiles or bicycles for short trips in dense urban zones (see Zipcar example in Chapter 12).

In the industrial arena, a common example of service innovation is "chemical management," whereby specialized contractors take over all of the functions associated with the procurement, storage, handling, and disposal or recycling of chemicals used in manufacturing operations such as semiconductor fabrication. This "turnkey" approach enables manufacturers to focus on their core business, while the service providers offer expertise in chemical handling and compliance with environmental, health, and safety requirements. Moreover, the service provider can seek greater efficiencies by pooling resources and taking a systems approach to service design.

The servicization concept has been embraced by UNEP, which defines a "product-service system" as *the result of an innovation strategy, shifting the business focus from designing and selling physical products only, to selling a system of products and services which are jointly capable of fulfilling specific client demands* [4]. Examples of such systems include

- Mobile services that provide on-demand delivery of products to a home or business
- Energy services that utilize innovative technologies to deliver heat and power
- Agricultural services that provide locally grown organic produce to consumers
- "Virtual office" services that provide on-demand resources for small businesses

These types of innovations have the potential to dramatically reduce the supply chain capital and resources needed to deliver value to the customer.

B. Design for Detoxification

Dematerialization reduces the total throughput of materials and energy, thereby conserving resources and reducing waste and emissions. However, simply reducing the mass of materials is insufficient. Some materials released into the environment can have significant adverse human health and environmental impacts even at very low levels; well-known examples include lead, mercury, and dioxin. Therefore, analysis of material flows alone is not sufficient for an understanding of product sustainability. A variety of methods for analyzing environmental impacts and the associated risks are described in Chapter 9.

Detoxification refers to the prevention or reduction of adverse human or ecological effects associated with materials use. It can include a variety of approaches, such as reduction in the volume of harmful wastes generated, restrictions on the use of specified materials, replacement of toxic or hazardous materials with benign ones, waste modification through chemical, energetic or biological treatment, waste containment or isolation to prevent human and ecological exposure, and *in situ* waste treatment.

B.1 Design for Release Reduction

Dematerialization, i.e., reduction of throughput, is the most direct way to reduce the releases of toxic or hazardous wastes. However,

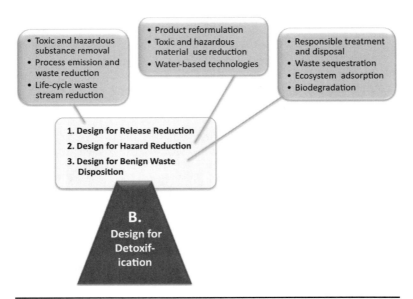

Figure 8.5 Design for detoxification.

even when production flow remains constant, it is possible to reduce the volume of harmful waste generated by modifying the product or the production process. As discussed in Chapter 3, the introduction of disclosure requirements and publication of the EPA Toxic and Hazardous Release Inventory (TRI) has been instrumental in motivating manufacturers to reduce such releases voluntarily. The guidelines below focus on reducing total releases, but another viable strategy is to convert toxic or hazardous wastes into valuable by-products (see Section C, Design for Revalorization).

Toxic and Hazardous Substance Removal

Removal of toxic or hazardous substances from pollution and waste streams is the traditional method for limiting human exposure and environmental dispersion. Manufacturing facilities may routinely release a variety of toxic or hazardous substances to the environment via stack emissions, fugitive emissions, wastewater discharges, and solid or hazardous waste disposal. Additional environmental concerns include ozone-depleting chemicals, such as perfluorocarbons, and greenhouse gas emissions (see Section D.2, Design for Natural Capital). Airborne emissions may be partially removed through pollution control devices, such as oxidizers, scrubbers, filters, or electrostatic precipitators. Liquid and solid wastes may be treated or disposed of on-site, sent to a municipal waste treatment facility, or transferred to a waste management contractor.

While requirements for control of workplace and environmental releases are typically regulated, there are many companies that choose to go "beyond compliance" and further limit releases in order to reduce potential liabilities or to anticipate more stringent regulations. In some cases, companies have requested that their suppliers adopt similar stringent measures if the regulations in their home country are more lax. The objective of these practices is to reduce the potential for chronic exposure to substances that may cause illness to plant workers, neighboring residents, or wildlife.

Process Emission and Waste Reduction

The largest source of toxic and hazardous substance releases is typically in manufacturing processes that involve physical or chemical transformations, as opposed to assembly of components. Process emissions and wastes can be broadly separated into internal releases within the production facility, which influence occupational exposures, and external releases to the environment. Examples of guidelines for process waste reduction include

- **Reduce process throughput**—the flow of materials through the supply chain can be reduced by identifying and minimizing non-product output; for example, by recycling pallets, solvents, catalysts, scrap materials, or process water.

- **Reduce excess inventory**—holding unnecessary inventory requires extra resources and can lead to product spoilage and waste. Lean process design and just-in-time techniques can help to streamline inventory, although going too lean can actually threaten business continuity (see Section D.3, Design for Economic Capital).

- **Reduce fugitive emissions**—Improved design of chemical containment, piping, and exhaust recovery systems such as ventilation hoods can significantly reduce the amount of losses or leakages of liquids and vapors. Often these emissions can be cycled back into the process, thus increasing efficiency while reducing human exposure (see Section D.1, Design for Human Capital).

Life-Cycle Waste Stream Reduction

Waste is generated throughout the life cycle of a product, from sourcing and manufacturing to customer use to end-of-life. The actual waste associated with the discarded product at end-of-life may be only a fraction of the waste generated by the manufacture, use, and disposal of a product. Thus, the investigation of opportunities for toxic and hazardous release reduction can encompass a variety of processes in the supply chain, many of which are beyond the control of the manufacturer. Practicing product stewardship means making an effort to collaborate with both suppliers and customers and to redesign supply chain processes in ways that enhance business and environmental performance. Examples range from supplier collaboration in chemical management to consumer initiatives such as battery recycling. By considering waste reduction opportunities during product design, engineers are able to minimize harmful wastes and/or maximize the value extracted from waste materials.

> **Example:** The Suppliers Partnership for the Environment is an innovative partnership between automobile original equipment manufacturers and their suppliers, co-sponsored by the U.S. Environmental Protection Agency (EPA). Through focused work groups, these companies collaborate to reduce their energy use and their environmental footprint, and to provide technical assistance to small and medium-sized businesses in the automotive supply chain.

B.2 Design for Hazard Reduction

An alternative to reducing the volume of toxic and hazardous substances released is to redesign products and processes so that the presence of such substances in waste streams is minimized. Applying DFE can help reduce the burden of compliance and reporting requirements, while also identifying opportunities for cost-effective design changes that reduce the potential exposure and risk pathways for humans or the environment.

One of the challenges in hazard reduction is the potential for excessive control of suspect materials. While the evidence for some environmental risks is overwhelming, others can be more speculative. The precautionary principle can lead to restriction of substances based on flimsy toxicological evidence, and it is not always clear that the public benefit in terms of risk reduction is commensurate with the economic burden of chemical substitution. In some cases, there are unintended adverse consequences; for example, the substitution of methyl tertiary butyl ether (MBTE) for lead in reformulated gasoline resulted in widespread contamination of underground water reservoirs.

The following are examples of guidelines for hazard reduction:

- **Product reformulation**—An effective way to avoid toxic releases is simply to eliminate them by redesigning the product. For example, brominated flame retardants in plastic materials have been eliminated from many electronic products due to evidence of potential chronic toxicity. Some companies have established lists of preferred and restricted materials in an effort to avoid specification of toxic or hazardous constituents in new products. The practice of "green chemistry" has expanded, as scientists explore ways to reformulate chemical products using alternative reaction pathways that are "cleaner" and safer (see Chapter 13).

  Example: SC Johnson, the consumer products manufacturer, has established a Greenlist™ program to classify all the ingredients that go into its products according to their impact on the environment and human health. For example, the company has made a considerable effort to eliminate chlorine-based packaging, including PVC bottles. In one case, the company reformulated a popular metal polish product so that it could be packaged in a non-PVC bottle (PET), and actually reduced overall life-cycle costs. The new formula uses fewer chemicals, matches the performance of the old product, eliminates the need for the E.U. "Dangerous for the Environment" hazard label, and can be warehoused together with other products.

- **Toxic and hazardous material use reduction**—While older plants can be retrofitted with pollution control devices, designing "cleaner" processes is a more effective means of pollution prevention. Avoiding the use of toxic or hazardous chemicals is the best way to eliminate the associated risks. As illustrated in Table 9.2, many companies have developed lists of materials to be avoided due to their regulatory status or known health and environmental hazards. For example, materials containing certain flame retardants (e.g., PBDO's) may be restricted from normal recycling channels due to toxicity concerns. As discussed in Chapter 3, European Union directives such as WEEE and RoHS have influenced global companies to work more proactively on material identification and screening.

> **Example:** The electronics industry has developed a Joint Industry Guide to assure that parts suppliers are not using certain restricted materials, including asbestos, heavy metals, ozone depleting compounds, and many other specific chemicals. This global, multiyear effort has produced an online registry where suppliers can provide their declarations.

- **Water-based technologies**—One common form of chemical use avoidance is the substitution of aqueous solvents for cleaning parts and other industrial processes. Through better control of mechanical and hydraulic factors, aqueous cleaning can be as cost-effective as traditional methods that use more hazardous chlorinated solvents. Similarly, advances in water-based paint technology have produced high-performance paints suitable for demanding applications.

> **Example:** Volkswagen, the German automobile manufacturer, decided to eliminate solvent-based paints rather than try to tighten VOC emission controls. They invested over $1 billion in water-based paint technology at their major plants and have virtually eliminated the use of solvents. As a result, beginning in 1992, they were able to reduce annual VOC emissions by a factor of about 5 and have sustained that performance over time. In addition to VOC reduction, VW's approach improves working conditions and enhances paint quality. (See the description of DuPont's water-based coating system in Chapter 13.)

B.3 Design for Benign Waste Disposition

While zero waste is a desirable goal for any industrial system, it is difficult to achieve in practice. Assuming that there are irreducible waste streams, potentially including persistent, bioaccumulative, toxic or hazardous substances, companies need to consider carefully how these wastes will be disposed of and whether there are ways to mitigate their ultimate impacts. A variety of waste disposition options are available, ranging from sanitary landfills to destructive incineration. Responsible treatment and disposal of wastes is a statutory obligation, but discretionary strategies can help to reduce costs, protect the environment, and avoid potential future liabilities. One example is *in situ* waste treatment, which can reduce the effective concentrations or adverse impacts of wastes that have previously been discharged into the environment.

Manufacturers are considered "point sources" of waste and pollution, and are easier to control than other more widely dispersed sources, including "mobile" sources such as vehicles and "non-point" sources such as pesticide application on farms. For example, fertilizer runoff containing nitrogen and other nutrients can cause *hypoxia*, or oxygen depletion in water bodies; an extreme example is the dead zone in the Gulf of Mexico. One way to combat this problem, according to Professor William Mitsch at The Ohio State University, is to build wetlands and riparian buffer zones that act as natural filters along waterways.

There are several available strategies for achieving benign waste disposition in the environment:

- **Waste sequestration**—It is possible to isolate wastes for varying lengths of time, so that they do not come in contact with humans or sensitive ecosystems. Examples include:
  - ○ Deposition of liquid wastes in settling ponds adjacent to industrial facilities
  - ○ Isolation of radioactive nuclear wastes for long periods of time in sealed containers
  - ○ Underground sequestration of carbon dioxide, currently being explored by the electric power industry and the U.S. Department of Energy (see Chapter 18).

  However, all of the above solutions are reversible and potentially vulnerable to unforeseen problems.

- **Ecosystem adsorption**—There exist a variety of ecosystem processes that can reduce the concentrations or impacts of wastes in an environment over time. These include processes that change wastes into less toxic forms, e.g., sorption to sediments, as well as processes that disperse and transport wastes in ways that dilute their impact. However, there is also a possibility that ecosystems will concentrate wastes into "hot spots" of relatively high waste concentrations, including bioaccumulation through the food chain. The assimilative capacity of an ecosystem to adsorb waste may be defined as the amount and rate of a given waste that can be added to an ecosystem before some specified level of detrimental effect is reached [5]. Given the current rates of material throughput, many terrestrial ecosystems are already stressed, and reliance on ecosystem adsorption cannot be taken for granted.

- **Biodegradation**—Many consumer products are disposed of rather than recycled, or may have unavoidable residuals that enter municipal solid or liquid waste streams. Examples include plastic bags, polystyrene pellets, batteries, detergent residues, and pharmaceuticals. While the concentrations of harmful substances in these wastes may be low, the cumulative volume and potential exposures may cause concern. A worthwhile approach toward risk reduction is to ensure that such substances are biodegradable in the environment.

Biodegradation involves breakdown of substances into harmless molecules by enzymes produced by living organisms. Most organic materials are biodegradable, and even synthetic materials can be biodegradable under the right disposal conditions. In some cases, companies have gained competitive advantage by designing products or packaging that are environmentally benign at end-of-life.

Example: BASF, a European chemical manufacturer, has developed a novel line of synthetic plastics, called Ecoflex®, that are completely biodegradable and will decompose in soil or compost within a few weeks. Introduced in 1998, it has become the world's leading synthetic, biodegradable material and is commonly used for trash bags or disposable packaging. Another product line, Ecovio® is a blend of Ecoflex® and polylactic acid made from corn, and is used in flexible films for shopping bags.

C. Design for Revalorization

Revalorization describes a variety of methods for recovering value from a product at the end of its useful life. It applies mainly to durable products that are placed into service for some period of time and then retired or discarded. However, in the case of consumable products such as foods, revalorization may be applicable to auxiliary materials, such as packaging and utensils that are traditionally treated as wastes. The fundamental purpose of revalorization is to prevent a product or material from becoming a waste by diverting it to an economically viable use.

C.1 Design for Product Recovery

From a product stewardship perspective, every product should be considered an asset, even after its ownership has been transferred to customers, and even after customers have terminated their use of the product. One way to improve the life-cycle utilization of products is to extend their longevity (see Section A.1, Design for Energy and Material Conservation), but this is often not feasible or economical. However, the residual value of discarded products can often be recovered for the benefit of the manufacturer and other supply chain partners.

In cases where product technology becomes obsolete rapidly, entire systems can be refurbished and resold. For example, there are

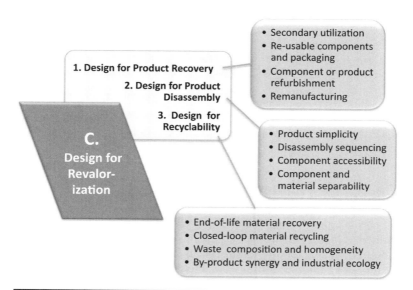

FIGURE 8.6 Design for revalorization.

thriving secondary markets for automobiles, cellular phones, and personal computers. If the entire system is either not operable or not marketable, the next priority is to disassemble it and try to recover valuable components (see Section C.2, Design for Product Disassembly). Although most salvaged components are resold to secondary markets, many companies are beginning to recover refurbished components for remanufacturing into new products (see the Caterpillar example in Chapter 10).

Design considerations are important in determining the end-of-life value of various components. Product developers can increase this value in a number of ways:

- Design components that are reusable for purposes of *closed-loop* remanufacturing

- Design components that are reusable for secondary applications due to their generic functionality, flexibility, or programmability

- Design reusable packaging—for example, Intel has saved millions of dollars annually by developing lighter-weight plastic trays that are used to move microprocessor units through the fabrication process and deliver them to customers

- Facilitate the nondestructive removal of components—for example, surface-mounted chips are difficult to recover because their tiny leads become bent

- Design components in a way that speeds diagnosis and refurbishment

Major computer manufacturers have established programs to recover value from their own equipment and, in a few instances, the equipment of other companies (see Chapter 11). Alternatively, component recovery can be accomplished via a turnkey arrangement with third-party salvage companies. For example, there are a variety of firms that specialize in recovering and reselling components from electronic products. With the advent of take-back legislation in Europe (see Chapter 3), most durable goods manufacturers have developed solutions for product and packaging recovery.

C.2 Design for Product Disassembly

The purpose of design for disassembly is to assure that a product system can be disassembled at minimum cost and effort. This is an important prerequisite to other end-of-life considerations, such as component separability and recyclability. If you can't get access to a component, no matter how valuable, then you cannot recover it. An added benefit of easy disassembly is that it generally contributes to product serviceability, which increases customer satisfaction.

Product Simplicity

Design elegance has long been a desirable attribute of products. For example, during the 1930s, the Bauhaus school of design formalized its criteria of elegance in terms of minimalism and functional orientation. With the advent of environmental consciousness, elegance in the form of simplicity has acquired a strong added motivation. Simplicity usually leads to lower manufacturing cost, lower material mass, greater durability, and easier disassembly for purposes of maintenance or asset recovery. There are several ways in which designers can try to achieve greater simplicity:

- Reducing the complexity of the product enclosures and assemblies in terms of their geometric and spatial design, as well as their functional operation.

- Reducing the number of distinct parts that are incorporated into a design; this is a well-known technique in the field of design for manufacture and assembly.

- Designing multifunctional parts that serve a variety of different purposes; e.g., using a single type of fastener for all assemblies.

- Utilizing common parts in a number of different designs, representing either different models in a product family or successive product generations.

- Using fewer different types of materials, which tends to lower costs by facilitating processes associated with procurement, manufacturing, and disassembly. Designing with fewer materials both facilitates identification and results in larger volumes of each material, potentially increasing the salvage value that can be obtained (see Section C.3, Design for Recyclability).

Disassembly Sequencing

The extent to which a unit, module, assembly, or component should be disassembled depends not only on the costs of disassembly, separation, inspection, sorting, and refurbishment, but also on the reuse, resale or salvage values. For example, it may not be cost-effective to disassemble products that contain many different, difficult to identify materials. Similarly, if components, assemblies and modules cannot be reused or refurbished, and if they contain largely nonrecyclable materials, little disassembly is warranted.

The sequence of assembly and the extent to which a product is disassembled can be represented hierarchically by a *disassembly tree*, as shown in Figure 8.7, which identifies all major modules and components in the product. There are a sequence of choices in order of priority to be made at each step in the sequence—reuse as is, refurbish and resell, disassemble, shred and recycle, or simply dispose of as waste.

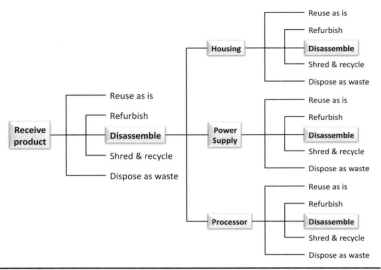

FIGURE 8.7   Example of a disassembly tree for electronic products.

Using mathematical optimization techniques, it is possible to maximize the net value recovered through product disassembly, based on information about the likely condition of the components, material and component salvage values, and average disassembly times.

Component Accessibility

At any stage of disassembly, the ability to remove a component or part is a key consideration. For example, parts that are embedded cannot be recovered easily for reuse. Moreover, if a part is embedded in an incompatible material, it makes it difficult to recycle the assembly. If the design does not allow for easily separable parts, then embedded parts should be constructed of recyclable and compatible materials. Likewise, separability of dissimilar materials is a key factor in the efficiency of disassembly operations (see Section C.3, Design for Recyclability). The following guidelines will contribute to speeding up the disassembly process and recovering a larger proportion of system components:

- Avoid springs, pulleys, and harnesses which complicate the disassembly process.
- Minimize the use of adhesives and welds between separable components or between incompatible materials. Adhesives introduce contaminants, can detract from quality due to the potential for bond failure, and increase the costs associated

with disassembly. If adhesives are required, try to use adhesives that are compatible with the joined materials.

- Use snap fits to join components where possible. Snap fits involve an undercut on one part, engaging a molded lip on a mating part to retain an assembly. Snap fits are relatively inexpensive to manufacture and have attractive mechanical properties.

- Avoid threaded fasteners (screws), if possible, because they increase assembly and disassembly costs.

- Use alternative bonding methods, such as solvent bonding or ultrasonic bonding. Such methods may be acceptable for bonding parts made from the same material and which will not be separated at end-of-life.

- Spring clips or speed clips can be an inexpensive and effective way of joining parts and materials. They permit easy assembly and disassembly, and do not introduce contaminants.

> **Example:** The End-of-Life Vehicle Directive, enacted by the European Union in 2000, imposed requirements on automotive manufacturers to design for disassembly and recycling (see Chapter 3). However, German companies such as BMW and Volkswagen had already developed disassembly methods at least 10 years earlier. For example, Volkswagen has operated a recycling plant since 1990, at which cars are dismantled, liquids are drained, wastes are separated for reprocessing, and some of the plastics are recycled into new automobile bumpers. Through design changes, such as elimination of threaded fasteners and part count reduction, they were able to significantly reduce disassembly time.

C.3 Design for Recyclability

A high priority in any DFE effort is to minimize waste by assuring that all by-products, materials, components, and packaging can be recovered and reused or recycled at the end of their useful life. While some companies have established "zero waste" as a goal, this is difficult to achieve in practice for both economic and pragmatic reasons. Therefore, design teams should make realistic assumptions about the proportions of end-of-life products that can be reused or recovered, and ensure that any residual wastes can be readily and safely recycled or disposed of.

To be recoverable with positive economic value, materials need to be as close as possible to the state of desired manufacturing feedstock. An important consideration in end-of-life material recovery is the ease of disassembly, which determines the residual economic value (see Section C.2, Design for Product Disassembly). Product materials or components that cannot be recovered economically become potential wastes but can still be recycled and thus converted into useful by-products.

Closed-Loop Material Recycling

Industrial processes can be designed or retrofitted to increase scrap utilization. With increasing attention to eco-efficiency, many companies have discovered that post-industrial scrap, such as trimmings, off-spec product, and even dust can often be recovered and reprocessed cost-effectively (see the Owens Corning example in Chapter 17). This is effectively a dematerialization strategy, because it directly reduces the quantity of materials required as process inputs (see Section A.2, Design for Source Reduction).

End-of-Life Material Recovery

When products are disassembled, materials need to be sorted into different categories for purposes of recovery and recycling. For this reason, using similar or compatible materials can greatly reduce the amount of end-of-life separation effort required. A key strategy for separability, and an inexpensive one, is to facilitate identification of materials by means of coding or marking. For example, the International Organization for Standardization (ISO) has developed a generic identification and marking standard, ISO 11469, to help identify plastic products for purposes of handling, waste recovery, or disposal.

Material homogeneity, purity, and reprocessability are important considerations in determining their recovery value. Recyclable materials include thermoplastics, engineering plastics, metals, and glass. As recycling technologies and materials science improve, we are reaching the point where recyclable materials can be found for virtually any application. Factors to consider in material selection include structural and aesthetic requirements as well as stability under varying conditions (e.g., temperature, moisture). For example, one of the factors inhibiting wider use of bio-based plastics is their limited tolerance for high temperatures.

Composite materials, such as carbon fiber composites used in tennis racquets, are prized for their light weight and superior mechanical properties; for example, the use of composite auto parts can improve fuel economy. However, composites can be problematic from an environmental point of view because they cannot easily be separated into their simpler and purer constituent materials. One viable approach is to grind composite materials for use as fillers and reinforcements.

In general, the recyclability of a material depends on a number of factors:

- The economic attractiveness of recycling the material and the existence of end-use markets
- The volume, concentration, and purity of the recycled material
- The existence of recycling and separation technologies and an adequate recycling infrastructure

While beverage containers are commonly recycled by consumers, other types of materials are more difficult to recycle because of the lack of qualified recyclers. Without a recycling infrastructure, it is pointless to specify recyclable materials. Many manufacturers have addressed this barrier by forming alliances within an industry sector; examples include the Vehicle Recycling Partnership created by USCAR and the European Recycling Platform.

Product packaging represents a universal opportunity for waste minimization, since it is generally discarded upon product use. Recycling of packaging materials is the next most desirable option once packaging mass has been minimized (see Section A.2, Design for Source Reduction). Great strides have been made in developing methods and technologies for the recovery and recycling of packaging materials. In the European Union, the Green Dot® program has been established to help companies comply with the Packaging Waste Directive (see Chapter 3).

Finally, it should be noted that recyclable materials may or may not originate from recycled materials. The issue of incorporating recycled content is addressed in Section A.1, Design for Energy and Material Conservation.

Example: One of the best examples of material recovery can be found in the aluminum container industry. According to the American Beverage Association, approximately 50% of aluminum cans are recycled in the United States, and, in other countries, the recycling rate is even higher. Within a few months of collection, the aluminum from a discarded can is incorporated into a new can, requiring only about 5% of the energy needed to produce an equivalent amount of "virgin" aluminum (see the Alcoa example in Chapter 17).

Waste Composition and Homogeneity

The following guidelines should be considered in order to ensure the recyclability of known waste streams:

- **Avoid material contaminants**—There are a number of potential contaminants that cannot easily be separated from product or packaging materials. Examples include adhesives, inks, paints, pigments, staples, and labels. Many manufacturers have begun to use integral finishes instead of painted finishes, which has advantages in terms of both lower manufacturing cost and recyclability.

- **Avoid discrete labels**—While discrete labels are sometimes necessary, integral labels are preferable from an environmental point of view. Discrete labels are often made of incompatible materials and introduce contaminants (e.g., adhesives, paper, inks). They must be completely removed before a part can be recycled, which greatly increases the material separation cost. Alternatives to discrete labels include molded-in or embossed labels. If discrete labels are necessary, it is best to avoid printed labels or to use compatible inks. Discrete labels can be manufactured from the same material as the base part and attached without use of adhesives.

- **Convert wastes into by-products**—with changes in both environmental awareness and economics, many materials that have traditionally been sent to landfill are being recycled. For example, as described in Chapter 18, utilities are increasingly marketing fly ash from boilers as an additive in construction materials. Regional networks are beginning to emerge where companies deliberately work together to find new uses for their manufacturing residuals (see Advanced Resource Recovery at the end of this chapter).

- **Design for waste incineration**—the last resort in the pollution prevention hierarchy, as an alternative to waste disposal, is conversion of waste to energy through incineration. Waste-to-energy facilities can be found in many parts of the United States that incinerate municipal solid waste, biomass (e.g., wood chips), or other wastes (e.g., automotive tires, railroad ties, utility poles) to generate steam and produce energy in the form of electricity. Hazardous wastes are typically not accepted for conversion to energy.

  Example: Discarded automobile tires, which used to sit in unsightly and hazardous refuse heaps, are now being recycled into a large variety of applications. In 2005, nearly 87% of the all scrap tires in the United States were utilized in one way or another. The major applications of scrap tires include fuel in cement production, power plants, and waste-to-energy incinerators; shredded tire applications in civil engineering, such as construction of lightweight backfill and leachate collection systems; and utilization of ground rubber in asphalt production, artificial turf production, molded products, and tire retreading.

D. Design for Capital Protection and Renewal

Capital protection and renewal is a fundamental requirement for the sustainability of an enterprise. This is an unfamiliar topic for designers accustomed to focusing on product quality and customer satisfaction. However, the broad scope of DFE encompasses many issues that are not visible to the customer, yet are vital to the success of the product and the company as a whole. Here, capital is used in the broadest sense, including human and natural capital as well as economic capital. For example, design decisions may be instrumental in improving process safety and reliability, thus maintaining continuity and productivity. Likewise, design decisions may facilitate access to raw material feedstocks based on natural resources, thus ensuring adaptability and sustainability.

D.1 Design for Human Capital

In the context of business enterprises, the term "human capital" refers to the skills and knowledge of management and employees. It is considered part of intellectual capital, which includes patents and proprietary knowledge, as well as relationships with customers and others in the supply chain [6]. Attraction and retention of talented, productive employees is one of the most important sources of competitive advantage for any business, as discussed in Chapter 4. Furthermore, human capital is influenced by the broader context of social capital. According to the World Bank, social capital refers to the institutions, relationships, and norms that shape the quality and quantity of a society's social interactions, and can be measured in terms of civil engagement and interpersonal trust.

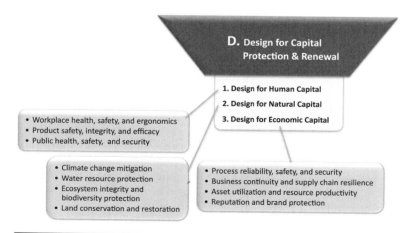

FIGURE 8.8 Capital protection and renewal.

When considering the environmental implications of product and process design, companies generally assign the highest priority to protection of human health and safety, including workers, customers, and the public at large. However, an additional important consideration is whether the products in question contribute to human well-being, dignity, and growth. A sustainable enterprise is concerned about human capital renewal—generating respect, commitment, and loyalty on the part of customers, suppliers, employees, and the broader stakeholder community. Just as the "customer experience" determines the success of a product, the "stakeholder experience" determines the long-term success of a company. The following are important guidelines for the design of products and processes from a human capital perspective.

Workplace Health, Safety, and Ergonomics

Traditional environmental health and safety practices include methods for monitoring and controlling routine releases of potentially toxic or hazardous substances, such as airborne emissions, liquids, and solid wastes. In the workplace, employees may be exposed daily to a variety of potential long-term health hazards, including:

- Process emissions such as fumes, vapors, mists, and dust
- Other airborne emissions, such as exhausts and cleaning fluid vapors
- Industrial chemicals, such as solvents, epoxies, or reagents
- Chronic disease agents, such as lead or asbestos
- Noise, radiation, and other potential hazards
- Repetitive motion and physical strain.

Accordingly, companies and regulatory agencies have developed extensive procedures for occupational health and hygiene, including protective equipment, ventilation hoods, exposure monitoring, and training. Material Safety Data Sheets (MSDS) are used routinely to communicate the toxic or hazardous properties of materials in the workplace. Ergonomic factors have been studied extensively, enabling the design of processes that prevent chronic injuries associated with lifting, bending, or other repetitive physical activities. Process safety management programs guard against fires, explosions, spills, and other life-threatening incidents (see Section D.3, Design for Economic Capital). Finally, the design of the workplace environment itself should be conducive to physical and mental health. There is increasing evidence that building occupants respond to improved comfort and well-being by working more productively, making fewer errors, and being absent from work less often [7].

Example: Herman Miller, a manufacturer of office furniture, is known for incorporating environmental design into high-quality products such as the famed Aeron chair. The company is also recognized as a leader in sustainable facility design. Herman Miller's headquarters was one of the first "green" office and manufacturing complexes built in the United States, and the enhanced workplace led to noticeable increases in employee satisfaction and productivity. The company has set ambitious goals for the year 2020: to eliminate solid and hazardous wastes as well as air and water emissions, to use 100% green electrical energy, to construct buildings to a minimum of LEED silver certification, and to have 100% of its sales from DFE-approved products.

Product Safety, Integrity, and Efficacy

A commitment to product stewardship implies making environmental, health, and safety considerations an integral part of designing, manufacturing, marketing, distributing, using, recycling, and disposing of products. This type of life-cycle commitment requires careful consideration of a broad range of possible scenarios and engagement with suppliers, customers, and other stakeholders. Points to consider during the design phase include the following:

- **Ensure supply integrity**—The globalization of material procurement has created vulnerabilities in the supply chain for many products, and there have been many examples of product contamination with harmful substances (e.g., melamine from China). To avoid such incidents, manufacturers should establish programs to screen and audit their suppliers, and ensure that all product constituents can be obtained from reliable sources.

- **Specify benign materials**—Designers should strive to avoid materials or ingredients that contain persistent, toxic, and/or bioaccumulative materials (see Section B.2, Design for Hazard Reduction). Substitution of benign materials that are non-toxic and biodegradable will reduce the burden of disclosure and exposure to potential liability, which is a risk even if no actual harm occurred.

- **Prevent product abuse or misuse**—Communication of proper procedures for product use is essential, including placement of warning labels where appropriate. Incorporation of safety features, such as child-proof closures for medicines and electrical surge protectors, has become a common practice.

- **Ensure responsible waste disposal**—Designers should strive to enable end-of-life recovery or recycling (see Section C.3, Design for Recyclability). However, they should also anticipate alternative customer behaviors and waste disposal scenarios, and try to ensure that the potential for improper disposal is minimized. For example, inkjet printer cartridges often come with a postage-paid envelope for mailing back the used cartridge.

Public Health, Safety, and Security

Company-owned facilities should be designed to protect the public from accidental exposure to health or safety threats, including hazardous materials, industrial processes, and moving vehicles or machinery. In an age of heightened concern about sabotage and terrorism, security precautions at industrial facilities have been tightened. However, there may be aspects of the product life cycle that are less carefully monitored, such as access to hazardous or flammable materials during transport. Enterprise risk management should include a review of the full product life cycle to ensure that all plausible scenarios involving public exposure and risk have been fully anticipated and appropriately addressed.

Many of the environmental issues addressed above, including process safety, product integrity, and waste management, have a bearing on public well-being. Stakeholder engagement should extend beyond the supply chain to include local communities, disadvantaged groups, and nongovernmental organizations that have an interest in public health and safety.

D.2 Design for Natural Capital

Natural capital refers to the ecological resources and services that make possible all economic activity, indeed all life. Ecological resource flows include edible organisms, sand, wood, grass, metals, and minerals, while ecological services include various forms of energy provided by the water cycle, wind, tides, soil, and pollination. Although ecosystem products and services are the foundation of our economy, they are excluded from typical energy and emissions accounting. For example, corn ethanol proponents overlooked the fact that land capacity is limited and that using crops for fuel could constrain our food supply. Even in the case of cellulosic biofuels, the use of agricultural wastes as a renewable energy source can hurt agricultural productivity by reducing the resilience of soil ecosystems.

As mentioned in Chapter 2, an understanding of potential threats to natural capital is essential for sustainable development, but industrial societies have tended to take these services for granted. Already, according to the Millennium Ecosystem Assessment, the majority of ecosystem services have been degraded due to unconstrained economic development. Climate change is causing polar ice to melt and sea level to rise, supplies of fresh water are dwindling, and irreplaceable ecosystems are being destroyed. Design strategies that can help to slow these effects and, thus, protect natural capital include the following:

- Radically increase resource productivity (see Section A, Design for Dematerialization)
- Mimic natural cycles by eliminating waste (see Section C, Design for Revalorization)

- Shift to a service-based economy (see Section A.3, Design for Servicization)

In addition, nations need to invest in the restoration and renewal of natural capital through improved agricultural practices, reforestation, and proactive conservation.

Climate Change Mitigation

The paramount environmental issue facing most businesses today is climate change. Driven both by stakeholder expectations and rising energy costs, companies are hastening to assess their carbon footprint and develop climate protection programs (see Chapter 3). Some mitigation of GHGs can be achieved immediately through energy conservation and more careful operations management, but the gains will be incremental. Energy conservation is attractive because it is both easy to implement and results in direct cost savings. According to the U.S. Environmental Protection Agency, which sponsors the Green Lights and Energy Star programs (see Chapter 3), every kilowatt-hour of electricity use avoided prevents the emission of approximately 1.5 lb. of CO_2, 5.8 g. of SO_2 and 2.5 g. of NO_x.

However, to fundamentally change the unsustainable trajectory of industrialized economies will take more than conservation—it requires disruptive innovation. Thanks to a flood of venture capital, a host of new technologies are emerging for reducing dependence on petroleum and generating energy from alternative, renewable sources (see Chapter 18). Many companies see an opportunity to gain competitive advantage through early adoption of energy-saving technologies. Some companies, particularly in the energy-related industries, are discovering new growth markets and building new businesses around climate change mitigation or adaptation. In all cases, companies need to build an awareness of energy implications into their product development processes (see Section A.1, Design for Energy and Material Conservation).

Greenhouse Gas Intensity Reduction

Energy conservation is certainly the first step in climate protection. A complementary strategy is substitution of *renewable* energy sources, such as solar and hydro power, for nonrenewable sources, such as fossil fuels. Natural resources are considered renewable if the rate at which they are replenished is sufficient to compensate for their depletion. However, as discussed in Chapter 9, life-cycle assessment studies have indicated that the full environmental impacts of some renewable sources, such as biofuels[†], may be greater than those of

---

[†]Renewable fuels based on biomass are effectively "borrowing" carbon dioxide from the atmosphere and then releasing it again through combustion, so there is no net increase in greenhouse gas concentrations.

nonrenewable sources, if one takes into account the energy and labor-intensive activities required to construct and operate the necessary equipment and facilities (see Figure 9.2).

Not all GHG emissions are due to fossil fuel combustion. There are many industrial and agricultural processes that emit GHGs; for example cement calcination (see Chapter 17). Herds of dairy cows and even landfills generate methane emissions. Therefore GHG reduction efforts should examine all emission sources and utilize "green chemistry" either to achieve reductions in GHG releases or to sequester GHGs so that they cannot enter the atmosphere. Chapter 9 provides additional information about carbon footprint assessment.

Finally, GHG emissions can be reduced by recovering and reusing waste energy, especially waste heat. It is a fundamental law of thermodynamics that some energy loss is necessary in order to create higher-quality energy, such as electricity. However, much of the energy consumed in our economy is wasted unnecessarily due to inefficiency and poor practices. Examples of energy recovery technologies at manufacturing facilities include combined heat-and-power systems, steam recovery from boilers, and synergistic activities, such as heating of adjacent fish ponds. One example of a waste heat recovery initiative at an Owens Corning plant is described in Chapter 17.

Water Resource Protection

The importance of water resources has been overshadowed by the climate change debate, but the global threats to water quality and availability are arguably more urgent. Over a billion people are without access to clean water, while the rate of depletion in freshwater resources continues to rise due to agriculture and other demands. Ironically, the "green revolution" enabled huge increases in crop yields to feed the world's population, but the new varieties of high-yielding crops are much more water-intensive—while food production has doubled, the corresponding water consumption has tripled [8]. Besides agriculture, other major consumers of fresh water are industrial activities, such as power generation and material processing, and, of course, municipal water supplies. Water is never depleted, of course, since it eventually returns to the earth, but water quality can be severely degraded through human or industrial contamination. Design strategies for protecting water resources include many of the same approaches used for dematerialization:

- Reduce the water intensity of the supply chain through elimination of water-intensive operations or through closed-loop recycling of process water. Note that the use of water-based technologies to reduce solvent emissions conflicts with this strategy.
- Reduce the water content of products by increasing their concentration or delivering them in dehydrated form. Note that

this strategy will significantly reduce the mass and volume of shipments, thus reducing transportation energy use.

- Reduce unnecessary flow-through of water by installing devices, such as automatic shutoff valves, flow reducers, and waterless urinals, and by shifting to low-water landscaping methods.
- Utilize alternative, lower-quality sources of non-potable water for industrial operations in order to reduce demand on municipal water supplies.
- Implement water treatment processes beyond compliance requirements in order to enhance the quality of wastewater discharges.
- Find beneficial secondary uses for process wastewater, such as floor maintenance.
- Collect rainwater as an alternative source of fresh water.

Many companies have emphasized water management programs as part of their overall sustainability commitment (see the Coca-Cola case study in Chapter 15). Going beyond facility water usage, life-cycle assessment methods are increasingly being used to assess the overall water "footprint" of a company's value chain from supplier to consumer (see Chapter 9).

Example: Intel Corporation uses ultra-pure water in its semiconductor fabrication plants, some of which are located in water-stressed areas such as Arizona and Israel. The Corporate Industrial Water Management Group supports the operating sites in implementing local strategies for sustainable water use. For example, at Intel's Chandler, Arizona facility, treated process water is sent to an off-site municipal treatment plant, brought up to drinking water standards, and reinjected into the underground aquifer at a rate of about 1.5 million gallons per day.

Ecosystem Integrity, Biodiversity, and Land Conservation

Ecological resources, including biomass, water, and genetic diversity, are nurtured in the world's ecosystems, which span a vast range of geographies and climates. Each ecosystem is unique, and contains a network of species that have adapted to the local or regional conditions. Instead of simply building facilities, occupying land, and ignoring the surrounding ecosystems, business entities can interact with them in several ways:

- Harvest resources, such as water, crops, or wood, without exceeding the capacity of the ecosystem to renew those resources.
- Protect resources by limiting vehicular traffic, creating buffer zones for sensitive habitats, and avoiding pollution or waste that would disturb the ecosystem.
- Restore damaged resources by reintroducing native plant and animal species, removing artificial barriers such as dams, and treating polluted water.

- Collaborate with regional governments and other stakeholders to implement ecosystem protection and renewal projects.

  Example: LaFarge, one of the world's largest cement companies, has partnered with the nonprofit WWF since 2000 to assure ecosystem sustainability and biodiversity in its worldwide quarry rehabilitation program. In Kenya, Lafarge's local subsidiary Bamburi Cement, has been developing expertise in landscape regeneration at Baobab Farm, a nature and wildlife park created by René Haller and Sabine Baer on a former limestone quarry site in Mombassa.

D.3 Design for Economic Capital

Economic capital, in accounting terms, refers to the productive assets of a company; and asset management typically focuses on assuring the productivity of company-owned facilities, process equipment, and working capital. Improving the utilization of company-owned assets is a fundamental approach toward increasing shareholder value because it reduces the capital investment required to generate the company's cash flow (see Chapter 4). Many companies have sought to improve their asset utilization by identifying underutilized assets, such as idle equipment, and recovering their value.

Process Reliability, Safety, and Security

The design and engineering of industrial operations typically involves trade-offs between cost, throughput, and reliability. Methods for increasing reliability, including redundancy, maintenance scheduling, and equipment monitoring, are beyond the scope of this book. However, one of the most important drivers of plant availability is process safety. A minor safety incident, even if it causes little or no damage or human injury, can disrupt operations and cause significant delays.

The historic mission of process safety management has been to prevent industrial accidents or releases of hazardous materials that may have adverse impacts upon human health, property, or the environment. Examples of potential acute hazards in an industrial facility include:

- Exothermic chemical reactions with potential for fire or explosion
- Handling and storage of flammable or corrosive gases or liquids
- Presence of multiple moving vehicles (e.g., at loading docks)
- Presence of high temperatures, pressures, or electric voltages
- Presence of dangerous process equipment, such as rotating machinery
- Requirements for entry into confined spaces (e.g., for cleaning tanks)

As in chronic risk management, many companies have gone beyond compliance to ensure safe working conditions and hazard minimization, and have instituted extensive training and emergency preparedness programs. A properly implemented process safety management program will not only reduce the likelihood and consequences of adverse incidents at operating facilities, but will also deliver direct economic benefits to the business. Examples of such benefits include improved reliability, reduced downtime, improved yield, reduced operating and maintenance cost, and, of course, reduced liabilities and insurance costs. To maximize the value of process safety, a company should consider the following guidelines:

- Minimize on-site hazards, such as storage of caustic or flammable materials and use of high-temperature or high-pressure vessels.

- Strive to recognize the underlying root causes of process safety incidents, and design systems and procedures to eliminate these root causes.

- Take advantage of capital investment opportunities, when building or upgrading facilities, to design them for *inherent* safety [9].

Business Continuity and Supply Chain Resilience

Business continuity refers to the ability of an organization to maintain operations despite disruptive events and to recover quickly from a business interruption. Recently, the supply chain management community has embraced the broader concept of *resilience*—the capacity of an enterprise to survive, adapt, and grow in the face of turbulent change (see Chapter 20). The trends toward globalization and outsourcing have resulted in global supply chains that are long, complex, and often beyond the control of the final manufacturer. Designing a resilient supply chain requires consideration of a variety of factors that can increase agility and flexibility or decrease vulnerability. For example, geographic dispersion of assets makes companies less vulnerable to natural disasters.

Many of the factors that influence resilience are connected with environmental issues. For example, the likelihood of supply chain disruptions may be higher when products and processes are sensitive to variations in climate conditions, constrained by the availability of scarce natural resources, dependent on a continuous energy supply, or susceptible to hazardous releases. Moreover, assurance of environmental and social responsibility will decrease the likelihood of deliberate supply chain interruptions by regulators, environmental groups, or local activists that are concerned about the environmental impacts of business operations.

Asset Utilization and Resource Productivity

Resource productivity generally refers to the rate at which value is derived from inputs to production, including materials, assets, and labor; for example, many companies measure sales per employee. In the case of facility operations, resource productivity can be measured in terms of annual production per dollar of invested capital. As it happens, this measure is closely linked with eco-efficiency, since the scale and complexity of plant and equipment assets determines the amount of energy, operating supplies, and maintenance required to support them. By following the maxim of "doing more with less" companies can develop more productive assets while reducing waste (see also Section A, Design for Dematerialization).

Many companies are beginning to apply Lean and Six Sigma techniques to environmental management (see examples in Chapter 14.) There is a wealth of literature on how to design leaner, more efficient manufacturing and logistics processes and systems, which is beyond the scope of this book. The emphasis here is on improving the environmental aspects of process design as part of integrated product development, thereby contributing to the overall performance and continuity of supply chain operations. Key considerations that drive resource productivity include:

- *Complexity*, i.e., the number of distinct unit operations and the characteristics of each unit, including set-up requirements, number of inputs and catalysts required, piping and instrumentation, and peripheral equipment such as pollution control devices.
- *Availability*, i.e., the fraction of time that a process is online and capable of operation, which is driven by set-up, inspection, and maintenance requirements as well as process reliability and frequency of interruptions.
- *Flexibility*, i.e., the range of different operations that can be performed by a specific process or production line as well as tolerance for variability. Greater versatility results in a smaller capital footprint and hence a reduced ecological footprint.

> **Example:** Velocys, Inc., founded in 2001, is commercializing a revolutionary chemical processing technology based on microchannel reactors, originally developed at Pacific Northwest National Laboratory. These small-scale, modular systems provide much higher throughput per unit volume and can be combined into large arrays, providing energy and chemical companies with substantial capital cost savings, improved product yields, and greater energy efficiencies (see Figure 8.9).

Reputation and Brand Protection

Besides contributing to the bottom line through reduced operating costs, insurance premiums, and capital costs, sustainable business

FIGURE 8.9 Microchannel reactors provide high throughput with small capital footprint.

practices contribute to shareholder value in a broader and more strategic way—by strengthening brand equity, reputation, human capital, alliances, and other important "intangibles" that can account for up to 90% of a firm's market value (see Chapter 4). It has been shown that proactive initiatives to address environmental issues can lead to new product innovation, development of new markets, and improved process technologies. For example, 3M and Bristol-Myers Squibb have incorporated product life-cycle review into their new product development processes, resulting in faster times to market and reduced compliance burdens (see Chapter 10). In addition, differentiation of a company through a reputation for corporate responsibility can enhance brand equity and strengthen its license to operate. For example, Dow Chemical and DuPont have been recognized as industry leaders through their initiatives to reduce air and water emissions in their global operations (see Chapter 13).

As discussed in Chapter 4, the most direct way to generate shareholder value through DFE is to develop innovative products and processes that address unmet needs in the marketplace. Examples include renewable energy systems, nutritional organically grown foods, and products designed for economically disadvantaged populations—the so-called "base of the pyramid." Even with conventional products and services, adherence to DFE principles can reduce the cost of ownership for customers and enhance customer loyalty. Many companies have succeeded in linking their brand with an "eco-friendly" image, improving consumer awareness and customer loyalty. However, if such marketing campaigns are not backed up with a genuine enterprise-wide commitment, they can be viewed as "greenwashing." Finally, transparency is critical; companies must live up to their social responsibility commitments by clearly disclosing their policies and practices, as well as their shortcomings.

Advanced Resource Recovery

Resource Cascading

One approach to Design for Environment that seeks to maximize the utility of resource usage is based on the theory of *resource cascading* [10]. The aim of this approach, illustrated in Figure 8.10, is to find a sequence of resource uses that extracts as much economic value as possible from a given resource as it evolves from higher-quality to lower-quality forms. Resource cascading incorporates many elements of the above guidelines, including dematerialization and revalorization.

For example, a batch of solvent used for degreasing in electronics manufacturing could, after a single use, be transferred to a metal cleaning operation. It could then be used in repeated cycles, until eventually being relegated to its lowest quality use, as a paint solvent. Finally, through purification and recycling, the spent solvent could again be shifted back up the cascade chain to a higher-quality use.

Similarly, in a closed-loop cascading approach, plastic materials can be used first for cosmetic parts, secondly for internal structural parts, and finally for base parts in a particular manufacturing process, before being recycled into a commingled stream. By the laws of thermodynamics, all materials and energy reach an equilibrium state of maximum entropy, but resource cascading enables us to capture as much economic value as possible during this decline.

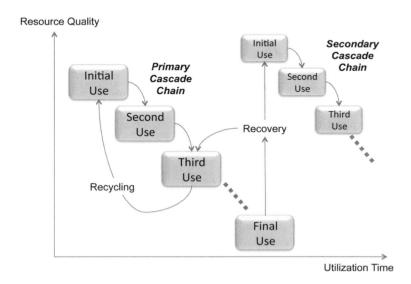

FIGURE 8.10   Illustration of resource cascading and relinking.

There are four basic principles associated with resource cascading:

1. **Appropriate fit**—reserve the highest quality resources for the most demanding uses.
2. **Augmentation**—increase the utility of a resource by extending utilization time or counteracting the decline of resource quality (e.g., through regeneration).
3. **Relinking**—at each consecutive link in the cascade chain, consider shifting the resource to a secondary cascade chain where its utility may be greater (e.g., by using waste or by-products as feedstocks for other processes).
4. **Sustainability**—ensure that the rate of resource consumption is balanced with the rate of resource regeneration.

Note that augmentation and relinking invariably involve some expenditure of resources, so that decisions about alternative pathways should take into account the associated net utility. Examples of resource utility maximization efforts include:

- Impeding quality decline (e.g., adding preservatives)
- Supplementing quality losses (e.g., chemical replenishment)
- Increasing use intensity (e.g., sharing among multiple processes)
- Improving durability (e.g., protective coatings)
- Separation of a substance into basic resources (e.g., solvent filtration)
- Recollection of dispersed material (e.g., cadmium reclamation from batteries)
- Regeneration of quality (e.g., thermal transformation).

While inorganic substances, such as metals, can often be relinked to the highest quality level through physical processes, organic materials must eventually be allowed to cascade through biochemical decomposition to the bottom of the chain, where they are regenerated into new life forms through solar energy.

The principles of resource cascading have already been applied in diverse areas, for example:

- Reuse of agricultural biomass to produce biofuels such as ethanol
- Repeated use of chlorinated solvents for degreasing operations
- Recovery of wood products for particleboard and pulp production
- Recycling of plastic materials used for packaging into textiles or roofing materials
- Energy cascading in multistep alcohol distillation.

Although it is tempting to consider "Design for Cascadability" as a product development practice, there are a number of challenges to this approach. It is difficult to anticipate the requirements of subsequent links in the chain when designing an initial product. Moreover, understanding these subsequent links may require collaboration among designers in different companies. The quality and reliability of relinked or augmented products is difficult to predict. Finally, and perhaps most challenging, there is little economic incentive to motivate the extra effort and expense of designing cascaded processes unless the benefits of cascading accrue directly to the original manufacturer (e.g., cascading of forest products in the pulp and paper industry). Nevertheless, with increasing interest in industrial ecology, resource conservation, and greenhouse gas emission reduction, new government incentive schemes and creative resource cascading partnerships among companies are certain to emerge.

Industrial Ecology

Industrial ecology, as mentioned in Chapter 6, advocates shifting from the traditional linear model of industrial systems to a closed-loop model that resembles the cyclical flows of natural ecosystems. In nature, there is no waste—one creature's wastes become another creature's nutrients. Rethinking conventional product and process technologies can lead to the discovery of innovative pathways for transformation of industrial wastes into economically valuable resources. This reduces the need for virgin natural resources, mitigates environmental pollution and greenhouse gas emissions, and alleviates the burden on limited landfill space [11].

The conversion of industrial process wastes to by-products, called *by-product synergy* (BPS), is a particular version of industrial ecology whereby companies collaborate to convert wastes into useful energy and materials, rather than operating as isolated entities. In simple terms, one facility's wastes can become another facility's feedstocks, leading to financial gains for both parties. The U.S. Business Council for Sustainable Development (USBCSD) originated the BPS concept in 1997 [12] and has helped launch BPS networks in several regions of the United States. One example is a state-wide Ohio By-Product Synergy Network, launched in 2009, that has engaged dozens of companies and will divert an estimated 30,000 tons of waste from landfill annually, saving about $3.5 million and avoiding about 660,000 metric tons of greenhouse gas emissions per year. In the United Kingdom, the National Industrial Symbiosis Program has attracted thousands of industry participants, diverted millions of metric tons from landfill, and generated significant economic and environmental benefits.

One of the earliest and best-known examples of industrial ecology sits on the coast of Denmark in the Kalundborg industrial region. Over a thirty-year period, beginning in the 1970s, a complex web of waste and energy exchanges has developed among a number of participants, including a Novo Nordisk pharmaceutical plant, Asnæs coal-fired electric power station, StatOil refinery, Gyproc plasterboard plant, Kemira, a sulfuric acid manufacturer, Biotek…nisk Jordrens, a soil remediation company, local farmers, and others (see Figure 8.11).

The Novo Nordisk plant produces 40% of the world's supply of insulin, as well as industrial enzymes. One million tons per year of sludge by-product from pharmaceutical processes are used as fertilizer for nearby farms. 5000 tons per year of surplus yeast from the pharmaceutical plant's production of insulin is shipped to farmers for pig food. The 1500 megawatt power station uses refinery gas from Statoil to generate electricity and steam, which is supplied to both Statoil and Novo Nordisk for heating of their processes. A local cement company uses 135,000 tons per year of fly ash from the power plant, while 200,000 tons per year of gypsum produced by the power plant's flue gas desulfurization process are shipped to Gyproc, which produces gypsum wallboard.

These synergistic relationships and efficient use of resources have led to numerous benefits including decreased water demand, reduced energy consumption, and reduced airborne emissions [13]. The water demand of the "big four" industry participants has dropped 20 to 25 percent, a reduction of 1.2 million cubic meters per year. At the same time, oil consumption has been reduced by 19,000 tons per year, coal consumption by 30,000 tons per year, CO_2 emissions by 130,000 tons per year, and SO_2 emissions by 25,000 tons per year.

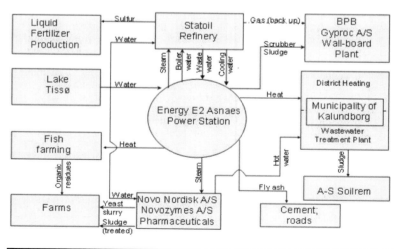

Figure 8.11   Industrial symbiosis in the Kalundborg region of Denmark.

Part 3 of this book includes numerous examples of companies that have adopted industrial ecology practices. These include:

- Recovery and recycling or refurbishment of used electronic components (Chapter 11).

- Utilization of renewable materials and conversion of wastes into by-products in automotive manufacturing (Chapter 12).

- Incorporation of recycled content into packaging, as well as recycling of packaging materials, in the pharmaceutical, food and beverage, and consumer products industries (Chapters 14–16).

- Diversion of waste streams from landfills into fuels or feed-stocks by companies in the materials and energy industries (Chapters 17–18).

References

1. E. U. von Weizsäcker, A. B. Lovins, and L. H. Lovins, *Factor Four. Doubling Wealth, Halving Resource Use* (London: Earths can. 1997).
2. F. Schmidt-Bleek, "Factor 10: Making Sustainability Accountable—Putting Resource Productivity Into Praxis," Factor 10 Institute, Provence, 1999.
3. Green-Blue Institute, *Design Guidelines for Sustainable Packaging*, Charlottesville, Va., 2006.
4. United Nations Environment Programme, Division of Technology, Industry, and Economics, *Product-Service Systems and Sustainability*, Paris, France, 2002.
5. Millennium Ecosystem Assessment, *Ecosystems and Human Well-Being: Current State and Trends* (Washington, D.C.: Island Press, 2005), Chapter 15, "Waste Processing and Detoxification," p. 429.
6. T. A. Stewart, *Intellectual Capital: The New Wealth of Organizations* (New York: Currency Doubleday, 1997).
7. J. Heerwagen, "Do Green Buildings Enhance the Well Being of Workers?" *Environmental Design+ Construction*, 2000, 3(4): 24–30.
8. F. Pearce, *When the Rivers Run Dry* (Boston: Beacon Press, 2006).
9. T. Kletz, *Plant Design for Safety: A User-Friendly Approach* (New York: Hemisphere, 1991).
10. T. Sirkin and M. ten Houten, *Resource Cascading and the Cascade Chain: Tools for Appropriate and Sustainable Product Design*, Interfaculty Department of Environmental Sciences, University of Amsterdam, Netherlands, 1993.
11. J. Fiksel, "Sustainable Development through Industrial Ecology," in R. L. Lankey and P. T. Anastas, Ed., *Advancing Sustainability through Green Chemistry and Engineering* (Washington, D.C.: American Chemical Society, 2002).
12. BPS examples and case studies are available at http://www.usbcsd.org/byproductsynergy.asp
13. N. Jacobsen, Industrial Symbiosis in Kalundborg, Denmark, *Journal of Industrial Ecology*, 2006, 10 (1–2): 239–255.

CHAPTER 9

Analysis Methods for Design Decisions

DFE Principle 5. *Develop and apply rigorous, quantitative tools to analyze the environmental, economic, and other consequences of design decisions and to weigh the trade-offs of alternative choices.*

Analyze This!

According to the integrated product development framework described in Chapter 5, *analysis methods* are required to enable systematic assessment and quantification of environmental performance for a new product or process design. Assuming that metrics have been defined and objectives have been set, analysis methods can be used to support design decisions in several ways:

- To evaluate the degree to which a particular design meets cost or performance objectives and requirements.

- To compare alternative designs and evaluate their relative merits.

- To identify potential design improvements and evaluate their expected benefits.

Although quantitative methods are preferable for purposes of continuous improvement and competitive benchmarking, in many cases quantification is difficult and qualitative methods can serve adequately.

The first principle of DFE is to consider all stages of the product life cycle, including raw material extraction, processing, transport, component manufacturing, product assembly, distribution, end use, service, disposal, and recycling. This makes rigorous analysis of environmental performance quite challenging. While product development teams need not analyze each life-cycle stage in exhaustive detail, they need to be aware of those environmental issues, at any stage of the life cycle, that are relevant to either customer needs or technical and financial constraints.

Table 9.1 lists the different types of methods available for each of the following purposes.

- **Screening** and comparison methods are used to narrow design choices among a set of alternatives; examples include threshold limits for chemical properties such as biodegradability, and material selection priority lists based on recyclability or other characteristics.

- **Performance assessment** methods are used to evaluate the expected performance of designs with respect to particular objectives; examples include calculation of expected material recovery rates and estimation of environmental concentrations associated with specified emission levels. Assessment methods may range from simple qualitative indices to life-cycle assessment studies to sophisticated numerical simulation models that assess environmental impacts.

- **Trade-off analysis** methods are used to compare the expected cost and performance of several alternative design choices with respect to one or more attributes of interest. Such methods may include quality-function matrices, "what-if" simulations, cost/benefit analysis, or other types of integrated analyses that draw upon various performance assessment methods described above. For example, parametric analysis

	Screening & Comparison	Performance Assessment	Trade-off Analysis
Tangible evaluation	Acceptance requirements	Functional testing	
Qualitative assessment	Criteria-based checklists	Qualitative indices	Scoring matrices
Environmental analysis	Footprint indicators	Life-cycle assessment	Predictive simulation
Risk analysis	Vulnerability assessment	Quantitative risk assessment	Integrated risk evaluation
Financial analysis		Life-cycle accounting	Cost/benefit analysis

TABLE 9.1 Overview of DFE Analysis Methods

can be used to show the relationships between environmental metrics and other cost or performance metrics, such as reliability and durability.

A full treatment of these many approaches is beyond the scope of this book. However, the following sections describe selected environmental performance assessment methods, along with examples of their application. The five categories of methods discussed below correspond to the rows of Table 9.1. Chapter 10 discusses the integration of these methods into a product life-cycle management process.

Tangible Evaluation

Tangible evaluation methods involve real-world physical or chemical procedures for assessing the characteristics of a product or component. Unlike the rest of the analysis methods discussed below, tangible evaluation provides observable results rather than theoretical predictions based on models and assumptions. Applications of these methods include:

- **Acceptance requirements** for design choices; for example, the criteria for material selection may specify physical properties, such as moisture resistance, strength, color, or other characteristics that can be evaluated directly. Relevant to DFE, the USDA has introduced a bio-based material standard that relies on carbon dating to identify the nonfossil fraction of carbon atoms in a given material.

- **Functional testing** of design performance; for example, electronic products are frequently subjected to actual use tests that simulate the wear and tear associated with repeated use or abuse, such as being dropped onto a hard surface.

Qualitative Assessment

Qualitative assessment methods have a number of obvious advantages over quantitative methods—they are easier to apply, require minimal data, and can be useful in spite of large uncertainties. For many companies, such methods are the logical first step in implementing DFE because they can provide value without requiring large resource expenditures. Qualitative methods are generally used to test whether product design choices satisfy certain environmental criteria, such as recyclability, energy efficiency, and absence of toxic constituents.

In practice, qualitative assessment methods have countless variations; the following illustrates two common types of methods: *checklists* and *matrices*.

Criteria-Based Checklists

The simplest and most widely used qualitative assessment tool is a checklist of criteria, stated in the form of questions or points to consider. The use of checklists is often one of the first DFE initiatives undertaken by product development organizations because they require only modest resources to update and maintain and are easy to understand and implement. There are several different varieties of checklists:

- **Material selection criteria**—Perhaps the most common DFE checklist approach is a list of materials to be consulted by design engineers when specifying parts or ingredients (an example is the Greenlist™ process developed by SC Johnson, described in Chapter 8). This can take two forms—a list of *preferred* materials based on environmental considerations, or a list of materials to be *avoided* because of regulatory restrictions and environmental concerns.

- **Supplier selection criteria**—The environmental "footprint" of a product or process is determined by the material and energy flows that characterize its life cycle. For example, automobiles and computers are assembled from a variety of components that typically are manufactured by one set of suppliers and recovered or recycled by another set of contractors. Companies that are highly leveraged in this way have begun to review the environmental performance characteristics of their suppliers and contractors. For example, as described in Chapter 11, worldwide electronics manufacturers have adopted a standardized supplier assessment tool that is based on the Electronic Industry Code of Conduct.

- **Product or process design criteria**—Design checklists are essentially a formalization of guidelines like the ones discussed in Chapter 8. They usually consist of a series of questions or criteria that address specific DFE considerations and are applied as a form of design review. The best time to use such checklists, of course, is during the concept development stage, but they are also useful as part of a milestone review to assure that the product team has considered relevant environmental concerns.

Despite their advantages, checklists do have important limitations:

- Checklists are qualitative in nature, even though it is possible to compute numeric scores. This means that they provide only crude measures of performance improvement. For example, a supplier checklist might pose the question "Do you have a waste minimization program?" A simple answer of "yes" does not convey much information. More detailed performance evaluation would take into account the baseline waste stream, the types of wastes, the difficulty of recycling, and the level of improvement achieved.

- Multiple checklists that reflect a large number of guidelines will often produce conflicts between different attributes of the design. For example, conflicts may arise between mass and recyclability (e.g., using polymers versus metals in automotive applications), reusability and energy consumption (e.g., refurbishing an old piece of equipment vs. acquiring a new energy-efficient one), and between toxic chemical use and energy consumption (e.g., using mercury-containing compact fluorescent lamps versus incandescent lamps) [1].

- Checklists provide no guidance to product developers regarding the relative importance of different issues or the degree of effort that is warranted in addressing a specific issue. For example, is it more important to reduce source volume or to assure recyclability? Is a 10% reduction in waste a reasonable goal? How much of the R&D budget should be committed to achieving these goals? These are challenging questions that can only be answered through a more rigorous trade-off analysis.

- Checklists can actually reduce creativity by encouraging a false sense of complacency. People who have worked through the checklist in a mechanical fashion may feel that they have done all that is necessary to consider environmental issues. Thus, they may fail to become sufficiently involved in DFE exploration and may overlook important opportunities or problems that are not covered on the list.

Nevertheless, checklists are an effective starting point for encouraging organizations to think about environmental issues and to begin taking positive actions.

Scoring Matrices

The use of aggregation and scoring techniques for interpreting environmental metrics was discussed in Chapter 7. Despite the limitations of these methods, qualitative scoring matrices can be a useful technique for trade-off analysis in design decisions. They involve creating a matrix diagram in which the rows represent competing options or objectives and the columns represent design attributes. Various indexing and scoring methods can then be applied, based on available data and/or subjective judgments, to derive categorical or numerical ratings. The assigned "scores" are seldom physically meaningful in an absolute sense but can be used to distinguish the *relative* environmental impact of alternative approaches. There are many variations on this basic technique, including scorecards, rating schemes, and "traffic light" signal charts. For example, Table 9.2 shows one component of a comprehensive scoring system used by Norm Thompson Outfitters to rate footwear and clothing designs [2]. Scoring matrices have many of the same advantages as checklists, and they can provide an easy means for evaluating trade-offs and representing more subtle interactions among design criteria.

Score	Examples of Materials	Criteria
3	Sustainably harvested/collected forest products, sustainable/organic agricultural products, (e.g., sustainably sourced vegetable-tanned leather or latex), pied au naturale, cork	• Materials without hazards across life cycle • Natural materials produced in sustainable manner • Biodegradable and/or easily recyclable • Recycled content for either natural materials or synthetics • No toxics in raw material processing
2	Recycled cotton, recycled metal, recycled plastic, hemp (natural retting) recycled polyester	• Materials with minor hazards in life cycle, but no carcinogens • Moderate recycled content • Biodegradable and/or potentially recyclable • Sustainably sourced materials • Toxics, if used, produced on-site and consumed or detoxified on-site; final product not toxic
1	Hemp (chemical retting), linen, common oak	• Materials with some hazards in life cycle, but no carcinogens • Recycled content • Downcyclable • Some ecosystem impacts

TABLE 9.2 Sustainability Scorecard for Material Selection: Soft Goods in Footwear Design

0	Vegetable-tanned leather, cold galvanized metal, ethyl vinyl acetate (EVA), conventional wool, polypropylene, polyethylene (LDPE, HDPE, LLDPE)	• Materials with moderate hazards in life cycle, but no carcinogens • Virgin/recycled content • Moderate ecosystem impacts
−1	Nylon, Delrin® (high molecular weight acetal, also known as polyacetal, polyoxymethylene, or polyformaldehyde), conventional cotton, polyester, plated metals (non-nickel, non-chrome), wood products—non-sustainably harvested	• Materials with moderate hazards in life cycle, but no carcinogens • Minimal recycled content • Moderate ecosystem impacts
−2	Chrome III (Cr3+)-tanned leather, stainless steel (300), PU and PU foam, Teflon/PTFE, spandex and spandex blends	• Materials with significant hazards in life cycle, but no carcinogens • Virgin content • Significant ecosystem impacts
−3	PVC and PVC coatings, nickel and nickel plate, chrome plate, chrome VI (Cr6+)-tanned leather, PU coatings (waterproof/breathable)	• Materials with extreme hazards (endocrine disrupters, carcinogens, high-level acute toxicity, etc.) in life cycle • Virgin content • Not biodegradable or recyclable • Extreme ecosystem impacts

TABLE **9.2**   Sustainability Scorecard for Material Selection: Soft Goods in Footwear Design *(continued)*

One popular approach that employs such matrices is Quality Function Deployment (QFD), which uses a "house of quality" model to make explicit the relationships between customer desires and product design parameters. This is essentially a matrix whose rows represent desirable properties of the product and whose columns represent controllable and measurable parameters of the design; interactions between these parameters can be represented in the "roof" of the house. The QFD analysis can be performed recursively by further analyzing and dissecting each of these parameters. A number of computer-based tools are available for creating, displaying, and printing QFD matrices.

Scoring and indexing schemes are particularly useful for capturing complex outcomes, such as ecological impacts, in cases where the absence of adequate data and validated models makes quantitative assessment impossible. However, by providing a mechanistic, repeatable algorithm, scoring methods can lead to false confidence in results that are at best approximate and occasionally outright wrong. It can be argued that such methods encourage "environmental illiteracy" on the part of design teams. For example, without delving into the logic of the system, it may be difficult for designers to understand what technical changes might result in an improved score.

> **Example:** An early example of a qualitative DFE matrix was adopted by AT&T for material selection [3]. The approach arrays alternative technologies (rows) against life-cycle stages (columns) and indicates both the magnitude and uncertainty of the impact in each cell of the matrix through graphic icons. Without requiring detailed data or laborious calculations, this approach offers useful insights into the environmental preferability of different technologies. For example, the matrix was used to assess three alternatives to lead solder in electronic devices—indium alloys, bismuth alloys, and conductive epoxies. The results suggested that, from a life-cycle perspective, lead is preferable to the alternatives because the latter tend to utilize more resources and generate more emissions in the extraction and processing stages.

> **Example:** A more recent example is a rating system developed by the Green Electronics Council (GEC) to help consumers choose products with environmentally preferable designs. The EPEAT (Electronic Product Environmental Assessment Tool) evaluates products on 51 environmental criteria in eight categories, including materials use, design for end of life, and packaging. GEC estimates that 2007 purchases of EPEAT-certified products have reduced the use of primary materials by 75.5 million metric tons (equivalent to the weight of more than 585 million refrigerators) and reduced the use of toxic materials, including mercury, by 3,220 metric tons (equivalent to the weight of 1.6 million bricks). In addition, the electricity savings are 42.2 billion kilowatt-hours—enough to power 3.7 million U.S. homes for a year.

Environmental Analysis

Not surprisingly, environmental analysis methods play an essential role in DFE. The purpose of these methods is to help designers understand the implications of alternative design choices in terms of their expected effects on ecological resources. However, ecological systems are complex, dynamic, and not well understood, so the available methods have inherent limitations. While these methods are based on scientific assumptions, they are generally impossible to validate, so the conclusions should be used with caution. This section focuses on the two most commonly used categories of methods—footprint indicators and life-cycle assessment. There is some overlap in that the calculation of an environmental footprint over the full product life cycle necessarily involves the use of basic life-cycle assessment methods.

Footprint Indicators

The term "footprint" has become popular in the environmental lexi-con, but it is used so loosely as to be virtually meaningless. Most practitioners conceive of the environmental footprint of a company, a household, or a community as being an aggregate measure of the total burden that it places on the environment. However, some have interpreted this in terms of a single metric, such as a "carbon foot-print" while others have interpreted it as a collection of indicators representing different environmental burdens (e.g., energy use, solid waste, air emissions). In the latter case, plotting these indicators on a "radar chart" enables a company to track its progress over time as the footprint shrinks toward zero.

A variety of different boundaries commonly used for footprint analysis are depicted in Figure 9.1. For example, an energy con-sumption footprint may include only nonrenewable energy sources (e.g., petro-fuels, coal, nuclear) or may include renewable sources (e.g., solar, wind, geothermal). A material footprint may analyze total mass throughput, may focus only on consumption of input materials, or may focus on wastes, which in turn may include solids, liquids, and/or airborne emissions. Note that a carbon footprint focuses on just one type of airborne emissions, greenhouse gases (GHG). Fur-thermore, a material footprint may include only products purchased within the economy, may include consumption of materials derived from ecological sources, such as biomass (e.g., grass, wood, fish), or may include ecological resources that are not consumed but can be degraded (e.g., water).

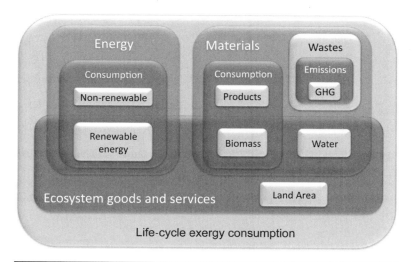

FIGURE 9.1 Alternative boundaries for scope of environmental footprint analysis.

A more challenging task is quantifying the environmental footprint of a company or other entity in terms of the ecosystem goods and services that are required to support its operations. For example, one technique below uses land area as an indicator, but it is difficult to capture the many important ecosystem services that we depend upon, such as climate regulation, water purification, and pollination (see Chapter 2). As discussed later in this chapter, new scientific methods based on **life-cycle *exergy* consumption** make it possible to quantify all of the indicators shown in Figure 9.1 within a common framework.

To complicate matters further, the scope of a footprint analysis can vary enormously based on the chosen life-cycle boundary. As illustrated in Chapter 1, Figure 1.1, a materials or energy footprint may be confined to the direct operations of a company or facility, it may extend to indirect activities associated with purchased goods or services, or it may encompass the full breadth of ecosystem goods and services. In published environmental reports, most companies are satisfied to quantify their environmental footprint in terms of direct consumption and direct generation of waste and emissions. As noted in Chapter 3, efforts have begun to include the full supply chain footprint in GHG inventories. While actual use of indicators to measure ecosystem service consumption is rare, many companies are beginning to track their water footprints.

The following briefly describes a number of methods for calculating different types of environmental footprints—carbon, materials, land area, and water.

Carbon Footprint

A carbon footprint can be calculated by taking an inventory of the total greenhouse gas emissions for a company, facility, product, community, family, or any other entity. The Kyoto Protocol identifies six greenhouse gases: carbon dioxide (CO_2), methane (CH_4), nitrous oxide (N_2O), sulfur hexafluoride (SF_6), perfluorocarbons (PFCs), and hydrofluorocarbons (HFCs). Each greenhouse gas has a global warming potential that can be expressed in terms of equivalent CO_2 [4]. Carbon footprints are typically organized in terms of three successively broader scopes, covering the following GHG sources:

- Scope 1: Fuel combustion in vehicles or facilities that are directly owned and/or controlled
- Scope 2: Purchased electricity from fossil fuel combustion (e.g., coal, oil, natural gas)
- Scope 3: Other indirect sources of GHG emissions (e.g., waste disposal, business travel)

The methodology for Scope 1 and Scope 2 assessment is straightforward—it involves enumerating emission sources and estimating their emissions based on standard "emission factor" coefficients. The results can either be expressed in absolute terms, i.e., CO_2-equivalent metric tons per year, or in normalized terms, e.g., CO_2-equivalent kg per sales dollar, per kg of product output, per employee, or per square foot of space. The accepted practice for Scope 3 is to allow considerable latitude in the inclusion of indirect emissions [5]. As mentioned in Chapter 3, establishment of the PAS 2050 international standard will help to achieve greater precision and uniformity in the estimation of Scope 3 GHG emissions using life-cycle inventory methods.

Materials Footprint

A mass-balance approach called *material flow analysis* (MFA) is widely used to estimate the total material and waste burdens generated by an economic system or a specific enterprise. MFA calculates the mass of materials entering and leaving a defined system boundary, as illustrated in Figure 9.2 [6]. This method is used in Europe, Japan, and other nations for purposes of material flow accounting, and provides several useful indicators for measuring the "mountain of waste" that was described in Chapter 1:

- Domestic Material Consumption (DMC), calculated by subtracting Exports from Direct Material Inputs, can be used to measure per capita material consumption. Studies have shown that to support the lifestyle of the average European resident requires a DMC of about 44 kg/day, or close to 100 lb/day. The majority of these materials are construction minerals, fossil fuels, and biomass from agriculture.

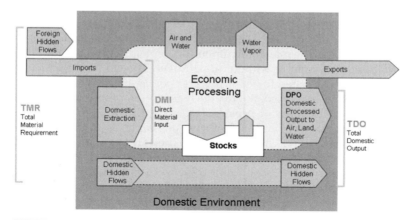

FIGURE 9.2 Overview of material flow accounting model (adapted from [6]).

- Total Material Requirements (TMR) measures the "rucksack" of indirect "hidden flows," including mining wastes and other discards, which are carried along with direct material inputs, but generate no economic value and may disturb the natural environment. For the average European, TMR is estimated to be about 220 kg/day. The size of the rucksack can be significant; for example a diamond ring weighing 10 g has a rucksack of about 6000 kg, while an average newspaper has a rucksack of 10 kg [7].

- Material intensity can be measured as the ratio of DMC to gross domestic product (GDP). MFA studies suggest that the E.U. economy has become more eco-efficient, since material intensity slowly declined from about 1.2 kg/€ in 1992 to about 1 kg/€ in 2000 [8]. However, the absolute DMC continues to increase due to economic growth.

Similarly, by drawing the boundary around an enterprise or a specific product system, MFA can be used to measure its material intensity. However, reliance on mass flow indicators can be deceptive for several reasons. First, not all materials are equal in terms of their environmental impacts, and MFA does not try to distinguish materials in terms of toxicity or other properties. Second, MFA often does not account for the hidden environmental burdens associated with imported materials. In an economy where global sourcing is increasingly the norm, the question of allocating accountability for these upstream material flows remains challenging.

Land Area Footprint

A technique called "ecological footprint" uses land area (hectares) as a metric for estimating the productive capacity needed to support both resource consumption and waste absorption for a specified economic activity such as power generation [9]. This footprint can be interpreted as the burden placed on the *carrying capacity* of ecosystems, which is the maximum amount of replenishment per unit time that they can support without impairment. The worldwide carrying capacity is estimated to be 2.1 hectares per capita, of which 1.6 hectares are land-based ecosystems, such as forests, pastures, and arable land and 0.5 hectares are ocean areas.

The average ecological footprint per capita is estimated to be 12.3 hectares in the United States, 7.7 hectares in Canada, and 6.3 hectares in Germany. Hence the oft-quoted statement that it would take three planet Earths to support the world's population if they all adopted the lifestyle of a "developed" nation. The average footprint for all nations is estimated to be 2.8 hectares per capita, suggesting that humanity has already overshot global capacity and is depleting the available stock of natural capital, rather than "living off the interest." While it has been used primarily at a national level, ecological footprint analysis can be applied at any level of granularity, from an entire nation to a single individual. Thus, it offers a method for estimating the ecological burden associated with economic enterprises, supply chains, or communities [10].

Water Footprint

The water footprint of a product or an entire enterprise can be defined as the total annual volume of fresh water that is used directly in operations and indirectly in the supply chain [11]. Some industries, such as food processing, use a great deal of water. Even businesses that use very little water for manufacturing may still have a substantial supply-chain water footprint if their raw materials come from agricultural sources; for example, the footprint for cotton garment manufacturers includes a large amount of irrigation water.* In addition, consumers that launder those garments use a considerable amount of water over the product life cycle, and this can be considered as part of the footprint from an extended producer responsibility perspective. In the United States, nearly 25% of all fresh water use is process water for steam turbines in electric power generation, and this is often the largest component of a company's water footprint. The water footprint of common foods can range from about 1000 liters per kg of grain to about 16,000 liters per kg of beef [12].

Water use can be measured in terms of water volumes consumed, i.e., evaporated, and/or polluted per unit of time. Thus, water that is simply "borrowed," as in hydroelectric power generation, does not count as usage. The water footprint can be split into three elements:

1. Blue water: the volume of fresh water that was evaporated from surface water or ground water resources.

2. Green water: the volume of fresh water that was evaporated from rainwater stored in the soil as soil moisture.

3. Grey water: the volume of polluted water, calculated as the volume of water that was required to dilute pollutant discharges in order to meet water quality standards.

Unlike greenhouse gases, the ecological or social impact of a water footprint depends not only on the volume of water use, but also on the geographic locations and timing of the water use. Water-stressed or arid regions are more vulnerable to water use, especially during dry seasons.

---

*The total water footprint associated with a product is sometimes called "virtual water."

Life-Cycle Assessment

Life-cycle assessment (LCA) methods are used to estimate the net energy or material flows associated with a product life cycle as well as the associated environmental impacts [13]. The Society for Environmental Chemistry and Toxicology was the first organization to develop a standard methodology for LCA in the early 1990s, involving the following steps:

1. **Goal and Scope.** Define the product, process, or activity to be assessed and the goal, scope, and system boundaries of the assessment.

2. **Life-Cycle Inventory.** Develop a system-wide inventory of the environmental burdens by identifying and quantifying energy and materials used and wastes released to the environment at each stage of the life cycle.

3. **Life-Cycle Impacts.** Assess the impacts of those energy and materials uses and releases upon the environment and/or human health.

4. **Interpretation.** Evaluate the results and implement opportunities for improvement.

The original LCA methodology has been updated and standardized through guidelines developed by the International Organization for Standardization (ISO 14040:2006 and ISO 14044:2006). These guidelines ensure that all assumptions are transparent, that the system boundaries and functional unit of analysis (i.e., product or service value delivered) are clearly defined, and that data quality, uncertainty, and gaps are clearly stated.

Traditional LCA studies have been based on "bottom-up" analysis of specific industrial processes along the supply chain, which can be burdensome. Although many companies have adopted LCA inventory methods, the use of impact analysis is more controversial. There are a number of limitations to the LCA methodology:

- Rigorous application of LCA requires specialized expertise and training, and can involve considerable time and expense.

- Process-level data are difficult to obtain and may have large uncertainties, especially with new technologies that have not been in widespread use.

- Drawing system boundaries is necessary but may omit important stages in the upstream supply chain or downstream product use chain.

- Inventory assessment alone is inadequate for meaningful comparison, yet impact assessment is fraught with scientific difficulties.

- Conventional LCA does not account for ecosystem goods and services and the impacts of renewable resource use

However, with appropriate definition of system boundaries, LCA can be useful for identifying the environmental advantages or drawbacks of various design options, thus supporting product development decisions [14]. Caution should be exercised in using the results of such analyses for external marketing and communication, such as comparative product claims. A good illustration of LCA application to consumer products is provided by the Kimberly-Clark example presented in Chapter 16. Another example of an industry-wide LCA for automobiles is shown in Chapter 12, Figure 12.2.

Impact Assessment

Within the life-cycle assessment framework described above, the third and most challenging step is assessment of the impacts associated with resource use and environmental emissions during each life-cycle stage—acquisition, manufacture, transport, use, and disposal of products. These impacts may include environmental, health or safety impacts upon humans and ecosystems, as well as economic impacts such as land use restriction and resource depletion. Moreover, impacts may be local, regional or global in nature. The assessment of impacts is problematic because we have a relatively poor understanding of the complex physical and chemical phenomena that determine the fate and effects of substances released to the environment. Despite a great deal of continuing scientific research, our knowledge remains fragmentary and largely theoretical. In some cases, such as greenhouse gas emissions or energy consumption, the impacts are cumulative and broadly distributed, but in other cases, such as mercury emissions or water resource consumption, the impacts are highly localized and dependent upon specific environmental conditions.

There is a vast literature on environmental impact assessment (EIA), mainly oriented toward the evaluation of proposed policies or projects that may affect the environment. In the United States and many other countries, EIA is a legal requirement prior to the initiation of major construction or development projects. However, most of the methods used in this field are not appropriate for product development purposes because they are detailed and site-specific; whereas LCA is applied at a broader system level. Instead, life-cycle impact analysis uses simplified models that provide *relative* measures of impact within broad categories. These categories reflect "midpoint" indicators of potential impact rather than final endpoints. For example, the TRACI tool developed by U.S. EPA uses the following categories [15]:

- **Stratospheric ozone** depletion due to airborne emissions of substances, such as chlorofluorocarbons (CFCs) and halons.

- **Global warming potential** due to the build-up of greenhouse gases, such as carbon dioxide, methane, and nitrous oxide.

- **Acidification** due to emissions that increase the acidity of water and soil systems.

- **Eutrophication**, meaning excessive plant growth (e.g., algae) in surface waters due to elevated nutrient levels caused by fertilizer runoff.

- **Tropospheric ozone formation** (i.e., smog) due to emissions of nitrogen oxides and volatile organic compounds, which is detrimental to human health and ecosystems.

- **Human health criteria pollutants**, including particulates that may cause respiratory diseases such as asthma.

- **Human health impacts** due to chemical releases that may cause cancer and other chronic diseases.

- **Ecotoxicity** due to chemical releases that may harm organisms either through direct toxicity or ecological damage.

Based on the above categories, using simplified impact assessment coefficients, it is possible to derive relative comparisons of design options in terms of their *potential* adverse effects on humans or the environment. To assess the actual risks of such effects requires more detailed environmental risk assessment methods, as described below under Risk Analysis.

Streamlined LCA

Because of the limitations mentioned above, many LCA practitioners have turned toward simplified LCA tools that provide results more quickly, with less effort, and also with less precision. Especially in the early stages of product development, such tools may be more appropriate for rapid design iteration. Alternative approaches have emerged that are more comprehensive and more streamlined, although less fine-grained than conventional LCA. For example, Carnegie Mellon University developed a tool that uses aggregate input-output data to model the entire economy from a top-down perspective [16]. Such methods provide a useful complement to detailed, bottom-up LCA. In particular, since streamlined LCA requires only basic data about resource inputs, it is helpful in assessing new products when emissions data are not yet available. Streamlined methods can also be combined with detailed methods through "hybrid" LCA studies, which embed a focused LCA for specific industrial processes within a broader "envelope" representing the rest of the economy.

Example: The Ohio State University has developed an online tool called Eco-LCA™ that combines an economic input-output model of 488 sectors of the U.S. economy with an ecological resource consumption model based on exergy analysis, as described below. This tool enables immediate assessment or comparison of proposed designs based on an approximate bill of materials and can display a variety of life-cycle indicators, ranging from a simple carbon or water footprint to a comprehensive ecosystem goods and services consumption profile [17].

Exergy Analysis

The latest advance in LCA involves modeling the material and energy flows in complex systems based on the laws of thermodynamics. *Exergy* is defined as the available work that can be extracted from a material; for example, the exergy content of a fuel is essentially its heat content [18]. More generally, exergy tends to be correlated with material scarcity and purity, since it measures the *difference* of a material from its surroundings. Bhavik Bakshi and his colleagues at The Ohio State University have shown that all of the factors of industrial production—energy, materials, land, air, water, wind, tides, and even human resources can be represented in terms of exergy flows. Therefore, exergy can be used as a universal indicator to measure eco-efficiency and sustainability in industrial-ecological systems [19]. This method has the unique capability to quantify the contributions of most ecosystem services and is particularly useful for analyzing new technologies when detailed process-level data are nonexistent. It is also useful for aggregation of environmental impacts, since it correctly accounts for the differences in quality among various resources (e.g., energy from sunlight is much lower in quality than electrical energy).

An example of an ethanol life-cycle assessment that incorporates exergy analysis is shown in Figure 9.3. This study uses a hybrid methodology, combining a detailed process model of corn ethanol production with the above-described Eco-LCA™ model of the U.S. economy (see below) to represent commodity flows from outside the process boundaries. Based on this approach, Figure 9.4 shows the results of a comparative life-cycle study of biofuels in terms of two sustainability metrics—renewability (percentage from renewable sources) and return on energy (mega joules delivered per megajoule required over the life cycle). This analysis indicates that the sustainability of fuel derived from municipal solid waste is far superior to corn ethanol, which requires energy-intensive harvesting. Gasoline has the highest return on energy, although it is not renewable [20].

Predictive Simulation

The methods described above are useful for assessing the performance of a design with respect to specific environmental indicators. However, at some point in the new product development process, there is a need to consider the trade-offs between environmental factors and other important objectives—cost, quality, manufacturability, reliability, and so forth. If environmental performance were independent of these other factors, they could be analyzed separately.

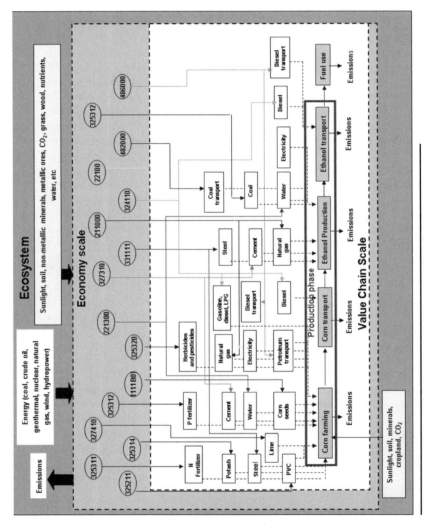

FIGURE 9.3 Hybrid multi-scale LCA model for corn ethanol.

182

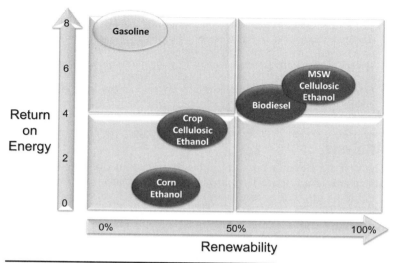

FIGURE 9.4   Life-cycle comparison of automotive fuels.

However, as seen in Chapter 8, environmental performance is often closely connected with other factors. Sometimes there are synergies, e.g., decreasing waste may result in lower costs; and sometimes there are compromises, e.g., increasing durability may interfere with recyclability. As mentioned above, scoring matrices provide a qualitative means of analyzing these trade-offs.

To fully understand complex trade-offs may require the development of predictive simulation models that capture the interactions among different design parameters. Detailed simulations can be extremely labor intensive, e.g., Gaussian plume dispersion models for airborne releases from production facilities. Only under exceptional circumstances would such tools be used for product and process development purposes. Nevertheless, with increased computing power, it is now possible to embed simulations into computer-aided design toolkits so that engineers can test new designs for compliance with important environmental constraints. One example of such a tool, still under development, is the Multiscale Integrated Models of Ecosystem Services (MIMES) system developed by Robert Costanza and his colleagues at the University of Vermont. This is a suite of dynamic ecological-economic computer models specifically aimed at understanding the impact of industrial activities on ecosystem services and human well-being.

Example: The Millennium Institute, a nonprofit based in Washington, D.C., has developed a sophisticated policy simulation tool called Threshold 21 (T21). The tool combines inputs from more detailed models to examine the interactions and feedback loops among economic, environmental, and social indicators (see Figure 9.5). T21's graphical human interface enables modelers to perform repeated iterations in order to explore the effects of key assumptions. Based on a powerful and flexible approach called *system dynamics*, the T21 modeling system has been applied over the last 20 years in dozens of countries around the world. In July 2007, Millennium Institute released a T21-USA model that is being used by lawmakers and advocacy groups to assess alternative national energy policies.

Risk Analysis

The risk management paradigm, as discussed in Chapter 4, predates the sustainability paradigm, but remains the basis for most environmental, health, and safety laws and regulations, as well as standard business practices. From a regulatory point of view, environmental risk reduction serves the public good. From a business point of view, it serves the interests of many stakeholders, including regulators, customers, local communities, and, of course, shareholders. Environmental risk management has been incorporated into the broader practice of enterprise risk management, which provides an integrated framework for examining all types of risks and opportunities faced by the corporation [21].

The term "risk" may be defined as follows: *a risk is the possibility of an adverse outcome associated with an event or activity.* There are generally two categories of environmental risk:

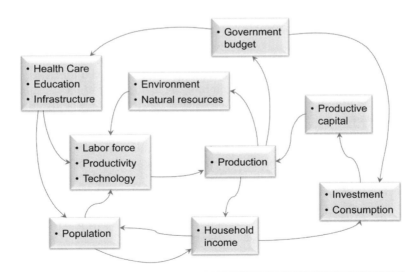

Figure 9.5   Structure and feedback loops of Threshold 21 system dynamics model.

1. Chronic risks are the result of long-term exposure to low levels of toxic or hazardous agents, which may cause adverse health effects in humans, or may harm fish, wildlife, and other environmental resources. An example might be traces of carcinogenic substances in drinking water.
2. Acute risks are the result of sudden, episodic exposure to substantial levels of toxic or hazardous agents that may cause immediate harm to humans or other living things. An example might be a hazardous chemical spill during transport.

Risk can generally be measured in two dimensions: 1. the *likelihood* that an adverse outcome will actually occur, expressed as a probability, and 2. the *magnitude* of the consequences expressed in terms that depend on the nature of the outcome (e.g. cost of ecological damage, incidence rate of disease). Acute risks are frequently assessed in terms of a *risk profile*, which assigns a probability distribution to the range of risk magnitude. Chronic risks are usually assessed in terms of the expected incidence of specific outcomes over a given time period. In principle, however, these two classes of risk are part of a continuum and can be assessed within a common framework.

Environmental risk analysis is a widely accepted approach whereby risks can be anticipated, understood and controlled. While there are many versions of this approach, a number of fundamental process steps are commonly practiced (see Table 9.1):

- *Vulnerability assessment* is the process of identifying and characterizing potential sources of risk (sometimes called *hazards* or *vulnerabilities*) in terms of their nature, mechanisms of action, and possible outcomes; this provides a basis for priority-setting to determine which risks merit greater attention.

- *Quantitative risk assessment* is the process of estimating the likelihoods and/or magnitudes of selected risks that have been identified; this generally includes assessment of the uncertainty associated with the "best" estimates of risk.

- *Integrated risk evaluation* is the process of assigning relative importance to risks that have been identified and/or quantified, based on regulatory, economic, social, or other factors that influence their acceptability to stakeholders.

- *Risk management* is the process of deciding how to avoid, mitigate, or otherwise control those risks that are deemed unacceptable; it generally involves a balancing of risks against the costs and benefits of mitigation alternatives.

- *Risk communication* is the process of understanding the concerns of stakeholders regarding identified risks and explaining the results of risk assessment, evaluation, and management decisions in terms that are meaningful to them.

Risk analysis is a complex subject, spanning a broad variety of risk sources, mechanisms, endpoints, and mathematical techniques. Despite the large amount of literature that has accumulated regarding environmental risk analysis, the methods are still evolving due both to theoretical advances and new empirical findings. Fundamental limitations on what is knowable (e.g., the "true" impacts on human populations of chronic low-level exposures) will continue to force reliance on predictive models, so that scientific debate over risk assessment methods may never be fully resolved. As a consequence, decision making regarding the mitigation of risks will continue to be challenged by the presence of significant assumptions and uncertainties in the available information, including the following:

- Types and magnitudes of risk *agents*, such as hazardous materials, waste and emissions, and ionizing radiation.

- Possible initiating *events* leading to unplanned releases, such as leaks, spills, fires, explosions, or deliberate human intrusion.

- Fate and transport *mechanisms* that describe how released agents are dispersed in the environment and partitioned among air, water, soil and other elements, as well as how they are chemically and physically transformed.

- Categories of *receptors* that may be exposed to released agents, including workers, community residents, sensitive populations (children, pregnant women, etc.), natural vegetation and wildlife, aquatic organisms, and domestic animals and crops.

- Exposure *pathways*, or routes, whereby humans and other biota may be exposed to released agents or their by-products, including inhalation, uptake through direct contact, ingestion in water, and bioaccumulation in the food chain.

Despite data limitations, it is still possible to develop and apply an objective decision-making process that takes into account the degree of uncertainty in risk estimates, as well as the relative importance of the corresponding endpoints from the perspective of different stakeholders. Moreover, it is possible to factor this type of information into a risk/cost/benefit analysis such that the associated trade-offs can be explicitly communicated to a variety of interested parties. To pursue such an approach requires great attention to the quality, adequacy, and credibility of the information upon which the analyses are based.

There are certain situations in which the traditional approach to risk analysis and management is not adequate. As the boundaries of the system become broader, several limitations emerge: increasing complexity, uncertainty, and interdependence with other systems make it difficult to identify risks, and the dynamic nature of these systems makes it difficult to acquire relevant and reliable data. Chapter 20 describes the complementary approach of resilience management, which enables enterprises to cope with unforeseen risks and adapt to a changing business environment.

Financial Analysis

The last category of DFE analysis methods, very important but sometimes overlooked, is the analysis of the financial implications of design decisions. Taking a product life-cycle perspective implies that designers must go beyond conventional cost accounting methods to consider the broader costs and benefits incurred either by the manufacturer, its customers, or other parties at various stages of the product life cycle.

Life-Cycle Accounting

Conventional accounting methods do not capture the costs or revenues associated with environmental improvements in a useful way. Because environmental budgets are usually assigned to overhead accounts, DFE initiatives in product or process engineering organizations cannot easily be credited for their monetary benefits. But if the cost savings associated with reduced energy use, reduced waste management expenses, and salvage values of recycled materials are taken into account, DFE investments become much more financially attractive.

> MOST ACCOUNTING SYSTEMS DO NOT PROPERLY INTERNALIZE THE ENVIRONMENTAL IMPACTS OF DECISIONS.

At a broader level, most enterprise accounting systems do not properly internalize the environmental impacts of product and process decisions. In particular, impacts upon resources, such as materials, water, soil, or energy are difficult to evaluate because market value based on classical supply and demand mechanisms fails to reflect the true societal value of these resources. As a consequence, under conventional accounting methods, it is usually difficult to justify the costs of environmental performance improvement because the benefits are not directly quantified.

To correct the above deficiencies, some companies have adopted a new approach called *life-cycle accounting*, also known as *environmental accounting* or *total cost assessment*. It can be defined as identification and quantification of direct, indirect, and other costs across the life cycle of a facility, product, or process, thus revealing cost-effective opportunities to simultaneously improve productivity and reduce waste, while encouraging business decisions that are both financially superior and beneficial to the environment.

Life-cycle accounting is a form of activity-based costing that captures the financial benefits of environmental performance improvement, including reduced operating costs, increased profitability, or economic value added (see Chapter 4). Life-cycle accounting is an extension of life-cycle costing methods used in the defense sector to manage large, multiyear weapon system programs where major costs are associated with deployment, logistical support, and decommissioning. Similar techniques have been used in the computer and other industries to capture the total "cost of ownership" for enterprise information systems.

To understand the full scope of life-cycle costs and benefits, it is helpful to divide them into the following categories [22].

- **Conventional:** Material, labor, other expenses, and revenues that are commonly allocated to a product or process (often called "direct" costs).

- **Potentially Hidden:** Expenses incurred by and benefits accrued to the firm that are not typically traced to the responsible products or processes, e.g., legal fees or safety training courses (often called "indirect" costs).

- **Opportunity:** Costs associated with opportunities that are foregone by not putting the firm's resources to their highest value use.

- **Contingent:** Potential liabilities or benefits that depend on the occurrence of future events, e.g., potential occupational health and clean-up costs related to a spill of a hazardous substance.

- **Good Will:** Costs/benefits related to the subjective perceptions of a firm's stakeholders, e.g., brand image, customer loyalty, or favorable relationships with regulatory agencies.

- **External:** Costs/benefits of a company's impacts upon the environment and society that do not directly accrue to the business, e.g., the benefits of reduced waste generation for product consumers.

As in LCA, life-cycle accounting requires the definition of a system boundary. Figure 9.6 illustrates different possible boundaries of analysis for product design, ranging from traditional unit cost to customer cost of ownership to total life-cycle costs. Similarly, Figure 9.7 illustrates alternative boundaries for facility life-cycle accounting. Table 9.3 gives examples of different types of costs and benefits that could be relevant to both product and facility accounting at each stage of a typical value chain.

Note that it is important to distinguish between environmental accounting "in the small," which provides an integrated corporate

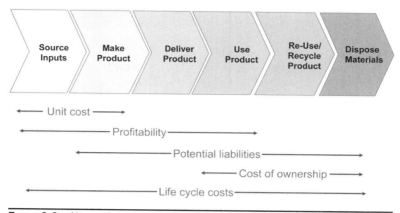

FIGURE **9.6** Alternative boundaries for life-cycle accounting in product design.

perspective on contributions to shareholder value, and environmental accounting "in the large," which seeks to develop appropriate prices for goods and services commensurate with their environmental footprint. The former is already being practiced in some form by many firms and is eminently feasible. The latter, also known as "full-cost pricing," is much more ambitious and controversial and will require broad political consensus to become realistic.

Cost/Benefit Analysis

For purposes of trade-off analysis, financial costs and benefits can be analyzed using conventional methods, such as discounted cash flow analysis or net present value. It is possible to introduce nonfinancial costs and benefits, such as greenhouse gas emission reduction, and

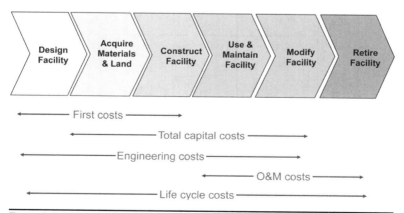

FIGURE **9.7** Alternative boundaries for life-cycle accounting in facility design.

	Purchasing	Material Handling	Storage	Material Recovery	Disposition	Product Take Back
Conventional	Production inputs	Reduced packaging material costs	Regulated material handling costs	Revenues from recovery of materials	Labor & fees for waste disposal	Reduced costs for purchasing components
Hidden	Activities to certify suppliers	Container use efficiency gains	Spill cleanup costs	Eco-efficiency gains, e.g., material usage	Decreased insurance premiums	Increased shipping costs due to returns
Contingent	Reduced risk due to supplier certification	Ergonomic and safety incident costs	Reduced worker claims of exposure	Decreased remediation liability due to reduced waste	Potential future liability for landfill risks	Preparedness for extended responsibility
Good will	Positive media coverage	Improved customer relationships	Improved employee satisfaction	Increased investor confidence	Enhanced brand image	Concerns about product quality
External	Ecosystem improvements	Reduction in accidental releases	Reduced emissions to environment	Reduced community medical costs	Decreased landfill burden	Decreased depletion of raw materials

TABLE 9.3 Examples of Life-Cycle Costs and Benefits [22]

to subjectively weigh the environmental benefits of a product improvement against the costs. For example, BASF has developed an eco-efficiency analysis tool that quantifies economic-ecological trade-offs (see Chapter 13). Some economists have also tried to assign monetary values to environmental outcomes, but these practices are rarely used for business decision making. The next section provides two straightforward examples of cost/benefit analysis for design decisions.

When the scope of decision making extends beyond an individual firm, cost/benefit analysis can become more challenging. For example, in a collaborative supply network or an industrial ecology network, companies may be considering a variety of alternative technologies and pathways for reducing their non-product output streams. Should a component manufacturer implement a reverse logistics system to recover shipping containers from its customers, or arrange for a third party to recover and recycle them? Should a company send its organic wastes to a composting facility or invest in an anaerobic digester to convert them into a biogas that could be sold as a by-product? Here the costs and benefits are distributed over multiple parties and may require formation of new business relationships. Advanced decision support tools have been developed to analyze multiple-agent design problems that involve material processing and transport among industrial clusters or networks. One approach is to use mathematical programming models to determine the network design and optimal flows among the processes that minimize total cost and/or environmental impacts.

Example: The Department of Public Utilities in Columbus, Ohio, used such an optimization model to conduct an analysis of its wastewater biosolids operations across multiple plants. It concluded that up to $2 million a year could be saved, representing nearly 25% of annual operating costs. In addition, the model identified opportunities for greenhouse gas emission reductions of up to 40% and energy use reduction of up to 64% [23]. The city is continuing to use the model as a means of evaluating new technologies and alternative operating policies.

Examples of DFE Decisions

The following two scenarios provide a simplified illustration of how life-cycle cost/benefit analysis principles can be applied to a product design decision and a process design decision respectively [24]. Although fictitious, they reflect realistic technological alternatives based on actual case histories.

Consumer Electronic Product Design

This example illustrates how DFE principles could be applied during the design of a consumer electronic product for global markets.

Because of product take-back regulations in the European Union, the design team must be concerned with the design issues that affect the end-of-life costs, including the product disassembly and recovery of high-value materials and components. The team has two primary design goals: 1. to minimize the product's total life-cycle cost and 2. to increase the recycled material content of the product.

The design team has identified several opportunities to accomplish their goals. First, they will reduce the number of different materials that are used in the product. This results in several benefits:

- Reduced raw material costs associated with purchases of higher volumes of fewer materials
- Reduced inventory requirements
- Lower material separation labor costs at end-of-life
- Increased material salvage value associated with higher volumes and reduced handling and transaction costs

The design team has also identified a recycled material for the product housing that satisfies all engineering requirements, although it is modestly more expensive than the alternative virgin material. As a related benefit, this new material can be again recycled, increasing the product's overall recyclability.

Comparing the life-cycle costs associated with the new design with those for the original design highlights the economic benefits (see Table 9.4). A cost saving of $30 per unit, or about 10% of total cost, translates to millions of dollars over the life of the product.

	Original Design	New Design
Costs per unit		
Material inputs	$100	$110
Manufacturing & assembly	$150	$150
Sales & service	$50	$50
Disassembly	$35	$15
Salvage value	($20)	($40)
Total life-cycle cost	$315	$285
Recycled content	25%	60%

TABLE 9.4 Comparison of Alternative Electronic Product Designs

Heat Transfer Process Design

This second DFE example applies similar concepts to the design of a process in a manufacturing facility. A company is reviewing options for replacing heat transfer fluids that are used in injection molding applications. Currently, heat transfer fluids are purchased on the basis of the requirements of the application, product availability, and initial purchase cost. Worker concerns about adverse health impacts, recent product price increases, and the availability of several new products have motivated the company to reevaluate the heat transfer fluids used in the manufacturing process. In selecting a heat transfer fluid, the company has the following goals:

- Minimize the total life-cycle costs,
- Decrease the volume of heat transfer fluids that are incinerated or landfilled, and
- Reduce adverse impacts to worker health.

Heat transfer fluids typically have different purchase costs, disposal costs, and useful lives. In addition, because of fundamental differences in their physical and chemical characteristics, the operating and maintenance costs (e.g., inspection costs, system repair costs, and worker training costs) can also differ.

Table 9.5 illustrates an analysis of several alternative types of fluid. While perfluorinated fluids are more costly initially, they have a longer expected lifetime (35 years), lower operating and maintenance costs, and much lower disposal costs (since they are recyclable) than other alternatives. These lower costs more than make up for the initial capital costs incurred to modify the heat exchangers. This life-cycle cost perspective shows that, on an annualized basis, perfluorinated fluids have significantly lower costs.

There are environmental benefits as well. Because perfluorinated fluids can be recycled, their use virtually eliminates the need to landfill or incinerate spent material. Furthermore, they are inert,

	Capital and Acquisition Costs	Operating & Maintenance Costs	Disposal Costs	Life-Cycle Costs	Annual Cost (NPV)
Mineral Oil	$275	$4,531	$130	$4,936	$3,705
Aromatic Hydrocarbon	$750	$6,023	$113	$6,886	$2,769
Perfluorinated Fluid	$18,500	$4,344	$1	$22,845	$2,369

TABLE **9.5**   Comparison of Alternative Heat Transfer Fluids

nonhazardous, and nontoxic, thus minimizing concerns about risks to workers' health and potential environmental liabilities.

The Challenges of Decision Making

The ultimate purpose of environmental performance assessment is to support decisions, whether business decisions by manufacturers or policy decisions by governmental and other organizations. It is important that such decisions be informed by a generic, life-cycle framework that recognizes the multitude of impacts that may affect stakeholders (see Chapter 10). Decisions based on considerations that are too narrow (e.g., considering only initial costs, or focusing only on recyclability) are bound to be flawed. On the other hand, decisions that attempt to factor in an overly broad range of outcomes (e.g., planetary health, societal welfare) are likely to become bogged down in uncertainty and confusion. A judicious middle point between these extremes is desirable.

One useful means of pruning the complexity of a decision problem is to identify first-order impacts that really matter to the decision-making organization and that are significantly influenced by the decision outcome. There are a number of key questions that need to be asked:

- What is the *minimal* set of environmental metrics that are adequate to represent the design's environmental performance?

- For each metric, can the degree of environmental improvement associated with each design option be assessed in quantitative terms?

- Do side effects of performance improvement need to be explicitly considered; for example, might an emission reduction be offset by an increase in risk associated with the substitute technology (e.g., waste incineration)?

- Can the analysis focus on the immediate impacts of the design change only, or do system-wide risks associated with materials and energy requirements need to be taken into account?

- Is it necessary and feasible, either implicitly or explicitly, to assign relative importance ratings or monetary equivalents to non-commensurate types of impacts?

- Is there a unifying conceptual model that defines the activities for which both cost and environmental performance estimates are developed? If not, are the underlying models for cost and environmental assessment compatible?

- How is uncertainty in cost or environmental assessments represented and managed? Does the range of uncertainty cast doubt upon the validity of the decision process?

- What is the sensitivity of the environmental performance estimates to specific design parameters? Are there particular features that account for the majority of the impacts?

These are challenging questions that will force analysts to look beyond the mechanics of their standard methodologies and adapt to the needs of decision-makers, who prefer simple insights to complex mathematical results.

References

1. S. J. Skerlos, W. R. Morrow, and J. J. Michalek, "Sustainable Design Engineering and Science: Selected Challenges and Case Studies," in M. A. Abraham (Ed.), *Sustainability Science and Engineering* (Amsterdam: Elsevier B.V., 2006).
2. Sustainability ToolKit and Scorecard, Norm Thompson Outfitters, Inc. and Michael S. Brown & Associates, 2002.
3. B. R. Allenby and D. J. Richards, *The Greening of Industrial Ecosystems* (Washington, D.C.: National Academies Press, National Academy of Engineering, 1994).
4. Intergovernmental Panel on Climate Change (IPCC), *Fourth Assessment Report: Climate Change.* (Cambridge, U.K.: Cambridge University Press, 2007).
5. "The Greenhouse Gas Protocol: A Corporate Accounting and Reporting Standard," World Business Council for Sustainable Development/World Resources Institute, 2004.
6. S. Bringezu, H. Schutz, and S. Moll, "Rationale for and Interpretation of Economy-Wide Materials Flow Analysis and Derived Indicators," *Journal of Industrial Ecology*, 2003, Vol. 7, No. 2.
7. H. Schutz and M. J. Welfens, "Sustainable Development by Dematerialization in Production and Consumption—Strategy for the New Environmental Policy in Poland," Wuppertal Institute for Climate, Environment, and Energy, Wuppertal, Germany, 2000.
8. Institute of Environmental Sciences (CML), Leiden; CE, Solutions for Environment, Economy and Technology; and Wuppertal Institute, "Policy Review on Decoupling: Development of indicators to assess decoupling of economic development and environmental pressure in the EU-25 and AC-3 countries." Draft final report, European Commission, DG Environment, October 2004.
9. M. Wackernagel, "Advancing Sustainable Resource Management: Using Ecological Footprint Analysis for Problem Formulation, Policy Development, and Communication," Report to European Commission, DG Environment, Feb. 2001.
10. OECD , *Indicators to Measure Decoupling of Environmental Pressure from Economic Growth*, 2002. www.olis.oecd.org/olis/2002doc.nsf/LinkTo/sg-sd(2002)1-final
11. P. W. Gerbens-Leenes and A. Y. Hoekstra, *Business Water Footprint Accounting*, UNESCO Institute for Water Education, 2008.
12. Water Footprint Network estimates. See www.waterfootprint.org.
13. M. A. Curran (ed), *Environmental Life Cycle Assessment* (New York: McGraw-Hill, 1996).
14. A good overview of life-cycle assessment methods and applications is: G. A. Keoleian and D. V. Spitzley, "Life Cycle Based Sustainability Metrics," in M. A. Abraham (Ed.), Sustainability Science and Engineering (Amsterdam: Elsevier B.V., 2006).
15. J. C. Bare, G. A. Norris, D. W. Pennington, and T. McKone, 2003. "TRACI—The Tool for the Reduction and Assessment of Chemical and Other Environmental Impacts." *Journal of Industrial Ecology*, 2003, 6(3-4), pp 49–78.
16. C. T. Hendrickson, L. B. Lave, and S. H. Matthews, Environmental Life Cycle Assessment of Goods and Services: An Input-Output Approach (Washington, D.C.: Resources for the Future Press, 2005).

17. With support from the National Science Foundation, a public version of the Eco-LCA™ tool is available at www.resilience.osu.edu.

18. R. U. Ayres, "Resources, Scarcity, Growth and the Environment" (Fontainebleau, France: Center for the Management of Environmental Resources, INSEAD, Fontainebleau, France, April 2001).

19. J. L. Hau and B. R. Bakshi, "Expanding Exergy Analysis to account for Ecological Products and Services," *Environmental Science & Technology*, 2004, 38, 13, 3768–3777.

20. A. Baral and B. R. Bakshi, "The Role of Ecological Resources and Aggregate Thermodynamic Metrics for Assessing the Life Cycle of Some Biomass and Fossil Fuels," *Environmental Science and Technology*, 2009.

21. Committee of Sponsoring Organizations of the Treadway Commission, *Enterprise Risk Management—Integrated Framework*, 2004.

22. J. McDaniel and J. Fiksel, *The Lean and Green Supply Chain: A Practical Guide for Materials Managers and Supply Chain Managers to Reduce Costs and Improve Environmental Performance*, U.S. EPA Office of Pollution Prevention & Toxics, EPA 742–R–00–001, January 2000.

23. K. S. Smith, K. A. Sikdar, R. E. VanEvra, D. Gernant, C. Bertino, and J. Fiksel, "Incorporating Sustainability Into Biosolids Master Planning," Proceedings of WEFTEC Conference, 2009.

24. J. Fiksel and K. Wapman, "How to Design for Environment and Minimize Life Cycle Costs," *Proceedings IEEE Symposium on Electronics and the Environment*, San Francisco, May 1994.

CHAPTER 10

Product Life-Cycle Management

DFE Principle 1. *When considering the environmental implications of product and process design, think beyond the cost, technology and functional performance of the design and consider the broader consequences at each stage of the value chain.*

Putting It All Together

Product life-cycle management is an integrated business process that brings together all of the principles, practices, guidelines, metrics, and methodologies described in previous chapters. Life-cycle thinking, as described in Chapter 5, provides a useful conceptual framework for life-cycle management. It is important to note the difference between life-cycle *assessment*, the analytical method described in Chapter 9, and life-cycle *thinking*, which is a paradigm for holistic understanding of enterprise sustainability and competitiveness. Life-cycle thinking does not necessarily require the application of any specific methods; it merely requires a broad awareness of how business decisions affect stakeholders across the entire value chain, from material and energy acquisition to the end of the product's useful life. Life-cycle thinking must combine an understanding of competitive opportunities in the marketplace with aspirations for sustainability and, thus, may give rise to innovative product or service concepts.

Life-cycle management (LCM) can be defined as: *Systematic consideration of all life-cycle stages in the evaluation, management, and improvement of an enterprise's products, services, processes, and assets.* Originally, LCM methods were developed in the defense industry as a means of understanding and managing the long-term financial implications of weapons system development, since the life-cycle costs of system maintenance and logistical support, such as replenishment of supplies, often outweighed the initial purchase costs. Gradually, the practice of LCM spread to the private sector as a means of managing the full enterprise-wide costs of durable products and productive assets, such as factories. The concept has also been adopted by the software community with reference to the development, maintenance, and support of enterprise computing assets. Since the mid-1990s, LCM practices have expanded to include concurrent engineering and the full spectrum of financial and nonfinancial impacts.

More specifically, product life-cycle management (PLCM) is an overarching business process that interfaces with all of the functional processes involved in designing, developing, producing, marketing, selling, supplying, and supporting a line of products. Not every company chooses to formalize this process, and some may give it alternative names, such as product stewardship or value chain management. But it is impossible to fulfill the aspirations of DFE without a cross-functional process that monitors the performance of products over their life cycle, ensures that the design intent is being realized, and provides feedback for purposes of product improvement. Under the umbrella of PLCM, the practice of DFE involves the following activities:

- Characterize the environmental burdens associated with a product, process, or service including energy and materials used and wastes released

- Assess the financial and nonfinancial impacts of those environmental burdens from the perspective of internal and external stakeholders

- Assess the balance between the above impacts and the value contributed to the enterprise and to society by the product, process, or service

- Evaluate opportunities for value improvements or reductions in adverse impacts, and establish goals

- Propose design or process changes that will address the goals

Figure 10.1 depicts PLCM as a continuous improvement cycle in which product development is followed by product deployment; and, eventually, products are retired, replaced, or upgraded.* Performance metrics play a key role in this cycle, from the initial establishment of design objectives to the product review and release decisions to the ongoing monitoring of product performance. The elements of the cycle include:

- **Life-Cycle Objectives**—Establishment of specific product objectives, requirements, and performance targets as a basis for innovation, whether for product enhancement or new product

---

*Note that the product development portion of the cycle is analogous to Figure 5.1.

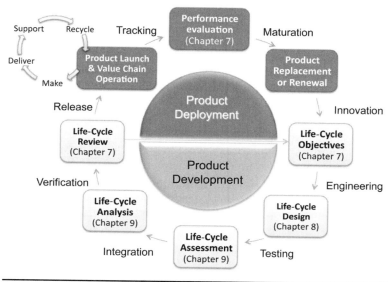

FIGURE 10.1   Product life-cycle management.

development. This requires the use of life-cycle metrics, as defined in Chapter 7.

- **Life-Cycle Design**—Product design, development, and engineering to achieve the product life-cycle objectives. DFE guidelines are presented in Chapter 8. However, as discussed in Chapter 5, DFE is just one of many elements in the concurrent engineering of product life-cycle performance.

- **Life-Cycle Assessment**—Testing of product performance characteristics and assessment of the expected financial and nonfinancial impacts, including socio-economic and ecological well-being. Methods for life-cycle cost and environmental assessment to inform design decisions are described in Chapter 9.

- **Life-Cycle Analysis**—Integrated analysis of total life-cycle costs and benefits associated with products, processes, or services. As discussed in Chapter 9, for planning or decision-making purposes financial outcomes need to be weighed against other trade-off factors such as environmental performance and customer satisfaction.

- **Life-Cycle Review**—Verification that the product design meets the stated life-cycle objectives, normally accomplished through a stage-gate review process. If necessary, the objectives may be refined or the product team may go through additional design iterations prior to release.

- **Product Launch & Value Chain Operation**—Ramp-up and implementation of value-chain operations and business processes, including product introduction, marketing and sales, manufacturing and distribution, and customer service. These activities correspond to the *physical* life cycle depicted in Figure 5.2.

- **Life-Cycle Performance Tracking**—Periodic monitoring of product performance indicators, and reporting of the results to management, shareholders, and external stakeholders. Indicators and metrics relevant to life-cycle environmental performance are discussed in Chapter 7.

- **Product Replacement or Renewal**—Periodic interventions to enhance, upgrade, extend, or replace the product, as it matures and as market conditions evolve. Eventually the product may be retired, although successful products have been sustained for many decades. Decisions to modify the product line will trigger a new cycle of product development.

Product life-cycle management encompasses an array of processes and services that are required to develop, manufacture, distribute, and support the product. For global manufacturers, this involves coordination of a complex set of supply chain capabilities around the world. At a more granular level, the same type of continuous improvement cycle can be applied to life-cycle management of individual processes, facilities, or services. As shown in Figure 10.2, there is an intersection between the facility life cycle and the product life cycle, but a specific product typically relies on multiple

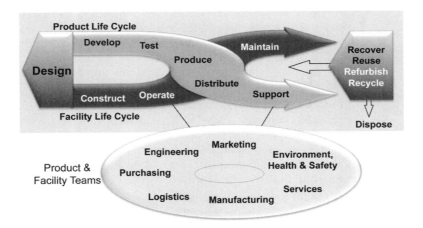

FIGURE 10.2 Life-cycle stages for management of products and facilities by cross-functional teams.

production and distribution facilities, while a specific facility typically handles multiple products over its lifetime. Each product and major facility should have a cross-functional team assigned to conduct the LCM activities depicted in Figure 10.1.

Companies that practice LCM are arguably more sustainable and competitive, since they are better able to understanding the full impacts of business decisions across the value chain, and can identify opportunities to enhance shareholder value by modifying their business processes or inserting innovative technologies to enhance performance, quality, and profitability. Many leading U.S. companies have established life-cycle management programs; for example:

- In the chemical industry, Dow Chemical, DuPont, and Solutia (formerly part of Monsanto) were early adopters of life-cycle environmental accounting to support their product stewardship and waste elimination efforts.

- In the pharmaceutical industry, which has long and costly product development cycles, most companies have instituted some type of life-cycle review process. For example, Bristol Myers Squibb estimates that each product life-cycle review costs about $25,000 and yields an average of $340,000 in benefits to the enterprise.

- Most major automotive firms have programs aimed at improving quality, productivity, service, price, and environmental performance in the supply chain. For example, Chrysler estimated that, over a 3-year period, LCM reviews avoided $22 million of hidden costs and potential liabilities.

This chapter highlights the LCM practices of two highly respected companies, Caterpillar and 3M, who have adopted life-cycle management as an essential component of their new product development processes.

Caterpillar: New Engines from Old
Extended Producer Responsibility

Caterpillar, headquartered in Peoria, Illinois, is the world's largest maker of construction and mining equipment, diesel and natural gas engines, and industrial gas turbines, with about 100,000 employees worldwide. Caterpillar views sustainability as intrinsic to its business interests, including providing reliable and efficient energy solutions, promoting responsible use of materials, enabling the mobility of people and goods, and developing quality infrastructure for society. Among its sustainability goals for 2020, Caterpillar is aiming to improve its customers' energy and material efficiency by 20% through improved products and services. By 2020, the company also aims to increase energy efficiency in its own operations by 25%, reduce absolute GHG emissions by 25%, use alternative and renewable sources for 20% of its energy needs, and eliminate waste to landfill [1].

A cornerstone of Caterpillar's sustainability strategy is product take-back, based on the concept of "extended producer responsibility." Caterpillar has been in the remanufacturing business for over 30 years and its remanufactured products include on-and-off-highway engines, engine components, transmissions, hydraulic components, and electronic components. Caterpillar also offers remanufacturing services to original equipment manufacturers that serve the rail, industrial, defense, and automotive industries. In 2006, Caterpillar acquired Progress Rail, one of the largest service providers in the rail industry, which remanufactures and recycles used railcars, locomotives, rail, and track.

Caterpillar's Remanufacturing Division, known as Reman, is a significant recycler of end-of-life Caterpillar products. The remanufacturing process reduces waste, minimizes the need for raw materials required to produce a new part, and helps ensure the recovery of end-of-life products through a closed-loop reverse logistics process. End-of-life components, called "cores," are taken back, inspected, and disassembled; and every part is remanufactured to print specifications incorporating all applicable engineering updates. The remanufactured parts, supplemented by new parts where required, are assembled into finished remanufactured products, tested, packaged for sale, and warranted the same as new products. If "Environmental Product Declarations" are provided to customers for new parts or components, a similar declaration would be provided for the remanufactured part or component.

Joe Allen, General Manager and Director of Sustainable Development for Cat Reman, explains it as follows: "Our remanufacturing program is based on an exchange system. Customers return a used component and receive a remanufactured replacement component with same-as-new quality, reliability, and warranty in return (see Figure 10.3). We then use the end-of-life returns in the remanufacturing process to produce like-new remanufactured products—or better-than-new if we are able to design and implement new engineering changes into that product."

Reman is Caterpillar's fastest growing division, due in large part to its profitability and its resilience during business downturns. The division's employee base has grown from 800 employees in 2002 to approximately 4,000 in 2008, distributed globally at 17 dedicated remanufacturing facilities in 8 countries. Some countries have restrictive trade policies on remanufactured products, and in expanding its international business Caterpillar has worked closely with the U.S. government and the World Trade Organization, among others, to help eliminate these barriers. Caterpillar opened its first Asian remanufacturing center in Shanghai, China, in 2006, and was the first

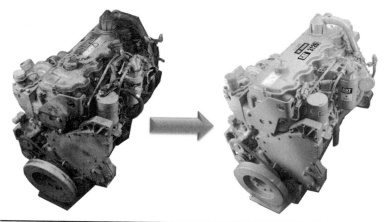

FIGURE 10.3   Caterpillar Reman restores old engines to warrantied like-new condition.

wholly owned foreign entity to receive a remanufacturing license in China. Caterpillar is also providing expertise to help Chinese research and development institutions pursue China's 4R initiative: reduce, reuse, recycle and remanufacture.

Benefits of Remanufacturing

By returning end-of-life components to same-as-new condition, Caterpillar is able to reduce waste and minimize the need for raw materials to produce new parts. Through remanufacturing, nonrenewable resources are kept in circulation for multiple lifetimes—supporting the company's goal for a zero landfill footprint by 2020. Moreover, by establishing a product take-back program, Caterpillar builds long-term relationships with customers through a series of product purchases and returns.

Remanufacturing continues to be a growing business for Caterpillar. In 2007, the company took back over two billion pounds of material—equivalent to over 56 billion aluminum cans. Core material received increased from 100 million pounds in 2002 to 141 million pounds in 2007. The current end-of-life return rate on a global basis is 93 percent, i.e., for every 100 remanufactured parts sold, 93 are returned. Nearly 70 percent of this recovered material is remanufactured and reused to produce Cat Reman products, as shown in Figure 10.4. The remaining material is recycled either by Caterpillar or by one of its approved foundries, mills, or recycling centers, with the ultimate goal of zero waste to landfill. Remanufacturing not only reduces material consumption and waste but also conserves a large portion of the labor and energy added to the original raw material. Caterpillar estimates that about 85% of the original energy value is preserved.

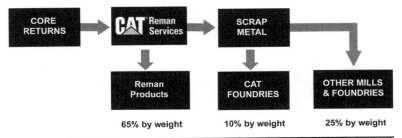

FIGURE **10.4**   Material flow in Caterpillar remanufacturing operations.

In 2006, Caterpillar conducted a limited study to compare the environmental footprint of the processes required to manufacture vs. those required to remanufacture an engine cylinder head. The study did not include the upstream impacts of extracting, transporting and processing raw materials to produce new parts. Even so, the results indicate that remanufacturing reduces GHG emissions by over 50%, water use by over 90%, energy use by over 80%, material use by over 99%, and landfill space by over 99%. These preliminary findings were encouraging, and Caterpillar is continuing to refine and expand on this study.

Aside from the direct environmental footprint reduction, Caterpillar's remanufacturing take-back business model allows the company to minimize the proliferation of restricted substances such as lead and chromium. It provides a convenient channel for customers to return used components such as lead-solder-containing electronic control modules or chrome-plated hydraulic cylinders, thus enabling them to responsibly dispose of these substances while extending the life of the product.

The Reman program has spawned a number of additional initiatives that are consistent with Caterpillar's commitment to sustainable development. For example:

- Caterpillar offers emissions retrofit kits, which allow engines to be upgraded from Tier 0 to Tier 1 emissions at the time of overhaul by replacing components, such as fuel pumps, nozzles, and turbo chargers. This results in reductions in nitrous oxide, particulate matter, carbon monoxide and hydrocarbon emissions.

- In addition to the standard core take-back process, Reman purchases cores directly from the marketplace. This not only creates additional core material for remanufacturing, eliminating the need to supplement with new parts, but also provides an avenue for dealers and customers to responsibly dispose of excess end-of-life products.

- The remanufacturing program includes repurposing products into completely different applications than originally intended. For example, one pilot project will take back diesel truck engines from North America, remanufacture them into gas configurations that burn methane (a potent greenhouse gas), package them into generator sets, and then deliver them to the developing world to help meet local electric power needs.

Thanks to its successful track record, the Reman Division has emerged as a key pillar of Caterpillar's future plans to focus on sustainable development as a business strategy. The division's commitment to sustainability is evident not only in its manufacturing operations but also in the office environment. From remanufactured office furniture to art sculptures and office displays designed with remanufactured parts and recycled content, sustainable development is an inherent part of the Cat Reman culture.

Design for Remanufacturing

Caterpillar strives to integrate life-cycle thinking into its product and process development activities. Accordingly, the company developed a business process called "Design for Remanufacturing," which was released enterprise-wide in 2006. The "Design for Reman" process ensures that, as part of the new product development process, all products are designed to optimize their remanufacture or recycling at the end of their useful life.

Using this process, Caterpillar design teams consider the entire product life cycle before the design phase even begins, considering such things as type of material used and ease of disassembly for repair, remanufacture, reuse, or recycling. By considering the product's remanufacturing potential in the product design phase, Caterpillar engineers have enabled some components to be in circulation for multiple life cycles. Thus, product life-cycle management principles have become a fundamental component of Caterpillar's equipment, engine, and service parts design. "Remanufacturability Design Guides" have been developed and are accessible via the Division's intranet website, allowing engineers throughout the company to readily view specifications and remanufacturing requirements when developing and designing new products, as well as modifying existing products.

As a complement to the Design for Manufacturability thrust, Caterpillar has introduced innovative business processes and capabilities to support its dealers, customers, and employees:

- Since Caterpillar's design focus includes the ability to reuse and rebuild at end-of-life, the company has developed the Cat Certified Rebuild Program to allow dealers to transform

Cat machines and engines back to their original, like-new condition at the end of their first useful life. Using this process, dealers reuse or recycle virtually all the original machine content, substituting Cat Reman and new parts for components that do not meet the Certified Rebuild criteria.

- In addition to remanufacturing and serviceability, Caterpillar is also focused on the upgradability and modularity of its products. For example, an Upgrade-to-New Program is available for specific products, allowing customers to purchase new products and return the core from old products. The core is then remanufactured and returned to the marketplace as a remanufactured component.

- Product take-back criteria for the remanufacturing program are provided to all dealers via simple and concise Core Acceptance Criteria. These guidelines explain how to inspect the cores for damage, such as cracks, and provide detailed instructions for dealers to check each core accurately and consistently. Dealer training is also provided to ensure that dealers are proficient in applying the criteria.

- Caterpillar has invested heavily in systems and processes to manage product take-back for its global remanufacturing program. The company's proprietary, web-based Core Management Information System provides real-time information regarding core receipts, core deposits, and logistics tracking for Cat's global distribution system and its dealers.

3M: Responsible Innovation
A History of Environmental Leadership

The 3M Company is a $24 billion diversified technology company serving a broad range of markets including consumer and office, display and graphics, electronics and telecommunications, health care, security and protection, and transportation. Headquartered in St. Paul, Minnesota, the company has over 75,000 employees in more than 60 countries and serves customers in nearly 200 countries. 3M is recognized as a perennial source of innovative technologies, many incorporated into its well-known consumer brands, such as Scotch, Post-it, Scotch-Brite, Scotchgard, Nexcare, and Filtrete.

According to Keith Miller, Manager of Environmental Initiatives and Sustainability, 3M has made impressive progress in reducing its corporate environmental footprint. Between 1990 and 2007, the company achieved a 95% reduction in absolute volatile air emissions, 95% reduction in absolute U.S. Toxic Release Inventory (TRI) releases, 62% reduction in absolute greenhouse gas emissions, and 61% reduction in solid waste indexed to net sales [2].

Long before sustainability became popular, 3M was regarded as a leader in environmental excellence, thanks partly to its pioneering Pollution Prevention Pays (3P) program launched in 1975 [3]. The program is based on the belief that preventing pollution at its source is more environmentally effective, technically sound, and economical than conventional pollution controls. By 2008, the 3P program had prevented more than 2.7 billion pounds of pollutants and saved the company an estimated $1.2 billion. 3P projects are conceived and managed voluntarily by 3M employees, and innovative projects are recognized with 3P Awards. There have been close to 7000 3P projects conducted to date, using a variety of DFE techniques, such as product reformulation, process modification, equipment redesign, or recycling and reuse of waste materials.

In 2000, 3M made a voluntary decision to phase out its current line of Scotchgard and other fluorochemical-based products that served markets such as water repellents, coatings for food packaging, and fire-fighting foams. The decision was based on the discovery of minute traces of a persistent, bioaccumulative by-product in the tissues of humans and wildlife. Although there was no evidence of harm, and 3M was not legally obligated to take action, the company decided to err on the side of caution and worked with regulators and customers to communicate its phase-out rationale. This decision demonstrated the company's commitment to a proactive product stewardship policy and, ultimately, enhanced its reputation. Since then, 3M has developed alternative chemistries that are considered safe for humans and the environment.

Life Cycle Management Program

In 1998, 3M introduced a corporate initiative called Life Cycle Management (LCM) to assure an awareness of potential impacts at every stage of the life cycle: from design and manufacturing to customer use and disposal. LCM is an integral part of 3M's sustainability strategy. Instead of only focusing on the manufacturing process to control environmental, health, safety, and energy effects, the spotlight is now on products throughout their entire life cycle, from manufacturing through customer use and disposal. 3M has adopted a Life Cycle Management Policy requiring all business units to conduct LCM reviews for all new products and for existing products on a prioritized basis.

With hundreds of new products introduced each year, the LCM process gives 3M a continuous flow of opportunities to significantly add to its environmental progress. Cross-functional, new product introduction teams use a Life Cycle Management matrix (see Figure 10.5) to systematically and holistically address the environmental, health, and safety (EHS) opportunities and issues over each stage of their product's life. The LCM process focuses on the broader impact of products and processes over their entire life cycle: from development and manufacturing through distribution and customer use to disposal. Initial life-cycle reviews may simply use the matrix as a qualitative "signal chart," while later-stage reviews may analyze individual cells of the matrix in great detail.

Inherent in 3M's Life Cycle Management Process is the characterization and management of both product risk and opportunity. *Risk* reflects the potential for exposure and the hazards of the materials associated with the product over its life cycle, as well as the degree of uncertainty and feasibility of controlling exposure. *Opportunity* lies in finding solutions to these issues, as well as environmental benefits for the customer. This proactive approach by 3M to product risk characterization and management complements the American Chemistry Council's Global Chemical Management Policy and the Responsible Care® initiative.

Factors that may be considered when reviewing a product include exposure, hazard, uncertainty and market opportunity:

- **Exposure** considerations may include the following: duration, concentration, distribution, frequency, pattern of product use, location, demographics, impurities and by-products, competency of user, and potential for abuse or misuse.

- **Hazard** considerations may include the following: type of outcome, potential severity of outcome, uniqueness of hazards, permanence of effect, and treatability.

- **Uncertainty** considerations may include the following: completeness of information, knowledge of product, product history and analogy to similar products, knowledge of customer, and knowledge of ultimate end-of-life disposition.

Life Cycle Stage / Impact	Material Acquisition	R&D Operations	Manufacturing Operations	Customer Needs	
				Use	Disposal
Environment					
Energy / Resources					
Health					
Safety					

FIGURE 10.5   Life Cycle Management process matrix used by 3M.

- Consideration of potential marketing **opportunities** based on a product's environmental, health, or safety benefits may include increased energy efficiency, reduced toxicity, reduced global warming potential, reduced energy use, recyclability, and reduced worker health and/or safety concerns.

The LCM process represents a DFE approach that has been integrated into 3M's New Product Introduction process, which is used by all business units for product development. In addition, 3M has integrated LCM into its Design for Six Sigma process, which includes tools for Voice of Customer as well as other tools for the early stages of product concept development and design.

Examples of Product Applications

One example of how 3M applied the Life Cycle Management process can be found in the wood adhesives business. Furniture companies turned to 3M for an environmentally improved, fast drying adhesive that would bond well with various materials used in the manufacture of their goods. 3M's Engineered Adhesives Division formed an international team to develop an easy-to-use, water-based adhesive that provides effective foam bonding and environmental advantages. 3M™ Fastbond™ Foam Adhesive 100, an entirely new product using new technology, was introduced. It eliminated an estimated 30,000 to 40,000 gallons of solvents in the first year following introduction and reduced waste disposal due to high coverage. One gallon of Fastbond foam adhesive provides about the same coverage as 3 to 4 gallons of a typical flammable foam adhesive.

Another example of LCM occurred in the fire protection market. When production of halons was banned in the early 1990s because of their high ozone depletion potential, several replacement products were rushed to the market in order to fill a void. Concerns continue, however, about the toxicity, regulatory restrictions, and impact on the environment of these "first generation" halon replacements. Specifically, most of them are hydrofluorocarbons (HFCs) that have high global warming potentials. 3M recognized that, as global climate-change policy continues to develop, these halon replacements would not be sustainable and that there would be a significant market opportunity for a sustainable halon alternative.

In fire protection, a "sustainable" technology can be defined as one that: extinguishes fires effectively; is economical to install and maintain; and, perhaps most important in today's business climate, offers a favorable EHS profile—allowing it to be used both today and in the foreseeable future with little or no regulatory restriction. Because fire protection systems are typically built into an infrastructure intended to last for years, there should certainly be a monetary value placed on the choice of a sustainable technology.

Several years ago, 3M embarked on an extensive research program that investigated hundreds of compounds to evaluate their potential as halon replacements. This effort began as a small-scale project championed by a divisional lab manager but quickly escalated into a major initiative. The team, comprising several disciplines including laboratory, environmental, and marketing, was assigned the challenge of determining the "best of the best" from the compounds that made the final list. The team used 3M's Life Cycle Management (LCM) process as one of its key tools in selecting a halon replacement, examining the EHS opportunities and issues at each stage of the product's life cycle.

The result was the development of 3M™ Novec™ 1230 Fire Protection Fluid. This new technology platform, based on fluorinated ketones, is superior in both extinguishing efficiency and safety. It provides a significant reduction in greenhouse gas emissions over HFCs, with a Global Warming Potential of 1 (one), the lowest for any halocarbon alternative to halon. It also has an atmospheric lifetime of 5 *days*, compared to years, decades, or even centuries for other halocarbon alternatives. In addition, it is low in acute toxicity and provides a significant margin of safety at design concentrations, making it ideal for use in occupied spaces. Unlike most halon replacements, it is a liquid at room temperature, so that handling and charging of fire protection systems is easier. Novec™ 1230 fluid is marketed by 3M Electronic Markets Materials Division.

In addition to being "climate-friendly," the product is a highly efficient extinguishing agent that offers both customer and environmental benefits, including:

- Is electrically nonconductive
- Is noncorrosive
- Rapidly vaporizes to gas during discharge
- Leaves no residue
- Does not damage electronics, electronic media, or delicate mechanical devices
- Can be safely used on energized equipment, helping to ensure continuity of operations during a fire emergency

Before introducing it into the marketplace, 3M met with representatives of the U.S. Environmental Protection Agency (EPA) to demonstrate the product and explain its benefits over first generation halon replacements. This meeting helped to increase EPA's comfort with the new product and to develop additional champions; the product was subsequently qualified for use by several major fire protection equipment manufacturers. One global pharmaceutical manufacturer, Roche, has specified Novec™ 1230 fluid as part of its worldwide directive to eliminate all ozone depleting substances. As climate-change policy continues to evolve and HFC-specific regulations are developed in the European Union and elsewhere, it is expected that Novec™ 1230 fluid will become a cornerstone in fire protection systems.

Toward Sustainable Supply Chain Management

While life-cycle management began as a company-centric practice that was concerned with costs, benefits, and risks to the firm, it has evolved into a broader-scope process that extends beyond the firm's boundaries to consider the needs of suppliers, customers, other key stakeholders, and society at large. As described in Chapter 4, business strategists have recognized the importance of environmental and social responsibility to the long-term success of the enterprise. LCM provides a consistent approach for implementing "sustainable" business practices, such as DFE, extended producer responsibility, and product stewardship. Of course, different types of companies may focus on different portions of the product life cycle; for example 3M's LCM program emphasizes the new product introduction process while Caterpillar's Reman program focuses on product take-back.

With increasing globalization, many companies have shifted their business models toward a greater reliance on partnerships and outsourcing. Leading companies increasingly view supply chain excellence as a source of competitive advantage, with the potential to drive performance improvement in customer retention, revenue generation, cost reduction, and asset utilization. Cross-functional teamwork is essential to orchestrate a company's supply chain business processes—managing relationships with suppliers and customers as well as managing the flow of goods, services, and information along the value chain. As a result, supply chain management has evolved from a traditional focus on purchasing and logistics to a broader, more integrated emphasis on value creation. Hence, the following definition [4]:

> *Supply chain management* is the integration of key business processes from end user through original suppliers, which provides products, services, and information that add value for customers and other stakeholders.

With the growing importance of environmental, health, safety, and security issues, there is a pressing need to proactively consider the full supply chain in both product development and life-cycle management. As exemplified by Caterpillar, manufacturers are beginning to take back products and packaging at the end of their useful life, so that designing for reverse logistics has become a strategic approach for converting wastes into assets and, thus, generating shareholder value.

In addition, companies are being held accountable for the content and quality of their products, and ignorance of supplier practices is intolerable. For example, the experience of the food and beverage industry with product contamination (see Chapter 15) has made clear the need for proactive supply chain integrity assurance.

Generally, there are two different supply chain perspectives that need to be balanced in LCM—upstream and downstream. As depicted in Figure 10.6, upstream LCM is concerned with maximizing operating efficiency, anticipating safety and security risks to ensure business continuity, and minimizing the environmental footprint in terms of resource utilization. Downstream LCM is concerned with ensuring that the product is used safely, delivers value to the customer, and is properly managed at end-of-life. There are a variety of mechanisms that companies have used to improve environmental practices in the supply chain; for example:

> **PERHAPS THE GREATEST OPPORTUNITY FOR DFE BREAKTHROUGHS IS COLLABORATION BETWEEN CUSTOMERS AND SUPPLIERS.**

- Electronics companies that depend on strategic suppliers to produce complex components and assemblies have emphasized supplier training, codes of conduct, and auditing (see Chapter 11).

- Chemical companies have emphasized product stewardship to ensure that their customers are handling, storing, using, and disposing of chemicals in a safe and environmentally responsible fashion (see Chapter 13).

- Consumer products companies that purchase natural resource commodities such as cotton or coffee have emphasized labeling, certification, and traceability (see Chapters 15 and 16).

Perhaps the greatest opportunity for DFE breakthroughs is *collaboration* between customers and suppliers to jointly explore redesign of products and processes. Collaborative innovation, sometimes called "co-creation," can enable solutions that might not have been possible

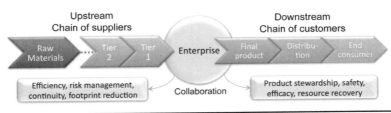

FIGURE 10.6 Life-cycle management across the supply chain.

if the parties worked separately. Individual companies inevitably are hampered by the constraints of existing markets and infrastructure. For example, electronic device manufacturers are limited by the performance characteristics of available materials and components, while chipmakers are constrained by the technologies of fabrication equipment. As a consequence, they may settle for "local optimization" of design features within their control. In contrast, by lifting constraints and pooling their talents, collaborating companies can seek "global optimization" and develop innovative technologies that benefit the entire value-added chain. An example is the introduction of entirely new fabrication processes that eliminate the use of chemical solvents.

Another type of collaboration that has flourished recently is the formation of joint sustainability initiatives among companies within an industry sector, often including direct competitors. Examples include the Beverage Industry Environmental Roundtable, the Electric Utility Sustainable Supply Chain Alliance, the Pharmaceutical Supply Chain Initiative, and the Electronic Industry Citizenship Coalition (see Chapter 11). In each case, the parties have decided that it makes more sense to work collectively on managing environmental and social performance in their upstream supply networks.

Integrated Business Decision Making

Practitioners of life-cycle management have come to believe that broader awareness of stakeholder concerns across the supply chain leads to better business decisions. In particular, design decisions made during new product development will have major impacts upon business performance for years to come and should be informed by a comprehensive life-cycle perspective. However, standard practices in business decision making are typically confined to financial analysis. Even among companies that are recognized as industry leaders in sustainability, integration of environmental or social issues into business case development remains largely an informal, qualitative exercise, and there is little use of systematic frameworks.

One exception is the Sustainable Business Decision Framework (SBDF), developed under the auspices of the World Business Council for Sustainable Development and validated through several applications with multinational cement companies [5]. The SBDF was motivated by a survey of global best practices across all industries, which revealed a widespread need for decision methods to account for sustainability-related trade-offs in business case development. Using a stakeholder value matrix, illustrated in Figure 10.7, the SBDF identifies linkages between sustainable development outcomes and the company's ability to reduce costs, increase profits, and build competitive advantage. Thus, it enables integration of sustainability issues into strategic and tactical decisions, including trade-offs

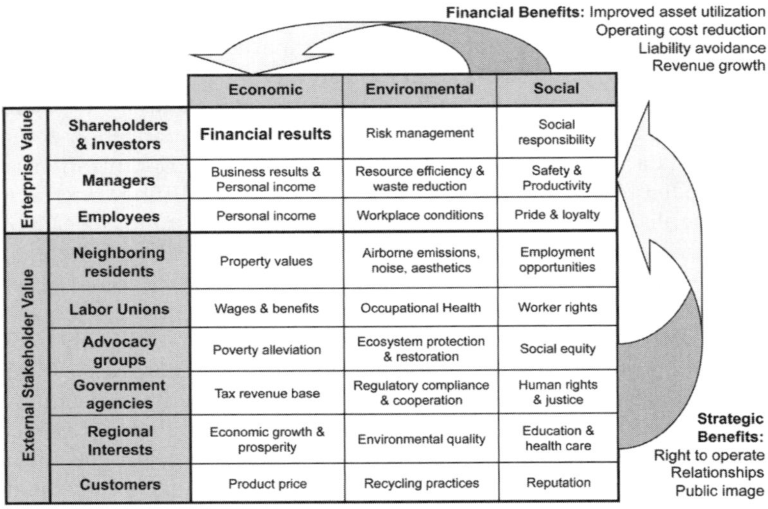

Financial Benefits: Improved asset utilization
Operating cost reduction
Liability avoidance
Revenue growth

	Economic	Environmental	Social
Shareholders & investors	**Financial results**	Risk management	Social responsibility
Managers	Business results & Personal income	Resource efficiency & waste reduction	Safety & Productivity
Employees	Personal income	Workplace conditions	Pride & loyalty
Neighboring residents	Property values	Airborne emissions, noise, aesthetics	Employment opportunities
Labor Unions	Wages & benefits	Occupational Health	Worker rights
Advocacy groups	Poverty alleviation	Ecosystem protection & restoration	Social equity
Government agencies	Tax revenue base	Regulatory compliance & cooperation	Human rights & justice
Regional Interests	Economic growth & prosperity	Environmental quality	Education & health care
Customers	Product price	Recycling practices	Reputation

The first three rows are grouped under **Enterprise Value**; the remaining rows under **External Stakeholder Value**.

Strategic Benefits:
Right to operate
Relationships
Public image

FIGURE 10.7   The Sustainable Business Decision Framework.

between financial gains and longer-term concerns such as company image and future barriers to growth. It also helps to ensure that voluntary sustainability initiatives will deliver maximum benefits in terms of shareholder value.

The SBDF can be used to support DFE in several ways: (a) by senior management to establish broad environmental objectives and decision guidelines; (b) by product development teams to include environmental considerations in product life-cycle reviews; (c) by marketing and communications staff to describe environmental and other benefits of the product; and (d) by external relations staff to support a balanced dialogue with stakeholders, including financial analysts. The important characteristics of the SBDF are as follows:

- Potential decision consequences are classified according to their impacts on *stakeholder value*. Internal company stakeholders are distinguished from external stakeholders such as local community residents, government agencies, and public interest groups.

- Decision consequences are further classified in terms of *economic, environmental, and social* impacts, including both positive and negative outcomes. Note that these three aspects often overlap (e.g., labor productivity, employee safety, and job satisfaction are closely linked) and should not be treated separately.

- Identification of important consequences for external stake-holders leads to assessment of strategic implications (positive or negative) for company interests, which in turn leads to evaluation of the financial implications of environmental and social impacts.

Conventional business case development is largely confined to the *upper left corner* of the SBDF matrix in Figure 10.7 namely, economic outcomes for the company. The SBDF prompts decision makers to expand their perspectives and consider other types of outcomes that are relevant to making good business decisions. For example, when selecting among competing technologies for a new product design, the development team can include additional evaluation criteria based on life-cycle environmental considerations and stakeholder priorities.

However, in order to implement this type of overarching framework successfully, companies need to carefully consider how it can be integrated with the established financial decision-making processes. To produce a rigorous, credible business case, stakeholder value analysis and life-cycle analysis methods need to be combined with traditional decision tools such as discounted cash flow and risk analysis (see Chapter 9). Finally, organizational characteristics such as management philosophy, leadership style, organizational structure, and behavioral norms will strongly influence the acceptance and adoption of any framework that represents a significant change in company practices.

References

1. Additional information about Caterpillar's sustainability programs and Remanufacturing Division is available at www.cat.com.
2. Additional information about 3M's sustainability programs is available at solutions.3m.com/wps/portal/3M/en_US/global/sustainability/.
3. T. W. Zosel, "Pollution Prevention Pays: the 3M Approach," Proceedings 1st International Congress on Environmentally Conscious Design & Manufacturing (Boston: Management Roundtable, 1992).
4. D. M. Lambert, M. C. Cooper, and J. D. Pagh, "Supply Chain Management: Implementation Issues and Research Opportunities," *The International Journal of Logistics Management*, 1998 ,Vol. 9, No. 2.
5. D. Fiksel, T. Brunetti, and L. Garvin. *Toward a Sustainable Cement Industry* (Geneva: Substudy 3, Business Case Development, Battelle Report to the World Business Council for Sustainable Development, 2002).

Walking the Talk: The Real-World Practice of Design for Environment

CHAPTER 11

Electronic Equipment Industries

Overview

Electronic equipment manufacturers include makers of computers and peripherals, telephone and telecommunications equipment, and a large variety of electronic devices used by consumers and industrial facilities. The global electronics industry was among the first to adopt DFE, as mentioned in Chapter 1. In the early 1990s, the American Electronics Association formed a DFE task force to develop basic principles and best practices and published a series of white papers on the topic. Contributors to this initiative included AT&T, HP, IBM, Xerox and the Microelectronics and Computer Technology Corporation (MCC). Subsequently, the Institute for Electrical and Electronic Engineers (IEEE) launched an annual Symposium on Electronics and the Environment, which continues to be an important forum for sharing research and applications in the field of DFE.

The electronics industry is relatively young compared to traditional "smokestack" industries but nevertheless has a significant environmental footprint. Semiconductor fabrication is one of the most resource-intensive of all industries. For example, it has been estimated that production of a 2-gram 32MB memory chip can require as much as 1,200 grams of fossil fuels, 72 grams of chemicals, and 32,000 grams of water. Chips are much more material-intensive than traditional products; the estimated weight ratio of production inputs to the final product is over 600 for a memory chip, compared to about 2 for an automobile or 4 to 5 for an aluminum can [1]. Moreover, electronic products tend to become obsolete rapidly, and recycling efforts have struggled to keep pace with the mounting flow of electronic waste. As discussed in Chapter 3, the European Union issued the WEEE directive to assign responsibility to manufacturers for waste recovery.

Most electronics companies have taken steps to embed DFE practices such as Design for Disassembly into their new product development processes. The next step has been shifting attention to the practices of their suppliers. Electronic supply chains are complex and difficult to manage, since many electronic components are manufactured in Asia or Latin America. In 2006, a global coalition of electronic industry companies entered into an unprecedented international collaboration to develop a set of supplier expectations, known as the Electronic Industry Citizenship Code. The code provides comprehensive principles for management of environmental releases, workplace health and safety, labor practices, and business ethics, and is supported by standardized supplier assessment and auditing procedures.

As shown in Figure 11.1, there are many opportunities for applying DFE practices in the electronic product life cycle. The following are selected examples of innovative DFE initiatives in the electronics industry:

- Motorola established a Green Design Project to develop and implement standards, methods, and tools for environmentally conscious product design. A customized software tool called the Green Design Advisor was developed to help Motorola engineers calculate the life-cycle environmental impact of a product and compare the environmental performance of different materials and processes used to create a product [2]. The tool includes a parts and connectivity model for analyzing the recyclability of devices, identifying design weaknesses and optimizing for end-of-life disassembly.

- Dell's packaging engineers are saving over 20,000 tons of packaging material annually thanks to a dematerialization program that resulted in reduction and elimination of corrugated, plastic foam, and wood materials. For example, Dell implemented slip sheets (three-pound, 0.03-inch thick plastic sheets) instead of wood pallets (which weighed 40 pounds and were 5 inches tall) for inbound chassis products.

- Panasonic developed a plasma display technology that cuts energy consumption in half, while maintaining the same brightness. The system uses new phosphors and cell design technology for improved discharge, and new circuit and drive technology to significantly reduce power loss. This double-efficiency technology will form the base for next-generation displays with thinner profiles, larger screens, brighter images, higher definition, lower power consumption, and lower weight.

- Intel's Core™ 2 Duo processor uses 40% less energy to do 40% more work than the previous Pentium processor. The total energy savings from all the Core 2 Duo products in the marketplace is equivalent to taking millions of cars off the road.

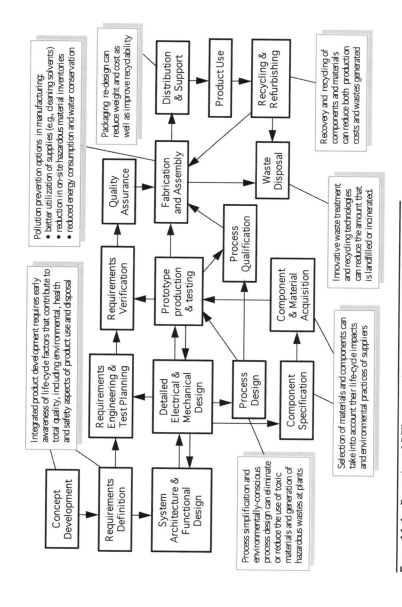

Pollution prevention options in manufacturing:
- better utilization of supplies (e.g., cleaning solvents)
- reduction in on-site hazardous material inventories
- reduced energy consumption and water conservation

Packaging re-design can reduce weight and cost as well as improve recyclability

Recovery and recycling of components and materials can reduce both production costs and wastes generated

Innovative waste treatment and recycling technologies can reduce the amount that is landfilled or incinerated.

Integrated product development requires early awareness of life-cycle factors that contribute to total quality, including environmental, health and safety aspects of product use and disposal

Selection of materials and components can take into account their life-cycle impacts and environmental practices of suppliers

Process simplification and environmentally-conscious process design can eliminate or reduce the use of toxic materials and generation of hazardous wastes at plants

Concept Development

Requirements Definition

System Architecture & Functional Design

Requirements Engineering & Test Planning

Detailed Electrical & Mechanical Design

Process Design

Component Specification

Requirements Verification

Prototype production & testing

Process Qualification

Component & Material Acquisition

Quality Assurance

Fabrication and Assembly

Waste Disposal

Distribution & Support

Product Use

Recycling & Refurbishing

FIGURE 11.1 Examples of DFE opportunities in the electronics industry.

221

The newer Xeon processor gives another 35–60% efficiency gain over previous products.

- Intel's Eco-Rack design for business data centers reduces power consumption by 16 to 18%. If it were used by all servers and data centers in the United States, this technology would save the equivalent of the energy used by 986,000 homes. Each Eco-Rack saves about $44,000 annually in electricity costs. Many data centers have adopted "virtualization" technologies to further lower energy costs by reducing the number of servers required for a given service load.

- Microsoft has adopted a commoditized manufacturing approach to make data centers modular, scalable, efficient, and as low-cost as possible. Its new Generation 4 design has a Power Usage Effectiveness—the fraction of energy actually used for computing—of 1.12, among the lowest yet achieved. Additionally, by working toward a chiller-free design, Microsoft hopes to eliminate the use of water.

- Building on IBM's experience in product stewardship and DFE, IBM Global Business Services introduced an Environmental Product Lifecycle Management service as part of its corporate social responsibility practice. The company employs more than 1,000 product life-cycle management experts worldwide, advising customers in a variety of different industries.

- EPA's DFE Program partnered with the electronics industry (see Chapter 3) to perform a life-cycle environmental impact assessment of alternatives to lead solder, helping to prepare for a phase-out of lead in compliance with the European Union's Restriction of Hazardous Substances directive [3].

This chapter highlights the DFE programs of three leading companies in the electronics industry, Xerox, HP and Sony, but there are many more stories worth telling.

Xerox Corporation: Reducing the Footprint of Printing
A Pioneering Commitment

Xerox Corporation was one of the earliest firms to adopt DFE, and enjoys a strong reputation as an industry leader in sustainability. With $17 billion in annual revenue and about 55,000 employees in over 160 countries, the company's famed brand name is synonymous with photocopying. Xerox introduced the first commercial copier system in 1959, and its legendary Palo Alto Research Center produced many breakthrough innovations, including the computer mouse. Xerox has adapted to the digital age by offering a broad portfolio of document products and services, including information management, business solutions, and imaging hardware and supplies.

Xerox has a legacy of regarding responsible corporate citizenship as a business and social imperative. Nearly 50 years ago Xerox pledged to "behave responsibly as a corporate citizen," as one of its six Core Values. That precept became the basis for Xerox's Environment, Health and Safety policy, adopted in 1991, which commits the company to "the protection of the environment and the health and safety of its employees, customers, and neighbors." The policy further states that it is applied worldwide and that protection "from unacceptable risks takes priority over economic considerations and will not be compromised" [4].

In the 1980s, Xerox was the first company to introduce power-down features to save energy and the first to make two-sided printing a standard. By 1999 the company was beginning to think beyond "eco-efficiency" and was looking toward methods to evaluate the life-cycle implications of its products, services, and manufacturing processes. Today, Xerox views sustainability not as a cost of doing business, but as a way of doing business. The company is responding to environmental sustainability challenges with four goals, elegant in their simplicity (see Table 11.1).

Xerox began its journey toward environmental sustainability with its goal of achieving waste-free products from waste-free factories to promote waste-free offices for its customers. That led to the development of pioneering remanufacturing technologies and industry-leading recycling expertise. These initiatives proved to be not only good for the environment, but they also saved the company hundreds

Challenge	Goal	DFE Accomplishments
Climate Protection and Energy	Carbon Neutral	18% reduction in CO_2 emissions from 2002–2006
Preserve Bio-diversity and the World's Forests	Sustainable Paper Cycle	Hardware innovations, new paper technology
Preserve Clean Air and Water	Zero Persistent, Bioaccumulative, Toxic Footprint	94% reduction in air toxics 91% reduction in hazardous waste
Waste Prevention and Management	Waste-Free Products, Facilities and Customers	Design for reuse/recycle New life to 2.8 million machines from 1991 to 2008

TABLE **11.1**   Xerox Corporation's Sustainability Goals

of millions of dollars. Xerox's greenhouse gas reduction project, Energy Challenge 2012, has likewise paid off environmentally and financially, as has a Zero Injury program to reduce workplace incidents.

Equally important has been Xerox's work to extend sustainability beyond its own walls. It has asked its materials and components suppliers to meet environment, health, and safety standards and eliminate toxic materials. Likewise, it has asked its paper suppliers to meet rigorous sustainability standards. It has worked with government agencies and industry organizations, including U.S. EPA on energy standards and the European Union on waste and materials initiatives. Most important, it has innovated to bring sustainability to customer workplaces.

Designing Total Systems

According to Patricia Calkins, vice president for Environmental Health and Safety, Xerox's DFE strategy was developed with a total systems perspective, focusing on the greatest opportunities to reduce environmental impacts along the value chain. In particular, since Xerox is one of the largest distributors of cut sheet paper under one brand in the world, the company has made a commitment to strive for a "sustainable paper cycle," which includes many elements:

- Designing hardware for efficient duplex capability and effective use of recycled content paper
- designing workflow tools (smart document technology) to reduce the hard copy requirements of document dependent business processes
- Introducing new paper technology, known as "high yield business paper," which requires half the wood input and generates 70% less greenhouse gas emissions (see below)
- Developing new print technologies (i.e., erasable paper) that could enable the reuse of paper multiple times
- Driving sustainability requirements up the supply chain through rigorous supplier requirements
- Achieving FSC, PEFC/SFI "chain of custody certification"
- Introducing FSC, PEFC, and SFI certified papers to the marketplace
- Partnering with The Nature Conservancy to drive improvements in sustainable forestry practices and enable more supply of sustainably sourced paper fiber

This effort involved several key innovations developed by Xerox scientists. For example, Xerox High Yield Business Paper™ is a *mechanical* fiber paper manufactured through a "greener" process than standard paper used with digital printers (see Figure 11.2). The sheet

FIGURE 11.2   Xerox High Yield Business Paper™ requires less wood, water, chemical, and energy inputs.

is produced by mechanically grinding wood into papermaking pulp instead of using the traditional chemical pulping process. As a result, 90% of the tree by weight ends up in the High Yield Business Paper versus only 45% in traditional digital printing paper. In addition, High Yield Business Paper requires less water and chemicals and is produced in a plant using hydroelectricity to partially power the pulping process. Optimized for digital printing, the new paper is lighter in weight, resulting in significant savings on postage and energy for paper shipments.

The key breakthrough was achieved at Xerox's Media and Compatibles Technology Center in Webster, N.Y., which was established in 1964 with the mission to ensure that Xerox papers are optimized for use in Xerox products, thus preventing waste and product downtime. Mechanical paper is widely used in offset printing, e.g., for newspapers, but there were two problems that prevented it from running reliably on printers and copiers—excessive dust contamination and curling of the paper due to heat. According to Bruce Katz, who led the Xerox research team, the mechanical paper curled because the back and the front of the paper shrank at different rates. Working with a paper mill and employing statistical techniques, the team developed a patented process that better distributed the fibers on both sides of the paper, reducing the curl. They also developed a surface treatment at the mill to minimize the paper dust.

Patty Calkins points out that, contrary to early predictions, the advent of digital technology and the Internet has helped to keep paper in the office—people print less of what they read, but they are reading a much larger stream of information. Xerox research shows that office workers discard about 45% of everything they print within a day. So, paper use will persist much longer than anticipated because paper is easier to access and annotate, easier to share or review with others, easier to compare with other documents, easier to scan, more pleasant to read, and less likely to be forgotten or lost. However, she says, "Even if the paperless office is a myth, Xerox believes that the office that uses paper responsibly is a goal that is within reach."

Closing the Loop through Asset Recovery

Xerox's hardware products exhibit a similar concern with total system performance. For example in the iGen3 commercial printing system over 97% of the components are recyclable or remanufacturable, and up to 80%, by weight, of the waste that it generates can be returned, reused, or recycled. Its all-digital technology means virtually no make-ready waste, and variable data printing means no obsolete inventory. iGen3 emits 80% less noise than a typical offset press. It uses dry inks that are nontoxic and have closed containers with a transfer efficiency rate near to 100%. This means that there are no chemicals requiring clean-up or degrading workplace air quality, no personal protective equipment needed, no regulated waste, and no air or water abatement capital or operating expenses.

Before sustainability was fashionable and before the term "reverse logistics" was invented, Xerox pioneered the practice of converting end-of-life electronic equipment into new products and parts. Xerox began a systematic "asset recovery" program in 1991, and by 2008 remanufacturing and recycling had given new life to more than 2.8 million copiers, printers, and multifunction systems, while diverting nearly two billion pounds of potential waste from landfills—111 million pounds (50,000 metric tons) in 2006 alone. Moreover, the program has saved more than $2 billion over that period. To accomplish this, Xerox developed a comprehensive process for taking back end-of-life products, including design methods for ease of disassembly and recovery as well as systematic processes for remanufacture, parts reuse, and recycling.

Xerox maximizes the end-of-life potential of products and components by considering reuse as an integral part of the design process. Machines are designed for easy disassembly and contain fewer parts. Parts are durable—designed for multiple product life cycles. Coded with instructions on how they can be disposed of, the parts are also easy to reuse or recycle. As a result, equipment returned to Xerox at end-of-life can be rebuilt to as-new performance specifications, reusing 70–90% of machine components (by weight), while meeting performance specifications for equipment with parts that are all new. Xerox also designs product families around modular product architectures and a common set of core components.

Thanks to these advances, a returned machine can be rebuilt as the same model through remanufacture, converted to a new model within the same product family, or used as a source of parts for next-generation models. Improved forecasting of equipment returns has allowed Xerox to rely on previous generations of equipment as a source of components for products in development. A Xerox product whose designs are based on previous models may have 60% of its parts in common with previous equipment. The practice of reusing parts reduces the amount of raw material needed to manufacture new parts, which generates several hundred million dollars in cost savings each year. Moreover, energy savings are significant; in 2006, energy savings from reused parts totaled six million therms (170,000 megawatt hours)—enough energy to light more than 136,000 U.S. homes for a year.

These capabilities have made it easy for Xerox to comply with the various country programs that implement the European Union's Waste Electrical and Electronic Equipment (WEEE) directive (see Chapter 3). The annual amount of waste diverted from landfill has declined since 2002, in part due to lighter-weight machines (see Figure 11.3). In addition, the number of office machines returned for remanufacturing has decreased since the WEEE legislation mandates the national collection and recycling of scrap office products.

FIGURE 11.3 Annual waste avoidance due to Xerox end-of-life asset recovery programs.

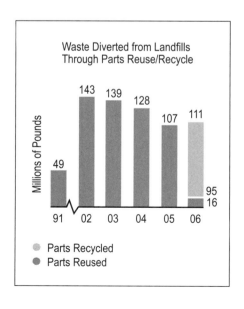

Customer Benefits

The ultimate justification for Xerox's sustainability efforts is that customers are realizing significant environmental and financial benefits from adoption of Xerox solutions. For example, by introducing Xerox multifunctional devices, Dow Chemical was able to reduce the number of printers by two-thirds and reduce its total annual printing costs by up to 30%. Northrop Grumman was able to cut in half the number of devices required and increase the efficiency of printing and copying operations, achieving over 25% energy and GHG reduction and 33% solid waste reduction. Xerox has truly discovered a sustainable path to customer satisfaction by reducing the environmental footprint of printing operations.

Hewlett Packard: A Green Giant

Stewardship the HP Way

With a portfolio that spans printing, personal computing, software, services, and information technology (IT) infrastructure, HP is among the world's largest IT companies, with annual revenue of more than $100 billion. For decades HP has worked to manage its environmental impact by adopting environmentally responsible practices in product development, operations, and supply chain. The company strives to be a global leader in reducing its carbon footprint, limiting waste and recycling responsibly. HP's efforts earned it recognition as one of Fortune Magazine's "Ten Green Giants" in April 2007.

In 1992, HP was among the first electronics industry companies to establish a DFE initiative. The initiative originated in a product stewardship program that was launched within the Computer Products Organization and later was expanded to HP's other businesses. By 1993, HP was measuring product performance improvements based on a set of DFE metrics that included material conservation, waste reduction, energy efficiency, and end-of-life recoverability. Today, the DFE program is coordinated by an Environmental Strategies Council that includes representatives from each global business unit and sales region, as well as from supply chain, operations and other corporate functions. A global network of product stewards works with design and development teams to incorporate environmental innovations into HP products [5].

According to Director of Sustainability Bonnie Nixon Gardiner, HP's approach to implementing a socially and environmentally responsible supply chain is based on early, frequent, and proactive involvement with key suppliers to develop a partnership for improvement. For example, HP was the first company to display the EPA SmartWay logo on its product packaging for a U.S. and Canadian surface transportation supply chain composed of 100% EPA

SmartWay compliant carriers, making it one of the region's cleanest. The SmartWay program is aimed at reducing fuel consumption, greenhouse gases, and other air emissions (see Chapter 3).

In response to customer interest, HP has certified its products to many eco-labels around the world, including a personal computer that was the first product to earn EPEAT gold status.* The company has also invested considerable effort in harmonizing environmentally related product standards. For example, HP was instrumental in the multi-stakeholder process that developed the environmental performance standard IEEE 1680, published by the Institute of Electrical and Electronics Engineers in 2006. This standard integrates a wide variety of existing regulations and standards, including U.S. Energy Star® and the European Union Restriction of Hazardous Substances (RoHS) and Waste Electrical and Electronic Equipment (WEEE) directives (see Chapter 3). It specifies 23 required and 28 optional criteria across eight areas of environmental impact covering all product lifecycle stages, providing a basis for buyers to assess the environmental performance of desktop and notebook computers and displays.

HP's strategy also includes growth through acquisition. For example, in 2006 HP acquired Voodoo PC, a manufacturer of high performance and personalized gaming computer systems, and in 2005 HP acquired Scitex Vision, a manufacturer of large format printers. When HP acquires such companies, it first ensures that their current products meet applicable regulatory requirements and then makes a transition to HP's more demanding DFE standards. This transition may take several product introduction cycles to complete. Until then, HP does not include those products in product goals or progress reports.

Design for Environment Program

HP defines Design for Environment as an engineering perspective in which the environmentally related characteristics of a product, process or facility are optimized. Together, HP's product stewards and product designers identify, prioritize, and recommend environmental improvements through a company-wide DFE program. HP's DFE guidelines derive from evolving customer expectations and regulatory requirements, but they are also influenced by the personal commitment of its employees. The Design for Environment program has three priorities:

1. Energy efficiency—reduce the energy needed to manufacture and use HP products.
2. Materials innovation—reduce the amount of materials used in HP products and develop materials that have less environmental impact and more value at end-of-life.

---

*EPEAT is a rating system developed by the Green Electronics Council (see Chapter 9).

3. Design for recyclability—design equipment that is easier to upgrade and/or recycle.

HP's DFE guidelines recommend that its product designers consider the following:

- Place environmental stewards on every design team to identify design changes that may reduce environmental impact throughout the product's life cycle.

- Eliminate the use of polybrominated biphenyl (PBB) and polybrominated diphenyl ether (PBDE) flame-retardants where applicable.

- Reduce the number and types of materials used, and standardize on the types of plastic resins used.

- Use molded-in colors and finishes instead of paints, coatings, or plating whenever possible.

- Help customers use resources responsibly by minimizing the energy consumption of HP's printing, imaging, and computing products.

- Increase the use of pre- and postconsumer recycled materials in product packaging.

- Minimize customer waste burdens by using fewer product or packaging materials overall.

- Design for disassembly and recyclability by implementing solutions such as the ISO 11469 plastics labeling standard, minimizing the number of fasteners and the number of tools necessary for disassembly.

The new TouchSmart PC is an example of how HP uses cutting-edge functionality as a driver to reduce the environmental impact of its products (see Figure 11.4). The touch-screen all-in-one PC uses HP power management technology to provide up to 86 kWh per

FIGURE 11.4   The HP TouchSmart PC was designed to reduce the environmental footprint.

year energy savings compared to PCs without power management enabled. The device is shipped in 100% recyclable packaging with more paper and less plastic foam for easier reuse, and the machine itself uses 55% less metal and 37% less plastic than standard PCs and monitors.

HP has also emphasized recycling of electronic products and operates recycling programs in more than 40 locations around the world. The programs seek to reduce the environmental impact of products, minimize waste going to landfills, and help customers conveniently manage products at their end-of-life in an environmentally sound fashion. In 2006 alone, HP recovered 187 million pounds of electronics globally, 73% more than IBM, its closest competitor. By 2010, HP is aiming to reach a cumulative total of 2 billion pounds of recycled electronics and print cartridges. Plastics and metals recovered from products recycled by HP have been used to make a range of new products, including auto body parts, clothes hangers, plastic toys, fence posts, serving trays, and roof tiles. In addition to recycling, HP offers a variety of product end-of-life management services including donation, trade-in, asset recovery, and leasing.

Greening the Data Center

The rapid growth of enterprise computing has led to the establishment of sophisticated data centers housing hundreds of servers and consuming significant amounts of energy for both powering and cooling. HP has responded with Thermal Logic, a technology designed to reduce power and cooling costs and increase data center capacity. The technology actively and automatically monitors and adapts power load and cooling capacity based on changes in demand and environment, via innovations that ensure the highest energy efficiency, redundancy, and scalability of power and cooling.

HP's new ProLiant blade server products were designed to use 25% less energy per watt, saving more than 700 watts per server enclosure, as well as a new BladeSystem enclosure that boosts the energy efficiency of server rack power supplies. By using a multizone architecture and multiple thermal sensors in each enclosure, Thermal Logic captures and analyzes power and temperature throughout the entire system and distributes power and cooling control where it is needed most. Customers can customize power and cooling thresholds for either the highest level of performance or the most efficiency; in addition, they can initiate cooling and automatically control cooling levels to react to and remove heat.

As part of the Thermal Logic portfolio, HP has developed a program called Dynamic Power Capping, which manages the power used by hardware to maximize performance while cutting power use as much as possible. HP says the program can increase a data center's capacity by as much as 300%. With that level of performance improvement, a one-megawatt data center could recover up to $16 million in capital expenditures and save $300,000 a year on energy bills. In addition, HP has launched a consultancy called Energy Efficiency Design and Analysis Services, which capitalizes on the company's experience in building and optimizing data centers. The service will suggest ways to reconfigure existing facilities or design new facilities with an eye toward peak performance at minimal cost, and can help data centers to achieve LEED certification.

Sony Electronics: Innovation in Design[†]
Green Management

The Sony brand name is synonymous with innovative products, and Sony has also been active in environmental design. Sony Electronics Inc. is dedicated to protecting and improving the environment in all areas of its business operations, including conserving resources, eliminating hazardous chemicals, increasing energy efficiency, reducing pollution, and offering convenient recycling opportunities to customers. Since 1991, the company has implemented a corporate-wide environmental policy to address waste minimization, waste management, and consideration of environmental impact when evaluating new products, projects, and operations. This led to the establishment of Sony's Environmental Vision and mid-term Green Management targets [6].

Green Management 2010 provides guidance, direction, and progress measurement systems for environmental initiatives for global Sony group operations. The plan creates a framework to uphold "green" policies and procedures in the creation, design, manufacture, packaging, transport, sales and service of products, and in all areas of operation in order to achieve these targets. All Sony North American manufacturing plants and all nonmanufacturing operations larger than 100 employees are ISO 14001 certified under the Sony's global environmental management system program.

The Sony Take Back Recycling Program was launched on Sept. 15, 2007 with Waste Management Inc. (WM) and is the first recycling initiative in the United States that connects a major consumer electronics manufacturer to a national waste management company and its physical network of recycling centers around the country. To encourage consumers to recycle and dispose of electronic devices in an environmentally sound manner, Sony is working with WM and its Recycle America locations to allow consumers, including businesses, to recycle all Sony-branded products for no fee at eCycling drop-off

[†]This section is based on Sony's website and a case study provided by the World Business Council for Sustainable Development.

centers throughout the United States. The program is also open to all makes of consumer electronics products, recycling any non-Sony product at market prices. Sony and WM Recycle America are also working toward having enough drop-off locations so there is a recycling center within 20 miles of 95% of the U.S. population. The company's overall goals are to make recycling of Sony products as easy and convenient as purchasing them and to recycle one pound of old consumer electronics equipment for every pound of new products sold.

Life-Cycle Innovation

Sony is taking a holistic approach to enhancing products and services to minimize environmental impacts. For example, Sony has been instrumental in using post-consumer resources including plastics in new products, thus supporting the recycling of materials. This initiative has led to the annual consumption of more than 15 million pounds of otherwise waste plastic. Innovations are not only improving production but also reducing environmental impacts in the product use phase. This has led to the development of a number of technologies that help reduce the environmental impact attributed to consumer behavior during product use, service, and disposal.

Life-cycle assessment points out that the major ecological burden of electronic products is due to the energy consumed in the use-phase of the product life. For example, European home consumer electronics consume 36 TWh (Terawatt-hour) annually and are forecasted to grow to 62 TWh by 2010. Sony's SDM-N50 liquid crystal display incorporates a unique set of user features that directly reduces the energy consumption of the product. The 12mm, 3kg display features an energy saving infra-red "user sensor" and an ambient light sensor. The user sensor automatically switches the screen to sleep mode (utilizing less than 3 watts) if no one is sitting in front of the screen. The light sensor adjusts the brightness (and therefore energy consumption) of the display according to natural "ambient" light conditions of the room.

Sony is also developing technologies and services that contribute to the extension of product lifetime. For example, the ProGlobe television is a European prototype design aimed at establishing technologies for improved serviceability. The prototype technologies being developed include a unique chassis design that allows service engineers to access the internal core of the TV at the consumer's home (potentially reducing the travel burden and cost of TV service); and a one-click, "eco-mode" function, giving the user the option of reducing the operational energy consumption of the TV by 25 watts (by adjusting the picture brightness). In addition, the ProGlobe's new power circuit layout provides a remarkable level of 0.5 watt standby power consumption.

Miniaturization and the shift to digital technologies have made product repair an increasingly difficult and specialized job. In Europe, Sony has introduced the "exchange and refurbishing program," an innovative service concept for the efficient repair of products with large sales volumes, such as the Walkman, Discman, PlayStation, and mobile phones. In the case of a defect, a customer can exchange their product at the dealer for a refurbished model, or even for a new one (if a defect is found within the three month guarantee period from date of purchase). Defective products are collected and refurbished, broken parts are repaired or exchanged, and housing parts are checked at a central European location. Since "factory-like" repair is more efficient, average costs have been reduced by 25%. This way more product parts and components get the chance for a second life. For example, Sony is achieving a 68% reuse rate for the plastic components of returned PlayStations, and virtually all of the remaining components that cannot be reused are recycled.

References

1. E. D. Williams, R. U. Ayres, and M. Heller, "The 1.7 Kilogram Microchip: Energy and Material Use in the Production of Semiconductor Devices," *Environmental Science & Technology*, 2002, 36, 5504.
2. More information about Motorola's Green Design Advisor is available at www.epa.gov/perftrac/events/design.pdf .
3. *Solders in Electronics: A Life-Cycle Assessment.* epa.gov/dfe/pubs/solder/lca/index.htm.
4. More information about Xerox's corporate citizenship is available at www.xerox.com/about-xerox/citizenship/enus.html.
5. HP's Global Citizenship report is available at www.hp.com/hpinfo/globalcitizenship/.
6. More information about Sony's environmental programs is available at news.sel.sony.com/en/corporate_information/environmental.

CHAPTER 12

Transportation Industries

Overview

The purpose of transportation systems is to satisfy the basic human need for mobility, which is essential to the functioning of the global economy and fundamental to human well-being. As democratic systems of government prevail and international trade barriers vanish, freedom of movement has become recognized as a universal human right. Each year, people travel trillions of miles to realize their personal and economic needs, while companies transport trillions of ton-miles of freight to achieve efficiencies of specialization and scale.

Transportation covers a broad range of industries, including the manufacture of vehicles and transportation equipment, the supply of fuel and replacement parts, and the operation and maintenance of roads, railways, airports, marine terminals, and transportation services. Transportation systems, and the automotive industry in particular, have an enormous impact on both the economy and the environment. As illustrated in Figure 12.1, the U.S. automotive industry consumes billions of dollars worth of goods and services and directly or indirectly creates about 13 million jobs [1]. The environmental impacts of transportation systems are extensive and highly visible, ranging from atmospheric pollution and greenhouse gas emissions to the accumulation of solid wastes, such as tires and scrap plastics from obsolete vehicles. Other adverse impacts include noise, traffic congestion, and highway fatalities.

The rising awareness of sustainability concerns has led government and industry leaders to address these environmental impacts. Transportation industries are among the most highly regulated, with requirements governing occupant safety, operator certification, equipment inspection, hazardous material transport, fuel efficiency, and many other issues. For example, the European Union's End-of-Life Vehicle Directive, described in Chapter 3, imposes responsibilities on original equipment manufacturers for disassembly and recycling or

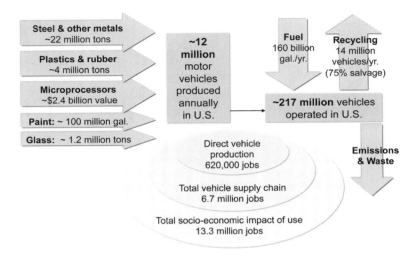

FIGURE 12.1   Aggregate resource flows and economic impacts of the U.S. motor vehicle industry.

disposition of vehicles. This has prompted manufacturers to implement design for recovery programs; as mentioned in Chapter 8, some companies like BMW, had anticipated these programs and developed advanced recovery technologies.

However, since 2001, a perfect storm of environmental, geopolitical, and economic realities have combined to drive home the fact that our world faces an increasingly uncertain energy future. Global climate change, regional conflicts in the Middle East, natural disasters in the Gulf of Mexico, and extraordinary economic growth in China and India are driving serious concerns about energy supply, the environment, sustainable growth, and even national security. Today, more than a third of the world's energy needs are met with petroleum. In 2008 the price of oil briefly exceeded $140 a barrel, and will likely do so again. Yet, the global automobile fleet is 96% dependent on fossil fuels [2]. Given these challenges, the automotive industry must do everything it can to design vehicles with environmental constraints in mind.

Sustainable Mobility Research

The continued growth in mobility demand has led to increasing concerns about potential impacts on the environment, public health and safety, and quality of life, raising questions about how existing transportation systems can meet today's mobility needs without compromising the welfare of future generations. This prompted the World Business Council for Sustainable Development to launch a program in 2002 called "Sustainable Mobility," sponsored by leading global companies and focusing on road transportation. Completed in 2004, the study proposed incremental solutions that assumed continued growth in private vehicle use but did not examine scenarios under which global mobility patterns might be fundamentally altered [3]. A follow-up study focused on solutions to mobility problems in cities around the world.

From a holistic perspective, the transportation industries have access to a broad portfolio of technologies that can be deployed to satisfy future societal needs for mobility. The choices include different modes of transport, such as air, sea, rail, and highway; different fuel sources—fossil fuels, biofuels, electricity, and hydrogen; and different infrastructure configurations. Yet we have only a vague understanding of the potential social, economic, and environmental conditions— for example, population density or use of digital communication—that will both drive the demand for mobility and constrain access to mobility. These conditions will vary enormously among developing and developed nations, between urban and rural settings, and across different geographic and climatic settings. Nor do we understand the full ramifications of technology choices upon economic vitality, ecological integrity, or community well-being under various future scenarios.

There is active ongoing research in sustainable mobility, addressing two main facets:

1. Technological innovation, including alternative materials, vehicle designs, energy sources, propulsion systems, and transportation networks that are safer, more effective, and more environmentally benign.

2. Technology assessment to determine the feasibility, eco-efficiency, sustainability, and resilience of alternative mobility technology combinations under various future scenarios, providing a sound scientific basis for public policy formulation and R&D priority-setting.

For example, the University of Michigan has launched a broad, interdisciplinary program in Sustainable Mobility and Accessibility Research and Transformation (SMART) that focuses on the growing challenges in urban regions of the world. The Center for Automotive Research at The Ohio State University has engaged both automotive and electric power companies in a new program focused on electric vehicles, including plug-in hybrids and intelligent charging. The benefits of plug-in hybrid electric vehicles are discussed in Chapter 18.

DFE in the Transportation Life Cycle

Investigation of life-cycle environmental issues in transportation systems is extremely complex for several reasons. First, many forms of travel, both business and personal, involve a mixture of different modes of transport. For example, freight container shipments routinely travel via a combination of rail, sea, and truck, while commuters routinely travel to work via a combination of private and public transportation. Second, as shown in Table 12.1, each mode involves integration of a variety of different systems with different life-cycle considerations—vehicle systems, energy systems, control systems, infrastructure systems, and associated maintenance systems. Third, the performance metrics that must be balanced are broad-ranging including satisfaction of customer needs for accessibility, affordability, speed, reliability, safety, and security. Lastly, redesign of transportation systems cannot be accomplished without an understanding of the relevant policy environment with regard to urban and regional planning and economic development.

Thus, genuine innovation in sustainable mobility will require extraordinary collaboration among a variety of stakeholders, both public and private. No single company can address all of the above issues in isolation. However, individual companies can make headway through the design of products, processes, and systems in a way that reduces environmental burdens or enhances environmental quality. Even better, companies can work with their supply chain partners to develop sustainable solutions. Indeed, understanding the environmental impacts of design requires life-cycle thinking, since the impacts occur at many points along the value chain.

For example, Figure 12.2 illustrates the results of a life-cycle study for a generic 4-passenger sedan, showing the percent of total environmental burdens contributed at each life-cycle stage. Energy use and

Transport Mode	Vehicle Systems	Energy Systems	Operational Systems	Infrastructure Systems	Life-Cycle Systems
Air	Aircraft, Dirigibles	Liquid Fuels, Gases	Air Traffic Control	Airports, Provisioning	Maintenance, Recycling
Water	Watercraft, Amphibious	Liquid Fuels, Wind Power	Harbor Control	Ocean Terminals, Provisioning	Maintenance, Recycling
Rail	Conventional, High-Speed	Liquid Fuels, Electric	Monitoring, Dispatching	Railways, Provisioning	Maintenance, Recycling
Road	Automobiles, Motorcycles, Moto-Rickshaws Trucks, etc.	Liquid Fuels, Fuel Cells, Electric, Hybrid, etc.	Monitoring, Congestion Management	Roadways, Bridges, Refueling Stations	Maintenance (Vehicle/Road), Reverse Logistics
Non-Powered	Pedestrian, Bicycles, Animal-Drawn	N.A.	N.A.	Roads, Dedicated Pathways	Maintenance, Animal Husbandry

TABLE 12.1 Design Considerations for Sustainable Mobility

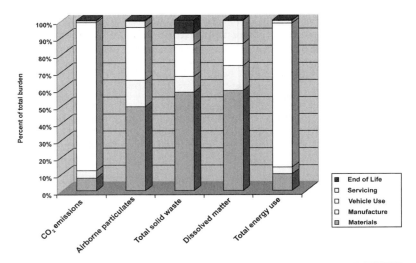

FIGURE **12.2**   Life-cycle environmental burdens due to production and use of motor vehicles [4].

greenhouse gas emissions are dominated by the vehicle use stage, whereas the largest contributions to waste and emissions occur in the materials acquisition stage. The contributions of the actual manufacturing stage are relatively small, since much of the final production involves assembly operations [4].

One of the most direct strategies for reducing energy consumption during vehicle use is the use of lighter-weight materials such as aluminum. For example, industry studies have shown that in automotive applications each kilogram of aluminum replacing mild steel, cast iron, or high strength steel saved between 13 and 20 kilograms of greenhouse gas emissions depending on the component. For metro or subway cars, the avoided GHG emissions are approximately 26 kilograms in Europe and 51 kilograms in the United States. [5].

There are numerous examples of DFE initiatives in the transportation industries that address energy and environmental goals at various points in the life cycle. The following are selected highlights, but there are many more similar stories of innovation:

- General Electric is designing a hybrid diesel-electric locomotive that will capture the energy dissipated during braking and store it in a series of sophisticated batteries. That stored energy can be used by the crew on demand—reducing fuel consumption by as much as 15 percent and emissions by as much as 50 percent compared to most of the freight locomotives in use today. In addition to environmental advantages, a hybrid will operate more efficiently in higher altitudes and up steep inclines. According to GE, the energy dissipated in braking a 207-ton locomotive during the course of one year is enough to power 160 households for that year.

- NetJets, the world leader in fractional jet ownership, introduced a multipronged Climate Initiative in 2007, including goals to improve energy efficiency and reduce carbon emissions from aircraft operations. To catalyze technological change in business aviation, NetJets is sponsoring the Next Generation Jet Fuel Project at Princeton University along with the University of California, Davis. The project is addressing technology and policy issues related to coprocessing coal and biomass along with carbon capture and storage, with the aim of producing an alternative jet fuel source with near-zero greenhouse gas emissions.

- When Airbus recently designed a new long-range airplane designed to carry up to 800 passengers, achieving the fuel consumption target posed a major challenge. To reduce the weight of the aircraft to meet fuel consumption requirements, engineers needed lightweight composite materials. Airbus worked with DuPont to develop a new lightweight, high-strength material made of Kevlar® and Nomex® aramid papers. Boeing is also optimizing weight, strength, safety, and fuel efficiency with this new material in its 787 aircraft. Reduction of greenhouse gas emissions can be as much as 12 million pounds per year based on the industry's annual fuel consumption.

- Pratt &Whitney, a division of United Technologies, launched a service business in 2004 called EcoPower® that offers environmentally friendly on-airframe engine washing. The service reduces engine fuel burn by as much as 1.2 percent, eliminates three pounds of carbon dioxide emissions for every pound of fuel saved, and decreases exhaust gas temperature by as much as 15 degrees Celsius, improving both performance and flying range. EcoPower® uses a closed-loop system with pure, atomized water, which is more effective and much faster than traditional engine washing processes and avoids potential contaminant runoff.

- The concept of car sharing, which began in Europe in the late 1980s, has now cropped up in many U.S. cities. This service offers a convenient alternative to car ownership and enables people to use the most effective combination of motor vehicles, walking, biking, or public transportation. The largest provider in the United States, Zipcar, claims that each of its cars replaces over 15 privately owned vehicles, thus relieving congestion and changing the urban landscape. Besides reducing fuel consumption and emissions, this reduces strain on the urban parking infrastructure, saving businesses, governments, and universities money.

- In 2007, Volvo achieved the first carbon-neutral automotive plant in the world in Ghent, Belgium, collaborating with the electric utility Electrabel. The plant uses hydroelectric electricity from the grid, a wood-burning heater to meet basic needs, an oil-based bio-heater for extra needs in winter and summer, three windmills, and 150 photovoltaic panels. These energy sources not only meet the operating requirements for electricity and heating, but also enable the sale of spare electricity to local customers.

- Researchers at Ford Motor have developed flexible, polyurethane foams that contain soy oil, replacing traditional, petroleum-derived polyols. Since 2007 Ford has introduced soy foam on the seat backs and cushions of the Mustang, Focus, Escape, F-150, Expedition, Navigator, and Mariner, reducing CO_2 emissions by about 5.3 million pounds per year (see Figure 8.4). Ongoing research on sustainable materials includes natural fiber-reinforced plastics to replace glass fiber and polymer resins made from high sugar content plants.

Even prior to the 2008 recession, General Motors and other U.S. automakers struggled to remain competitive in the face of burdensome retiree benefit programs and declining profits as sales of larger vehicles plummeted. While the recent economic recovery is encouraging, Toyota and other Japanese competitors remain better positioned in terms of cost structure, vehicle design, and propulsion technologies. This illustrates that environmental responsibility is not enough—companies must also learn to be resilient in the face of the turbulent changes that are sweeping the transportation industries. Enterprise resilience is discussed in Chapter 20.

> ENVIRONMENTAL RESPONSIBILITY IS NOT ENOUGH— COMPANIES MUST ALSO LEARN TO BE RESILIENT.

General Motors: Product and Process Innovation
A Roadmap for Sustainability

General Motors (GM), founded in 1908, is one of the world's largest automakers, employing about 250,000 people in 34 countries around the world. GM established its environmental policies and principles in the early 1990s and was the first Fortune 50 manufacturing company to endorse the Ceres principles in 1994. The company reports its progress with respect to global environmental metrics, established in 1999, which include energy use, water use, greenhouse gas emissions, and recycled and nonrecycled waste [6].

GM's laboratories historically have been leaders in automotive innovations that have provided tangible benefits to society. For example, GM developed the catalytic converter to reduce hydro-carbons, nitrous oxides, and carbon monoxide from vehicle tailpipe emissions. In 1974, GM was the first manufacturer to install the cata-lytic converter in all cars sold in the United States, and, in 1989, GM was the first to provide catalytic converters on all vehicles sold in Europe. GM allowed other manufacturers to use the technology without paying a patent fee, and today millions of vehicles around the world use catalytic converter technology developed by GM.

Like the rest of the U.S. auto industry, GM experienced declining market share and severe financial challenges during the early 2000s. In 2008, GM was obliged to request a bail-out package from the fed-eral government, and in 2009 declared bankruptcy in order to exe-cute a rapid restructuring of its operations. The new, leaner GM has restored profitability, and is continuing its long-standing emphasis on environmental excellence. GM and its competitors are placing a renewed emphasis on designing vehicles for an energy-conserving and environmentally conscious marketplace. The Hummer era is over.

There are already almost 900 million vehicles in a world of 6.6 billion people, and GM predicts that this number will grow to a bil-lion vehicles by 2020. This presents a clear challenge to sustainability in terms of energy, environment, safety, congestion, affordability, customer satisfaction, and the success of the automotive industry. Achieving sustainable mobility will require game-changing technol-ogies to address the complex societal, environmental, and economic interactions of the twenty-first century.

Recognizing that alternative sources of propulsion will be essen-tial, GM has adopted a strategy focused on displacing petroleum through energy efficiency and energy diversity. GM's Advanced Pro-pulsion Technology Strategy (GMAPTS) provides a roadmap of tech-nologies needed to address the societal, environmental and economic challenges faced by the auto sector. Figure 12.3 shows that many options are required; there is no "silver bullet," but there are key breakthroughs that will be developed along the way. The objectives of the GMAPTS are to

- Establish sustainability through diversified sources of energy
- Displace petroleum in order to alleviate the mismatch between supply and demand and reduce U.S. dependence on foreign imports subject to uncontrollable risks (i.e., geopolitical, oil refinery capacity, natural disasters, wars, etc.)
- Reduce GHG emissions.

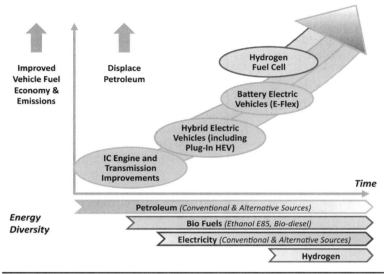

Figure 12.3 GM's Advanced Propulsion Technology Strategy.

GM's strategy incorporates the following key elements:

- Continue to improve conventional powertrains, recognizing that gas engines, diesel engines, and transmissions will be the principal propulsion systems for the foreseeable future.

- Develop alternative fuels, including biofuels blends such as E85 ethanol and B20 biodiesel as well as synthetic fuels derived from coal, natural gas, and/or biomass.

- Increased electrification of the automobile through expansion of the hybrid portfolio, including plug-in hybrids, EV range extenders, and ultimately, hydrogen fuel cells.

Chevrolet Volt: A New Automotive DNA

The latest example of GM's intent to break the automobile's dependence on petroleum is the development of the electrically driven Chevrolet Volt. According to former GM Chairman and CEO Rick Wagoner, "We believe this is the biggest step yet in our industry's move away from our historic, virtually complete reliance on petroleum to power vehicles."

Since its introduction at the 2007 North American International Auto Show in Detroit, the Chevy Volt sedan, powered by GM's Voltec electric propulsion system, has energized car enthusiasts in the United States and abroad with its potential to greatly reduce trips to the gas station for many commuters, as well as greatly reduce CO_2 emissions. A production version of the Volt is due in showrooms by 2010.

The foundation for GM's electrically driven vehicles is the new Voltec electric propulsion technology. The Voltec system can produce electricity using a generator powered by an internal combustion engine, which can be operated with a variety of fuels, including gasoline, diesel and ethanol. It can produce electricity using a generator powered by an internal combustion engine, which can be operated with a variety of fuels including gasoline, diesel, and ethanol. Or, it can generate electricity using a hydrogen fuel cell. Finally, it can operate on electricity that was stored in a battery when the car was plugged into the utility grid.

The Voltec electric propulsion system enables energy diversity because electricity and hydrogen can be generated from a wide range of energy sources. This allows GM to tailor the propulsion system to meet the specific needs and infrastructure of a given market. The electrification of the vehicle also enables the creation of a new automotive DNA, which exchanges the internal combustion engine for electric propulsion, petroleum for electricity, and mechanical linkages for electrical and electronic controls.

The Chevrolet Volt is a front-wheel drive, four-passenge Extended Range Electric Vehicle (E-REV) that uses electricity as its primary energy source and gasoline as its secondary energy source for propelling the vehicle. Energy is stored in a "T"-shaped lithium-ion battery pack. The battery pack powers the electric drive motor, which will propel the vehicle electrically for up to 40 miles without using a drop of gas. For longer trips, the Volt's on-board range-extending engine is used to drive an electric generator when the battery's energy has been depleted. The range extender, which can be powered by gas or E85 Ethanol, is able to generate additional electricity to power the car for hundreds of miles.

More than 75 percent of drivers in the United States commute fewer than 40 miles a day. For these drivers, a fully charged Chevy Volt will use no gasoline and produce no tailpipe emissions. The Chevy Volt is easily recharged by plugging it into a common 110-volt or 220-volt electrical outlet. At 10 cents per kilowatt hour, GM estimates that the cost of an electrically driven mile in a Chevy Volt is about one-fourth that of a conventional vehicle, assuming a gasoline price of $2.40 per gallon. That amounts to a reduction in running costs, including the cost of electricity, of about $1,200 per year for an average driver. At a cost of about 80 cents for a full charge that will deliver up to 40 miles of driving, the Volt will be less expensive to drive for most owners than purchasing a daily cup of coffee.

By-Product Management: No More Landfilling

In addition to vehicle design, GM has been introducing DFE principles into its production and supply chain processes. One example is

GM's Global By-Products Management Strategy, aimed at establishing a single global process to leverage all manufacturing by-product materials, maximize values, and reduce the environmental footprint of GM. By-products are defined as any manufacturing output materials that are not part of the intended product shipped to commerce. The new strategy was implemented in 2007, leveraging a Resource Management system that had been in place for ten years.

In the past, GM had multiple systems in its plants with several responsible entities managing by-products. The new program establishes a single point of contact within each facility whose specific mission is to manage all by-products. This entity has to interface with all stakeholder business groups, such as Manufacturing, Purchasing, Product and Materials Engineering, and others. This allows manufacturing personnel to focus on their core mission instead of managing scrap by-products. With extensive knowledge of environmental and other related regulations, as well as an understanding of end use markets, the dedicated by-product managers are able to maximize product disposition efficiencies and revenues.

Thanks to this strategy, GM was able to announce a commitment to make half of its major global manufacturing operations landfill-free by the end of 2010. Already, 33 of GM's global operations have reached landfill-free status, meaning that all waste has been converted to by-products. Gary Cowger, GM group vice president of global manufacturing and labor, comments: "As we develop new solutions in vehicle propulsion, GM is also making significant progress in reducing the impact that our worldwide facilities have on the environment."

At GM's landfill-free plants, over 96% of waste materials are recycled or reused, and about 3% is converted to energy at waste-to-energy facilities. As a result of the company's global recycling efforts, approximately $1 billion in revenue will be generated annually from recycled metal scrap sales. Additionally, in North America alone, GM will generate about $16 million in revenue from the sale of recycled cardboard, metal, wood, oil, plastic, and other recycled materials.

Part of the challenge in reaching landfill-free status is finding uses for recyclable materials. At GM's landfill-free plants, even the smallest piece of waste is put to a good use. Waste aluminum generated at GM facilities is sent to GM foundries to be reused to produce engine and transmission components. Steel, alloy metals, and paper are sent to recyclers to be made into a variety of products. Used oil is reconditioned for reuse in GM facilities. Wood pallets are reused, rebuilt ground into landscape chips, or sent to waste-to-energy facilities. Empty drums, totes, and containers are refurbished and reused again and again. Cardboard is collected, compacted, and sold for making new cardboard materials.

In the aggregate, over 3 million tons of waste materials were recycled or reused at General Motors plants worldwide in 2008. An additional 50,000 tons were converted to energy at waste-to-energy facilities. Some of the materials recycled at GM's zero landfill sites include 630,000 tons of scrap metal, 8,000 tons of wood, 7,500 tons of cardboard and 1,200 tons of plastic. As a result of these efforts, 3.65 million metric tons of carbon dioxide emissions were avoided. Additionally, using recycled by-products to make new products reduces energy use and manufacturing costs, compared to using virgin materials.

Toyota: The Future of Propulsion

Headquartered in Japan, Toyota employs about 300,000 people in 52 manufacturing companies worldwide, with annual sales of approximately 9 million vehicles under the Lexus and Toyota brands. Toyota was an early pioneer in envisioning alternative forms of propulsion that would satisfy both consumer expectations and environmental concerns. As a result, the company introduced the first gasoline-electric hybrid vehicles in the late 1990s, and by 2008, had sold over 1.5 million hybrid vehicles worldwide [7].

Toyota identified environmental issues as a management priority in 1992, the same year as the Rio Summit, and, in 1993, it launched the Global 21 project, which aspired to set the standard in automobiles for the twentieth-first century. In 1995, a group led by Akihiro Wada, a Toyota executive vice president, established a goal to double the fuel efficiency of existing vehicles. It was clear that meeting this goal would require hybrid technology, and, by that time, Toyota engineers had developed power trains that featured both high fuel efficiency and low emissions. At the Tokyo Motor Show in October 1995, the company unveiled a vehicle that combined an efficient gasoline engine with an advanced electric motor and required no charging. This was the forerunner of the Toyota Hybrid System for passenger vehicles, which was first released in Japan in the 1997 Prius Sedan.

The second-generation Prius aimed to improve both fuel efficiency and driving performance in order to encourage broader acceptance by mainstream customers. This required a complete redesign of the power train, including adoption of a high-voltage power-control circuit to increase motor output. The new system required exhaustive testing and went through seven prototype iterations. It further boosted fuel efficiency by about 25%, to over 50 miles per gallon. Toyota also decided to install the hybrid technology in sports utility vehicles (SUVs), including the Lexus RX400h and the Toyota Highlander Hybrid. By 2008, Toyota was firmly established as the market leader in design and marketing of alternative-propulsion vehicles.

Not content to rest on its laurels, Toyota continues to explore the design of "eco-cars" that minimize fossil fuel consumption and greenhouse gas (GHG) emissions. Toyota does not view hybrids as a transitional stage towards electric or fuel cell vehicles; rather, hybrid technology can be used with diesel and alternative fuel engines, as well as fuel cell technologies. Toyota is developing a fuel cell vehicle with hybrid technology and is also working on plug-in hybrids, which require higher battery capacity but could dramatically reduce GHGs. Looking to the future, Toyota is beginning to collaborate with other automakers to share technologies and promote the widespread adoption of hybrid systems.

DuPont: Eco-Efficient Automotive Paint

DuPont is one of the world's leading chemical companies and has incorporated sustainability principles into its product development efforts (see Chapter 13). One example is in the automotive paints and coatings business. Traditional coating processes use significant amounts of solvent and energy to apply paint to automotive vehicles, and the resulting volatile organic compound (VOC) emissions are regulated toxic substances. DuPont introduced a breakthrough, water-based coating system called EcoConcept®, which combines both product and process innovations to reduce production costs and energy usage, increase speed and productivity, and lower VOC emissions. These benefits have been confirmed by Volkswagen at its Puebla, Mexico and

FIGURE 12.4 Robotic application of EcoConcept® primerless automotive paint.

Pamplona, Spain assembly plants where the technology has already been used to produce more than 250,000 vehicles for consumer sales (see Figure 12.4). DuPont received the prestigious Automotive News PACE award for this innovation and a second PACE award was presented to VW for the best OEM-supplier cooperation while implementing this technology.

Paint line equipment and engineering can represent from one-third to one-half of the initial capital investment in an assembly plant, often ranging from $300–$400 million. Much of an assembly plant's investment in emissions abatement equipment is required in operating the paint shop. Additionally, while paint line operating costs and inherent multistep processing have been viewed as essential to vehicle quality and sales, the paint line has also been identified as a bottleneck to reductions in assembly time, requiring approximately 12 hours of manufacturing time per new vehicle. Paint itself represents just 20% of the operating cost in the coatings operation.

The basic idea of EcoConcept® is to add another paint component which modifies the characteristics of the basecoat layer so that it adopts the function of a primer-surfacer. The "primerless" system uses multiple basecoats in combination with a reactive crosslinker as a blending component. This eliminates one complete coating layer, the primer-surfacer that is usually a solvent-borne product, and combines it into a modified, environmentally compliant water-borne basecoat. As a result, a spray booth and curing oven are eliminated, producing a number of environmental benefits:

- Process energy and solvent emissions are reduced by approximately 25%.
- Material consumption is lowered by approximately 20–25%.
- Greenhouse and acidification gases are reduced by approximately 30%.
- Paint shop total time and costs are reduced by approximately 20–25%.
- The overall supply chain footprint of automotive paint acquisition is reduced by about 20% due to lower mass throughput.

Furthermore, DuPont estimates that adoption of this technology will save customers about $60 million in plant construction costs and from $3 million to $12 million in annual operating costs.

References

1. Center for Automotive Research, Ann Arbor, Michigan, 2002.
2. Personal communication, Terry Cullum, Director, Global Design for Environment, Public Policy Center, General Motors.
3. "Mobility 2030: Meeting the Challenge to Sustainability" (Geneva: World Business Council for Sustainable Development, 2004).
4. J. L. Sullivan et al. "Life Cycle Inventory of a Generic U.S. Family Sedan, Overview of Results, USCAR AMP Project." SAE Total Life Cycle Conference, paper #982160, 1998: 1–14.
5. *Improving Sustainability in the Transport Sector Through Weight Reduction and the Application of Aluminium* (London: International Aluminium Institute, 2007).
6. More information about General Motors' environmental programs is available at www.gm.com/corporate/responsibility/environment/.
7. "Toyota: Toward the Ultimate Eco-Car" (Geneva: World Business Council for Sustainable Development, 2008).

CHAPTER 13
Chemical Industries

Overview

Chemical products are fundamental building blocks in the supply chain of virtually every manufacturing or service industry. Therefore, every company that practices DFE must consider the implications of its chemical acquisition and use. The chemical industry has undergone enormous transitions since the advent of coal tar and petroleum-based chemistries in the 1930s and 1940s, which resulted in advances in polymers, pharmaceuticals, and agricultural chemicals. Synthetic chemistry was once hailed as a miraculous technology for converting industrial wastes into useful products, but by the 1970s suspicion was mounting about the potential adverse effects of chemical residues in the environment. In 1984, the accidental release of isocyanide gas from a Union Carbide plant in Bhopal, India marked a tragic turning point in public perception of chemical risks. The Chemical Manufacturers' Association, later renamed the American Chemistry Council, worked hard to repair the image of the industry, yet mistrust of chemical companies has lingered.

Meanwhile, regulation of chemical substances has continued to grow more stringent, as mentioned in Chapter 3. Right-to-know legislation has forced disclosure of chemical release information, while many voluntary programs have emerged to encourage pollution reduction. At the same time, the industry has diversified into biotechnology and "cleaner" biobased processes, while tightening the stringency of environmental, health, and safety management for existing processes. The Responsible Care® code adopted by the chemical industry represents a global standard of excellence for proactive risk management and product stewardship practices, and has been emulated by many other industries. As shown in Figure 13.1, during the ten-year period between 1996 and 2005, toxic and hazardous releases from chemical production dropped by over 60% [1]. Today, chemical industry leaders such as Dow and DuPont are in the vanguard of companies that have integrated sustainability into their business practices.

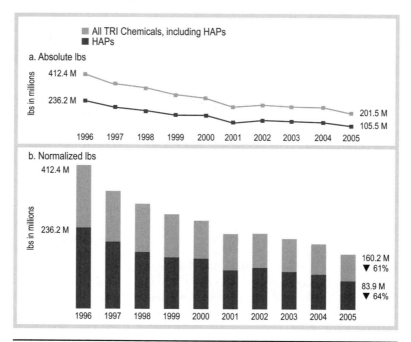

FIGURE 13.1 Decline in chemical industry toxic releases, including hazardous air pollutants [1].

Green and Sustainable Chemistry

The U.S. EPA and the American Chemical Society have been instrumental in fostering a "green chemistry" movement, which has encouraged industrial chemists and researchers to develop chemical products and chemical reactions that are inherently benign in terms of environmental impacts. Also known as "sustainable chemistry" this practice involves a variety of techniques including

- Use of alternative synthetic pathways involving biocatalysis, natural processes, such as photochemistry, and more innocuous and/or renewable feedstocks
- Use of alternative reaction conditions, involving lower temperatures and pressures
- Use of solvents with minimal human health and environmental impacts, such as aqueous-based solvents
- Increasing selectivity and reducing wastes and emissions from industrial processes
- Designing chemicals that are less toxic than current alternatives or are inherently safer with regard to accident potential

The following set of fundamental design principles for green chemistry was articulated by Paul Anastas and John Warner [2]:

1. **Prevention**—It is better to prevent waste than to treat or clean up waste after it has been created.

2. **Atom Economy**—Synthetic methods should be designed to maximize the incorporation of all materials used in the process into the final product.

3. **Less Hazardous Chemical Syntheses**—Wherever practicable, synthetic methods should be designed to use and generate substances that possess little or no toxicity to human health and the environment.

4. **Designing Safer Chemicals**—Chemical products should be designed to effect their desired function while minimizing their toxicity.

5. **Safer Solvents and Auxiliaries**—The use of auxiliary substances (e.g., solvents, separation agents, etc.) should be made unnecessary wherever possible and innocuous when used.

6. **Design for Energy Efficiency**—Energy requirements of chemical processes should be recognized for their environmental and economic impacts and should be minimized. If possible, synthetic methods should be conducted at ambient temperature and pressure.

7. **Use of Renewable Feedstocks**—A raw material or feedstock should be renewable rather than depleting whenever technically and economically practicable.

8. **Reduce Derivatives**—Unnecessary derivatization (use of blocking groups, protection/deprotection, temporary modification of physical/chemical processes, etc.) should be minimized or avoided if possible because such steps require additional reagents and can generate waste.

9. **Catalysis**—Catalytic reagents (as selective as possible) are superior to stoichiometric reagents.

10. **Design for Degradation**—Chemical products should be designed so that at the end of their function they break down into innocuous degradation products and do not persist in the environment.

11. **Real-time Analysis for Pollution Prevention**—Analytical methodologies need to be further developed to allow for real-time, in-process monitoring and control prior to the formation of hazardous substances.

12. **Inherently Safer Chemistry for Accident Prevention**—Substances and the form of a substance used in a chemical process should be chosen to minimize the potential for chemical accidents, including releases, explosions, and fires.

Every year, chemical companies and collaborating researchers vie for the coveted Presidential Green Chemistry Challenge Awards, which recognize outstanding chemical technologies that incorporate the principles of green chemistry into chemical design, manufacture, and use (see Figure 13.2). Examples of past award winners include the following:

- **Biobased toner**. Laser printers and copiers use over 400 million pounds of toner each year in the United States. Traditional toners fuse so tightly to paper that they are difficult to remove from waste paper for recycling. They are also made from petroleum-based starting materials. Battelle and its partners, Advanced Image Resources and the Ohio Soybean Council, developed a soy-based toner that performs as well as traditional ones, but is much easier to remove. The new toner technology can save significant amounts of energy and allow more paper fiber to be recycled.

- **Climate-friendly foam**. Foam cushioning used in furniture or bedding is made from polyurethane, a man-made material. One of the two chemical building blocks used to make polyurethane are polyols, which are conventionally manufactured

FIGURE 13.2 Presidential Green Chemistry Challenge Awards.

from petroleum products. Cargill's BiOH™ polyols are manufactured from renewable, biological sources, such as vegetable oils. Foams made with BiOH™ polyols are comparable to foams made from conventional polyols. As a result, each million pounds of BiOH™ polyols save nearly 700,000 pounds of crude oil. In addition, Cargill's process reduces total energy use by 23% and carbon dioxide emissions by 36%.

- **Low-emission paints.** Latex paints require coalescents to help the paint particles flow together and cover surfaces well. Archer Daniels Midland developed Archer RC™, a new biobased coalescent made from plant oils, to replace traditional coalescents that are volatile organic compounds (VOCs). Instead of evaporating into the air, the unsaturated fatty acid component of Archer RC™ oxidizes and even cross-links into the coating. This new coalescent has other performance advantages as well, such as lower odor, increased scrub resistance, and better opacity.

- **Recyclable carpet backing.** Conventional backings for carpet tiles contain bitumen, polyvinyl chloride (PVC), or polyurethane. EcoWorx™ carpet tiles have a novel, 100% recyclable thermoplastic backing that uses less toxic materials and has superior adhesion and dimensional stability. Because EcoWorx™ carpet tiles can be readily separated into carpet fiber and backing, each component can be easily recycled.

- **Nontoxic wood adhesive.** Adhesives used in manufacturing plywood and other wood composites often contain formaldehyde, which is toxic. Oregon State University, Columbia Forest Products, and Hercules Incorporated developed an alternate adhesive made from soy flour. Their environmentally friendly adhesive is stronger than and cost-competitive with conventional adhesives. During 2006, Columbia used the new, soy-based adhesive to replace more than 47 million pounds of conventional formaldehyde-based adhesives.

Another stimulus for environmental awareness in the chemical industry is U.S. EPA's Design for Environment (DFE) Program (see Chapter 3). EPA forms voluntary partnerships to collaborate with businesses, trade organizations, and other interested parties to design or redesign products, processes, and environmental management systems to be cleaner, more cost-effective, and safer for workers and the public. One key initiative is the Formulator Program, which labels products that EPA has reviewed and found to be safer for human health and the environment. EPA currently allows use of its DFE label on more than 600 products, and tens of millions of DFE products have been sold to consumers and institutional purchasers. In 2008, active partnership projects within DFE reduced more than 200 million pounds of chemicals of concern.

Chapter 8 describes some of the DFE strategies adopted by various industries to ensure environmentally responsible utilization of chemicals in their supply chain processes, including chemical management services, chemical property screening, and toxics elimination. The following case studies of Dow, DuPont, and BASF illustrate leading edge DFE practices in the chemical industry.

Dow Chemical: Raising the Bar

The Journey toward Sustainability

With annual sales of $54 billion and 46,000 employees, Dow is a diversified chemical company with global operations that encompass 175 countries worldwide. Dow was one of the first companies to establish a product stewardship program and has consistently been an outspoken proponent of sustainable business practices. All of Dow's business divisions are encouraged by management to connect chemistry and innovation with the principles of sustainability; and the resulting products and services address a variety of human needs, including fresh water, food, pharmaceuticals, paints, packaging, and personal care products. Dow strives to combine the power of science and technology with the "Human Element" to constantly improve what is essential to human progress [3].

Since the 1980s, Dow has achieved significant improvements in environmental performance as measured by conventional indicators, such as energy, emissions, and waste reduction. Dow was one of the first companies to understand the link between environmental excellence and profitability, and established a pioneering Waste Reduction Always Pays (WRAP) program in 1986. In the early 1990s, Dow formally adopted sustainable development principles, joined the World Business Council for Sustainable Development, and began applying life-cycle assessment to its products and processes. At the same time, Dow was among the leaders in the establishment of community advisory panels to foster stakeholder engagement.

In 1996 Dow announced a set of ambitious Environmental Health and Safety Goals for 2005, including goals for energy, waste, wastewater, emissions, priority compound emissions, illness and injuries, and process safety incidents. By the end of 2005, Dow had surpassed a number of the goals and had made significant achievements on all the goals. A rigorous assessment of the value derived from this effort indicated that Dow spent $1 billion to save about $5 billion during the time period of the goals.

In 2006, Dow's CEO and Chairman, Andrew Liveris, personally announced the company's ambitious sustainability goals for 2015.

He committed to achieving at least three breakthroughs to help solve the worldwide challenges of affordable and adequate food supply, decent housing, sustainable water supplies, or improved personal health and safety. In his words: "We will reach beyond the fences of our company. Sustainability begins at home, but its destiny is to engage the problems of the world." He also announced a commitment to improving the company's energy efficiency, developing alternative sources of energy, and addressing the challenge of global climate change that has been driven by the consumption of fossil fuels. Specifically:

- Between 1996 and 2005, Dow reduced its consumption of fossil fuels per pound of product by more than 20%. The new goals call for further improvements in energy efficiency of 25% over the next decade.

- In the area of climate change, the new goals call for reducing greenhouse gas (GHG) emissions intensity by 2.5% annually between 2006 and 2015. The company plans to achieve this reduction via a combination of energy efficiency, greater use of alternative, low GHG emission energy sources, and carbon offsets.

In addition, Dow's 2015 sustainability commitments include

- Continued improvement in employee health and safety at Dow locations

- New mechanisms for collaborating with communities where Dow has a major presence to spearhead efforts to address community concerns and support community goals

- Increased transparency of Dow products with risk assessments written for the layperson

- External evaluation and assessment of Dow product safety and product stewardship processes

- A renewed commitment to "sustainable chemistry," including new or enhanced products or services with additional sustainability benefits

Examples of Dow's sustainable chemistry innovations are described below.

Polyethylene from Sugar Cane

Brazilian sugar cane has already proved to be an energy-efficient source of ethanol for automotive fuels. In 2007, Dow announced plans to develop an energy-efficient, world-scale facility to manufacture polyethylene in Brazil. Polyethylene is the most widely used of all plastics and can be found in many everyday products such as food packaging, milk jugs, plastic containers, pipes and liners.

The new facility will have a capacity of 350,000 metric tons, serving both Brazilian and international markets. The facility will use ethanol derived from sugar cane, an annually renewable resource, to produce ethylene—the raw material required to make polyethylene. Ethylene is traditionally produced using either naphtha or natural gas feedstocks, both of which are petroleum products. The new process, using sugar cane both as a feedstock and as a process fuel, is estimated to reduce greenhouse gas emissions by over one million metric tons per year when compared to a similarly-sized conventional polyethylene plant, based on a critically-reviewed life-cycle assessment study.

The new facility will use Dow's proprietary technology to manufacture DOWLEX™ polyethylene resins, which combine toughness and puncture resistance with high performance and processability. These materials are used in a range of different applications, including pipes, films, membranes, and food and specialty packaging. At a molecular level, the sugar cane-based product will be identical to the resins manufactured at other Dow facilities. Thus, unlike most renewable resource-based plastics, industrial customers will be using a drop-in replacement made with a renewable resource, not a different polymer altogether. Also, like the traditional product, the sugar cane-based polyethylene will be fully recyclable using existing infrastructure.

Green Chemistry at Dow AgroSciences

Dow AgroSciences LLC, a wholly owned subsidiary based in Indianapolis, is a top-tier agricultural company with global sales of $4.5 billion. It was originally formed as a joint venture between Dow and the Plant Sciences Division of Eli Lilly and Company. The company provides innovative technologies for crop protection, pest and vegetation management, seeds, traits, and agricultural biotechnology to serve the world's growing population.

Many industries, including crop protection, have been under considerable pressure to move toward greener chemistry. Green chemistry, also known as sustainable chemistry, encompasses all aspects of and types of chemicals and processes that reduce negative impacts on human health and the environment including reduced waste, safer products, and reduced use of energy and resources. Accordingly, Dow AgroSciences has invested considerable R&D in developing better and more environmentally sound products with equivalent or superior performance to many accepted standards.

These "green insecticides" are able to effectively and selectively control pests without posing significant risks to the applicator, consumer and ecosystem as a whole, including mammals, birds, earthworms, plants and aquatic organisms. The company has earned several awards for green and sustainable chemistry innovations, including four Presidential Green Chemistry Challenge Awards.

One example of a green insecticide is *Spinetoram™*—an insect control technology derived from the fermentation of a naturally occurring soil organism followed by chemical modifications. The technology is based on *spinosyns*, a novel class of natural products that are active against certain types of insect pests. The development of *spinosad* began with a soil sample taken by an Eli Lilly and Company scientist on vacation. Upon analysis, a biological activity organism (see Figure 13.3) was discovered that had insect control potential. Full-scale trials were conducted and *spinosad* received EPA approval for use in numerous crops.

In searching for ways to further enhance the natural product, Dow AgroSciences was able to achieve improved efficacy and an expanded spectrum of pest applications by using artificial neural networks to investigate structure-activity relationships [4]. *Spinetoram™* offers a unique combination of characteristics, including:

- Provides long-lasting control of a broad spectrum of insect pests in a variety of crops
- Can be applied at low rates and has low impact on most beneficial insects
- Favorable toxicological profile as it relates to mammals, birds and aquatic organisms
- Safety to beneficial insects, providing an excellent fit with Integrated Pest Management (IPM) programs
- Unique mode of action makes it an ideal fit for resistance management programs

FIGURE 13.3 S. spinosa, the producing organism of the spinosyns, magnified 20,000 times.

Another award-winning technology is The Sentricon™ System, which eliminates termite colonies while replacing the need for widespread applications of pesticides. Traditionally, homeowners have combated termites through fumigation and the use of chemical barriers. With The Sentricon System approach, pest control companies use monitoring stations to detect the presence of termites, and then insert bait laced with a slow-acting termite growth inhibitor. The amount of pesticide required to eliminate termite infestations is as little as one ten thousandth of the amount used in traditional fumigation processes [5].

DuPont: Realizing Sustainable Growth

DuPont, with revenues of $30 billion, employs about 60,000 people in 70 countries around the world. Founded in 1802 as an explosives manufacturer, the company eventually became one of the world's largest producers of chemicals and materials. DuPont has established a reputation as an industry leader in the practice of sustainability and has continually transformed itself to address a changing business environment. In the 1970s and 1980s the company's focus was on safety and environmental compliance, and so diligently did it pursue these practices that its EH&S programs were emulated by many other companies. In the 1990s, DuPont was among the first companies to embrace a broader view of sustainability and established voluntary "stretch" goals, such as zero incidents and zero wastes. Again, these practices influenced broad adoption of sustainability goals among global industry leaders. DuPont was the first to articulate a value proposition around eco-efficiency—increasing shareholder value added while decreasing raw material and energy inputs and reducing emissions. DuPont coined the term "sustainable growth" to assert the compatibility of sustainability with business prosperity [6].

Today, DuPont views itself as a science company, using biotechnology, chemistry, and other scientific disciplines to develop sustainable products that address some of the world's most pressing challenges. The company is in a third phase of sustainable growth, characterized by a holistic approach that is fully integrated into its business models and has become a market-driven business priority throughout the value chain. As evidence of DuPont's commitment to cutting dependence on fossil fuels, the company that invented nylon, the miracle substance of the twentieth century, divested its nylon division in 2003. The company's new 2015 Sustainability Goals go beyond traditional footprint reductions to link business growth directly to the development of safer and environmentally improved products. The goals are divided into two categories:

1. Market-Facing Goals

- Double the investment in R&D programs with direct, quantifiable environmental benefits for our customers and consumers along our value chains.

- Grow annual revenues by at least $2 billion from products that create energy efficiency and/or significant greenhouse gas emissions reductions for our customers (estimated at least 40 million tonnes of CO_2 equivalent reductions).

- Double revenues from nondepletable resources to at least $8 billion. DuPont has introduced a Renewably Sourced initiative as a key pillar of new product development.

- Enhance the focus on protecting people, by introducing at least 1,000 new products or services that help make people safer globally. Examples include protective clothing materials and disease-fighting disinfectants.

2. Footprint Goals

- Further reduce GHG emissions at least 15% from a base year of 2004. (Since 1990, DuPont has reduced global GHG emissions by 72%.)

- Reduce water consumption by at least 30% at sites where fresh water is either scarce or stressed, and hold water consumption flat at other sites.

- Ensure that 100% of fleet vehicles represent leading technologies for fuel efficiency and fossil fuel alternatives.

- Further reduce air carcinogen emissions at least 50% from a base year of 2004. (Since 1990, DuPont has reduced global air carcinogen emissions by 92%.)

- Ensure that 100% of global manufacturing sites have completed an independent third-party verification of their environmental management goals and systems.

According to CEO Chad Holliday, "As an ingredient supplier in countless value chains, we have a broad and deep impact on global industries and therefore on global society." The company is forming partnerships around the world to speed the transformation toward sustainability. For example, DuPont is working with BP to create a new generation of biofuels that can be produced from locally grown crops containing sugar, such as corn in North America, sugar beets in Europe, and sugar cane in Brazil. The goal is to reduce dependence on fossil fuels without threatening the food supply. According to Chief Innovation Officer, Tom Connelly, DuPont will be growing in a variety of new directions that emphasize biobased and renewable materials.

As a result of DuPont's commitment to environmental excellence, the company has produced an impressive series of innovations that benefit their customers in different industries. One example is Eco-Concept® automotive paint, described in Chapter 12. Another example is triexta, a renewably sourced polymer carpet fiber that is used by Mohawk Industries for residential carpeting (see Chapter 16).

BASF: Beyond Eco-Efficiency

BASF is the world's largest chemical company, with annual sales of approximately $70 billion in 2006. Headquartered in Germany, the company employs about 95,000 people on five continents. BASF has placed great emphasis on developing a world-class approach toward managing environmental sustainability. In 2008, BASF was the first chemical company to publish a life-cycle carbon balance. Based on this extensive analysis, BASF claims that the use of its products can save three times more greenhouse gas emissions than the entire amount caused by the production and disposal of all BASF products. Examples of products that provide climate protection include building insulation materials, lightweight automotive plastics, emission-reducing catalysts, nitrification-inhibiting fertilizers, and biodegradable plastics [7].

In the area of plastic films, Ecoflex® is a line of biodegradable synthetic polymers, sometimes blended with thermoplastic starch, which decomposes on compost heaps within a few weeks without leaving any residues behind. Introduced in 1998, Ecoflex® has become the world's leading synthetic biodegradable material. In addition, BASF has developed a biodegradable plastic called Ecovio® which is a blend of 45% by weight of polylactic acid from corn, a renewable raw material, and Ecoflex. These products can be used for consumer applications, such as disposable packaging and grocery shopping bags.

To support the development of sustainable products, BASF has developed a rigorous eco-efficiency analysis tool that quantifies the economic and ecological benefits of a product or process over the complete life cycle. This tool is routinely used by BASF design teams to

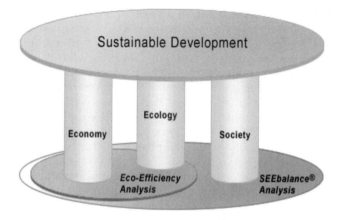

FIGURE 13.4 The SEEbalance® method developed by BASF.

help identify products and processes that consume less energy and generate less waste and emissions than alternatives, while maintaining or improving the products' commercial value [8]. The tool has been applied to over 100 different products and processes, such as asphalt microsurfacing, nylon fiber, building materials, automotive coatings, plastics, and adhesives. For example, BASF's eco-efficiency analysis for dyeing of blue denim revealed that electrochemical application of a vat solution was preferable to the use of powdered, granular, or biotechnologically produced indigo.

Recently, BASF has expanded the scope of its analysis to consider the three dimensions of sustainability—economy, environment, and society. The company's new SEEbalance® tool (SocioEcoEfficiency Analysis) enables integrated assessment not only of environmental impacts and costs but also of the societal impacts of products and processes (see Figure 13.4). The societal impacts are grouped into five stakeholder categories: employees, international community, future generations, consumers, and local and national community. For each of these stakeholder categories measurable indicators are considered, such as number of occupational accidents, and these are summarized in the form of a "social fingerprint."

References

1. U.S. Environmental Protection Agency, 2008 Sector Performance Report.
2. P. Anastas and J. Warner, *Green Chemistry: Theory and Practice* (New York: Oxford University Press, 1998).
3. More information about Dow Chemical's sustainability commitments is available at www.dow.com/commitments/.
4. T. C. Sparks, G. D. Crouse, J. E. Dripps, P. Anzeveno, J. Martynow, C. V. DeAmicis, and J. Gifford, "Neural network-based QSAR and insecticide discovery: spinetoram," *Journal of Computer Aided Molecular Design*, 2008.
5. N. Su and R. H. Scheffrahn, "A review of subterranean termite practices and prospects for integrated pest management programmes," Integrated Pest Management Reviews, 1998.
6. More information about DuPont's sustainability programs is available at www2.dupont.com/Sustainability/en_US/.
7. More information about BASF's sustainability programs is available at www.basf.com/group/sustainability_en/index.
8. P. Saling, A. Kicherer, B. Dittrich-Krämer, R. Wittlinger, W. Zombik, I. Schmidt, W. Schrott and S. Schmidt, "Eco-Efficiency Analysis by BASF: The Method," *International Journal of Life Cycle Analysis*, 2002.

CHAPTER 14

Medical and Pharmaceutical Industries

Overview

The medical and pharmaceutical industries are among the highest in terms of R&D spending as a percentage of revenue and, therefore, invest a great deal of effort in product and process development. From an environmental perspective, pharmaceuticals, along with semiconductors, have a high proportion of non-product output, thus offering a large target for resource intensity reduction. It has been estimated that the total mass of materials required to deliver one kg of product is 25 to 100 kg for pharmaceuticals, compared to 5 to 50 kg for fine chemicals, and only 1 to 5 kg for bulk chemicals [1].

Another emerging issue of concern is the presence of trace levels of pharmaceuticals in the environment. When pharmaceuticals are administered to patients, some of the active ingredients may not be completely metabolized. These residuals are generally excreted through the urine and find their way into sewage systems, where they are transported to waste water treatment systems that remove most of the pharmaceutical residues. However, extremely low concentrations may pass through the waste water treatment plant and be discharged to the environment. Despite low concentrations, low-level effects have been observed in aquatic organisms.

Not surprisingly, medical and pharmaceutical companies are among the leaders in the application of green chemistry, which offers ways to reduce environmental impacts while cutting material and energy costs (see Chapter 13). In 2005, several pharmaceutical firms, along with the American Chemical Society's Green Chemistry Institute, established a Pharmaceutical Roundtable to promote the integration of green chemistry and green engineering in the industry. For example, solvents used in production can account for up to 80% of the life-cycle impact of a pharmaceutical product; using green engineering Pfizer was able to cut the amount of dichloromethane solvent used at its three main research sites by almost two-thirds, while use of di-isopropyl ether was eliminated entirely.

One example of systematic application of DFE is the Eco-Design Toolkit© developed by Glaxo-Smith-Kline to integrate eco-design principles and practices into new product development. The toolkit includes material selection guides, a green chemistry and technology guide, and several life-cycle-based tools for packaging, materials, and process assessment. These tools are being integrated into R&D, Global Manufacturing, and appropriate business-level or corporate communities across the company [2].

The following sections describe selected DFE programs implemented by leading medical and pharmaceutical companies, including Johnson & Johnson, Baxter International, and Eli Lilly, but certainly do not provide a complete picture of the global activities in these industry sectors.

Health Care Without Harm

A nongovernmental organization called Health Care Without Harm (HCWH) has had a substantial influence on DFE in the medical field. It consists of an international coalition of hospitals and health care systems, medical professionals, community groups, health-affected constituencies, labor unions, environmental and public health organizations, and religious groups. The mission of the organization is to transform the health care sector worldwide, without compromising patient safety or care, so that it is ecologically sustainable and no longer a source of harm to public health and the environment [3]. HCWH has focused on a number of issues related to product safety and waste management in health care:

- **Mercury**. HCWH promotes the use of alternative products that do not contain mercury, which is a potent neurotoxin that can affect the human brain, spinal cord, kidneys, and liver, and can have adverse ecological effects as well. Examples of medical devices that may contain mercury are thermometers and sphygmomanometers, or blood pressure devices. In addition hospitals and medical research facilities may utilize mercury-containing fluorescent lamps, batteries, or fixatives in the laboratory. HCWH has urged pharmacies to phase out the manufacture, distribution, and sale of mercury-containing devices and, as a result, virtually all the major chain pharmacies have stopped selling mercury fever thermometers.

- **PVC plastic**. Polyvinyl chloride (PVC) is the most widely used plastic in medical devices, but many health advocates are opposed to its use because (a) dioxin, a known human carcinogen, can be formed during manufacture of PVC or during the incineration of PVC products, and (b) DEHP, a phthalate used to soften PVC plastic, may leach from PVC medical devices, and has been linked to reproductive birth defects and other illnesses. Despite the lack of definitive evidence of harm, PVC has become a controversial material, and many government agencies are urging caution in its use. A wide variety of PVC-free products have been developed (see Johnson & Johnson below).

- **Waste disposal**. In the past, many hospitals simply incinerated all their solid waste, which resulted in the release of dioxin, mercury, lead and other dangerous air pollutants. HCWH and other organizations have promoted responsible waste management through waste minimization as well as segregation and treatment of infectious wastes. Many innovative treatment technologies have been developed; for example, the University of Sydney, Australia, has developed a portable, solar-powered autoclave system for rural areas that can be operated in any weather conditions.

Another important movement is the application of "green building" principles to health care environments. HWHC and other groups have worked with The U.S. Green Building Council to develop a Green Guide for Health Care (GGHC), which builds on the principles of the well known sustainable design standard, LEED® (Leadership in Energy and Environmental Design). Green hospital buildings are healthier for the patients, doctors, and nurses; use less energy and water; and have less of an impact on the environment. According to Rick Fedrizzi, President of the U.S. Green Building Council, "Patients in green hospitals have greater emotional well-being, require less pain medication and other drugs, and have shorter hospital stays." In addition, studies have shown substantial increases in the job satisfaction, performance, and productivity of people who work in green buildings. By 2008, over 100 facilities had registered for GGHC certification.

Johnson & Johnson: A Matter of Principle
Healthy Planet Initiative

Johnson & Johnson (J&J) is the world's most comprehensive and broadly based provider of health care products and services for the consumer, pharmaceutical, and medical devices and diagnostics markets. The company employs approximately 116,000 people in 57 countries, with about $61 billion in annual sales. Well-known brands include Johnson's Baby Shampoo, Band-Aids, and AcuVue contact lenses. J&J has long prided itself on being a leader in environmental performance and sustainability [4].

The J&J Credo, originally framed in 1943, is one of the earliest examples of a corporate responsibility commitment. It identifies a variety of stakeholder groups and affirms the company's responsibilities for product quality and affordability; employee safety, dignity, and equal opportunity; community citizenship; environmental protection; and, of course, stockholder returns. Adherence to these principles helped the company to respond effectively to the Tylenol poisoning crisis in 1982 and to rapidly redeem the brand and recapture market share.

Today, the company's sustainability program is built around a threefold vision—Healthy People, Healthy Planet, Healthy Business. By 2005, J&J had largely achieved or exceeded its environmental goals, most of which were based on typical operational metrics, including reductions in waste, emissions, and resource consumption. A more unusual goal, already achieved, was 100% implementation of a systematic Design for Environment (DFE) process by all divisions. As shown in Table 14.1, the company identified business benefits for all of these goals, including over a half billion dollars in direct cost savings.

J&J's 2010 Healthy Planet goals, announced in 2006, are predicated on the notion that "the environment is the ultimate human health and safety issue." The goal categories include compliance, environmental literacy, biodiversity, transparency, product stewardship, water use, paper and packaging, greenhouse gas reduction, waste reduction, and external manufacturing standards. J&J's environmental footprint goals are not normalized in terms of sales volume—they call for absolute reductions regardless of growth.

By 2007, the company had conformed 93% of all its packaging with either 30% post-consumer recycled and/or certified content, and had eliminated virtually all polyvinyl chloride (PVC) from packaging. In addition, the company obtained 38% of its electric power from renewable sources, including a solar photovoltaic tracking system at its headquarters in New Brunswick, NJ. To help achieve its goal of 30% reduction in CO_2 per mile (from a 2003 baseline), J&J has acquired the largest corporate fleet of hybrids in the United States— approximately 1,600 as of 2008. Finally, J&J has established electronic product take-back programs for its used medical device products, recovering over 340 metric tons in 2007.

DFE Practices

As mentioned above, DFE is routinely applied as a business practice in 100% of J&J's business units. To support adoption of DFE throughout the company, J&J has established a corporate-level Design for Environment group. The group provides training and support to all

Environmental Performance Indicator*	5 Year Goal (%)	Actual Results (%)	2005 Status	Business Benefits (MM=$million)
Water Use	10	14 ✓	Exceeded	$9 MM
NPO*: Nonhazardous Waste	10	14 ✓	Exceeded	$77 MM
NPO*: Hazardous Waste	5	27 ✓	Exceeded	$47 MM
NPO: Toxic Waste	5	24 ✓	Exceeded	Included in $47 MM Hazardous Waste
CO_2 Emissions (absolute reduction from 1990 level)	4	11.5 ✓	Exceeded	$172 MM
Raw Materials	5	7 ✓	Exceeded	$262 MM
Packaging	10	8	Strong Progress	$163 MM
External Manufacturing Assessments	100	93	Strong Progress	Risk Reduction
ISO 14001 Registration	100	96	Strong Progress	Risk Reduction
Design for Environment	100	100 ✓	Achieved	Risk Reduction
Community Outreach/ Forums	100	96/91	Strong Progress	Reputation

*NPO denotes Non-Product Output. Indicators of waste and resource use reduction, unless noted, represent "cumulative avoidance" since the beginning of the 5-year period. The last four goals represent 100% implementation across all J&J divisions.

TABLE 14.1 J&J Corporate Environmental Goals, 2001–2005, and Business Results

business units and has developed a customized DFE toolkit to help design teams evaluate proposed product concepts. The available tools include a DFE assessment spreadsheet for products, packaging, and processes; a Chemical Regulation ColorGuide for scanning global chemical regulations; a Watch List for chemicals of concern; and an Environmental Product Design Guideline. According to J&J, these tools have enabled product developers to consider the environmental impacts of the materials that they choose and the end-of-life impacts of their products up front in the research and development process.

An outstanding example of DFE achievements at J&J is LifeScan, Inc., a J&J subsidiary. LifeScan pioneered the modern era of blood

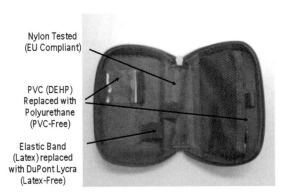

Nylon Tested
(EU Compliant)

PVC (DEHP)
Replaced with
Polyurethane
(PVC-Free)

Elastic Band
(Latex) replaced
with DuPont Lycra
(Latex-Free)

FIGURE 14.1 Environmentally friendly design of the One-Touch® meter carrying case.

glucose monitoring with the introduction of OneTouch® metering technology, which eliminated wiping and timing procedures. The blood glucose meter device itself was reengineered to meet the Restrictions on Hazardous Substances (ROHS) directive of the European Union, even though medical devices are exempt. In addition, the carrying case was redesigned to eliminate PVC, latex, and phthalate plasticizers, as shown in Figure 14.1. These changes enabled easier access to European and Asian markets and are a clear demonstration of the company's green commitment. Finally, J&J offers to take back the blood glucose meters at their end-of-life as part of the company's global product stewardship efforts. In 2007 over 700,000 LifeScan meters were taken back for recycling.

There are numerous other examples of DFE achievements across J&J's many divisions, including raw material source reduction, packaging avoidance, asset recovery, solvent reduction, and toxics elimination. The company's leadership in corporate responsibility has earned it considerable recognition, including membership in the Dow Jones Sustainability Indexes and selection as one of the Global 100 Most Sustainable Companies.

Baxter: Saving and Sustaining Lives
A Culture of Sustainability

Baxter International had annual sales of approximately $11.3 billion globally in 2007, with more than half its sales coming from outside the United States. The company employs approximately 46,500 people in 26 countries around the world. Baxter's main businesses are in therapeutic medicines, including recombinant and plasma-based formulations, medication delivery equipment and supplies such as intravenous systems, and renal dialysis products. Baxter's commitment is to deliver these life-saving and sustaining products in a way that also sustains the environment.

The environmental programs at Baxter are part of a broader sustainability approach that addresses "our people, our operations and products, our world." The company has laid out specific 2015 goals addressing each of these priority areas [5]. Baxter has a long history of focusing on environmental and occupational health and safety. The company was among the first to measure the business benefits of environmental excellence and continues to be an industry leader at integrating sustainability into its corporate culture. For example, to reduce its carbon footprint Baxter is not only reducing energy consumption at its facilities but also applying green technologies, such as cogeneration and increasing the use of renewable energy. In addition, the company is pursuing supply chain initiatives that reduce the carbon footprint of its automotive fleet and is continuing its involvement in carbon trading programs, such as the Chicago Climate Exchange. Baxter is also striving to improve the efficiency of its natural resource use and to reduce waste, especially plastics waste, which accounted for approximately 31% of its nonhazardous waste in 2007.

In early 2008, Baxter was confronted with a crisis similar to the J&J Tylenol incident. Baxter staff detected an unusual increase in allergic-type reactions in the United States associated with heparin sodium injection, a commonly used anticoagulant (blood thinner). The company subsequently recalled all of its U.S. heparin vial products and identified a sophisticated, heparin-like contaminant. Following the highly publicized recall, more than a dozen other companies in nearly a dozen countries reported similar heparin contamination issues. The problem was attributed to deliberate tampering with the raw materials. Baxter has responded with stringent new testing protocols and precautionary measures for improved supply chain visibility, and is working with international authorities to assure future product integrity.

Product Sustainability Review

Baxter includes a Product Sustainability Review (PSR) during the early stages of the product development process, since the design stage offers a unique opportunity to influence a product's environmental, health, and safety performance across the entire life cycle. It is at this time that decisions can be made regarding materials selection, components that impact energy use, and other factors. The Baxter PSR is a two-step assessment of the environmental, health, safety, and other sustainability-related impacts of a product throughout its life cycle. An initial screen reveals high-level sustainability risks and opportunities at the product development concept phase, in areas such as regulations and customer and other stakeholder requirements.

The second step is a comprehensive review that identifies improvement opportunities across the life cycle. It includes life-cycle impact assessment and computer modeling that details the environmental impacts of possible products. This assessment can also compare these impacts to those of existing products, assist with material choices, and assess disposal options. Baxter uses these results to confirm product feasibility, help establish product requirements, and influence design to minimize potential impacts to human health and the environment.

Beginning in 2005, Baxter has used PSR to evaluate all new medical devices, ranging from intravenous bags to renal machines that reach the concept stage of development. The PSR addresses both products and packaging issues, aiming to eliminate hazardous substances wherever possible and maximize product reuse, recycling, and service life. Several reviews have influenced materials selection; for example, heavy metals were eliminated from a new machine design, except for some that are exempt from the European Union Restriction of Hazardous Substances directive. In one case, the assessment led to redesigned, lighter product packaging.

One example of DFE accomplishments is Baxter's AVIVA premium line of intravenous solutions containers, launched in 2006. While providing similar functionality and benefits as the company's VIAFLEX flexible container systems, AVIVA containers are lighter in weight, made of non-PVC (non-polyvinyl chloride) film, contain no latex, and offer a DEHP-free fluid pathway to patients. AVIVA is designed to better accommodate certain newer therapies that are more complex and have the potential to be incompatible with existing container technology. The product also helps clinicians better meet the needs of sensitive populations, such as neonatal, pediatric, and oncology patients. Finally, Baxter purchased Green-e Certified Renewable Energy Certificates to offset the electricity used to produce AVIVA.

Baxter is continuing to improve the PSR process through enhanced life-cycle impact assessment computer modeling. In 2008, the company plans to align PSR with the external standard ISO 14040 (which describes life-cycle assessment principles and frameworks). Baxter is also piloting the international standard IEC 60601-1-9 (for environmentally conscious design) with one of its devices. Effectively tracking the materials and chemical substances used in Baxter products and manufacturing is complex, since a product may contain thousands of components from hundreds of suppliers worldwide. To better meet this challenge, Baxter is developing a new product stewardship database to manage environmental and other information related to all new products. This database will interface with other company information systems related to products and allow Baxter to better understand, manage, and optimize product environmental performance.

Product End-of-Life

The responsible treatment of electronic healthcare products at the end of their useful life is an emerging issue worldwide. Baxter has had programs for several years to refurbish and return products to use as appropriate, and to recycle products when reuse is not feasible. Baxter has introduced even more comprehensive product take-back in Europe, in accordance with the European Union Waste Electrical and Electronic Equipment (WEEE) directive (see Chapter 3).

Many of the electronic medical products Baxter sells, such as intravenous (IV) infusion pumps, are well suited to repair and refurbishment after the original customer has finished using them. In some countries, Baxter leases most of these products to customers, which helps ensure that products will be returned to Baxter after a set period of time. Repair and refurbishment extends a product's useful life and decreases the environmental impacts associated with product disposal and the manufacture of new products. For example, in Europe in 2007, Baxter conducted about 38,300 maintenance events and repairs on products (an increase of 9% from 2006) with a cumulative weight of approximately 1,600 metric tons (an increase of 7% from 2006). In the United States, Baxter refurbished a total of about 18,300 renal machines, blood collection devices, and nutrition compounding machines in 2007.

When customers return products to Baxter that contain batteries, such as infusion pumps, or when Baxter repairs those products onsite, Baxter sends the batteries to a recycler whenever feasible. As part of its global audit program covering all the regulated or medical waste recycling or disposal sites that Baxter uses, trained Baxter auditors assess battery recycling sites before using the recycling vendor. These sites are then reassessed at least once every four years to ensure they comply with Baxter's requirements and conduct their operations in an environmentally responsible manner. These audits examine all aspects of operations, including site history, possible contamination sources, regulatory compliance, financial conditions, insurance, and other factors.

In addition, Baxter works with customers and end-users to facilitate recycling. For example, Baxter is a charter member of the Chicago Waste to Profit Network, a public-private partnership launched in 2007 through which companies and other organizations work to identify and develop potential ways to convert waste from one company into value for another.

In Ireland, Baxter launched a program in 2007 with local waste management contractors to provide pick-up services at patient residences for home-use oncology products, such as vials, needles, and drugs that cannot be processed with regular municipal waste, as well as packaging and other materials that might be recycled. The contractors collect, process, and recycle or dispose of the products, while protecting patient confidentiality and privacy. This provides a valued service to patients and communities, while ensuring environmentally responsible waste treatment. Building on the original initiative, Baxter recently launched a similar pilot program for renal products supplied to home patients. Due to the success of this pilot, beginning in 2008 the company will offer waste-collection service to all renal home patients in Ireland. The two programs will benefit approximately 600 customers and recover about two metric tons of waste annually.

Integrated "Lean and Clean" Engineering

Baxter has been a leader in applying "lean" principles and Six Sigma statistical engineering tools to its environmental programs. Baxter's approach is to integrate environmental, health, and safety (EHS) expertise into Lean manufacturing initiatives to prevent negative environmental consequences *and* to identify opportunities for environmental improvement—known as "Lean and Clean." The approach includes

- Modifying traditional Lean tools, such as value-stream maps, by adding environmental aspects
- Applying Lean tools to such EHS processes as wastewater treatment to drive efficiencies
- Integrating traditional pollution-prevention techniques into Lean and Clean tools to provide a systematic way of finding opportunities.

For example, at Baxter's facility in Waluj, India, a Lean exercise that focused on water usage helped reduce wastewater effluent by 40%. The U.S. Environmental Protection Agency (EPA) has requested Baxter's help to help build a Lean/Six Sigma toolkit for businesses to use in integrating Lean principles into EH&S initiatives.

Baxter also is focused on the Lean and environmental performance of its suppliers. Through its activities with the Green Suppliers Network, U.S. EPA, and the U.S. Department of Commerce collaboration with private industry, Baxter urges its suppliers to be trained in Lean and Clean manufacturing, which creates opportunities for greater efficiency, cost savings, and waste- and resource-reduction.

Eli Lilly and Company: Reducing the Environmental Footprint

Environmental Achievements and Goals

Eli Lilly and Company (Lilly), founded in 1876, is a leading pharmaceutical and biotechnology company with products focused around diabetes/endocrine/osteoporosis care, neuroscience, oncology, and cardiovascular/critical care. In 1923 Lilly introduced the first commercial insulin product. With nearly $19 billion in sales in 2007, Lilly has about 40,000 employees providing products in over 143 countries.

Lilly is committed to continuously improving the environmental aspects of its operations, including employee health and safety, environmental protection, and the efficiency of energy and materials use. During the five-year period from 2003 to 2007, Lilly

- Cut hazardous materials purchases in half
- Reduced energy intensity (energy used per dollar of sales) by more than a third and reduced absolute energy use by 2.4% (see Figure 14.2)
- Cut greenhouse gas emissions per dollar of sales by 33%
- Reduced water use intensity by nearly a third and cut overall water use by 14%
- Reduced solid waste generation by 22%
- Cut emissions of volatile organic compounds by 40%

Lilly adopted next-generation environmental goals in 2008 aimed at achieving best-in-class performance within the industry [6]. These goals, with a base year of 2007, include:

- Improve energy efficiency and reduce associated greenhouse gases 15% by 2013.
- Reduce waste-to-landfill 40% by 2013, with the ultimate goal of "zero landfill."
- Reduce water intake 25% by 2013.

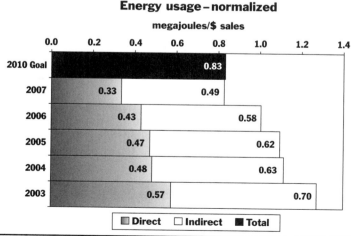

Energy usage – normalized

FIGURE **14.2**   Energy intensity reductions achieved by Lilly from 2003 to 2007.

Leading-Edge Environmental Practices

Energy Efficiency

Lilly has focused on improving the energy efficiency of the nine plants that account for 86% of its total energy use. These operations are relatively energy and water intensive since they use biological processes such as fermentation to produce natural and biotechnology products. Since 2006, the company has invested nearly $18 million in 44 energy conservation projects that are delivering $10 million in annual savings. Examples of successful projects include cogeneration, lighting management systems, and high-efficiency chiller systems. One plant in the United Kingdom purchased imported electricity (to meet about 82% of its needs) from a renewable energy source (hydroelectric) with zero carbon emissions. The other 18% was provided by an efficient on-site combined heat and power plant that ran primarily on natural gas.

Design for Process Safety

Pharmaceutical bulk manufacturing typically uses reactive chemistry involving quantities of hazardous materials and may also have recovery operations for solvents that are flammable and/or toxic. Therefore, process equipment designers and engineers need to ensure adequate safeguards against the loss of containment. Lilly developed a Globally Integrated Process Safety Management system, a set of practices that go beyond compliance by addressing hazardous processes from the design stage, during start-up, and throughout their operating life. Since 2001, additional practices, such as Inherently Safer Design and Layer of Protection Analysis have been implemented. The result has been a five-fold reduction in process safety management issues since 2002.

Product Stewardship

Lilly developed a Product Stewardship Standard to identify and prevent potential health, safety, and environmental (HSE) disruptions along a product's value chain, which might interfere with meeting patient needs. The Standard considers the full product life cycle including discovery and development, procurement, packaging, distribution, marketing, and sales. It also considers the needs of customers, suppliers, contract operations, and alliances, and incorporates business continuity planning. The Standard distributes responsibility for managing HSE aspects across each value chain functional group. This decentralized approach strengthens the organization's ability to integrate HSE into its business processes.

Applying Green Chemistry for Source Reduction

Lilly believes that the largest improvements in its environmental and safety profile will be driven by new production processes that are inherently safer, use fewer resources, and result in less waste. The company strives to discover and develop such processes by applying "green chemistry" principles during process development. For example, chemists use electronic lab notebooks that include green chemistry assessment tools, providing chemists with immediate feedback on process efficiency and the availability of less hazardous alternative materials.

According to Steve Gillman, Executive Director, Corporate Health, Safety and Environment, "The most significant health, safety, and environmental (HSE) improvements result from our ongoing efforts to design new products and processes to minimize HSE impacts from the start. Applying green chemistry and inherently safer design principles to our process development efforts helps us 'get it right the first time'."

At key milestones in its "stage-gate" development process, Lilly verifies material use efficiency based on metrics, such as *process mass intensity* (PMI), which is the ratio of material used per unit of active pharmaceutical ingredient produced. For example, a recent process improvement for a product designed to treat anxiety and depression would reduce hazardous material usage by more than 80%, saving more than 3 million kilograms of raw materials and 6 million liters of water per year at peak production. Figure 14.3 illustrates the reduction in PMI due to improved synthetic chemistry from an early development route to a process ready for manufacturing.

Lilly scientists and engineers are continually working to apply the latest scientific knowledge when designing pharmaceutical production processes. Product development teams evaluate attributes that predict the future HSE burden of a process, along with more traditional criteria, such as yield, quality, cost, and equipment needs. In

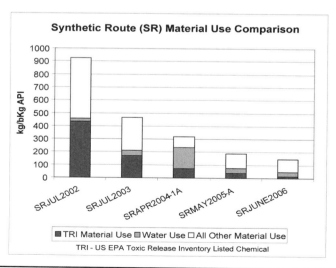

FIGURE 14.3 Process mass intensity reduction due to improved synthetic chemistry.

addition, Lilly is leveraging new technologies to make medicines more efficiently and with less waste. For example, in the case of one potential product, a treatment for sleep disorder, the team has been able to reduce raw material use by a factor of 100 using continuous reactors instead of conventional batch reactors.

Lean Six Sigma for Environmental Performance

Lean Six Sigma has become a core business tool at Lilly for improving productivity and eliminating process wastes, which has positively influenced environmental performance. Other projects have reduced the amount of laboratory wastes generated, improved the use of continuous emission monitoring systems, established common energy-tracking systems to drive reduction in energy use, and established a container recycling process. The following are examples of HSE benefits:

- **Waste Elimination.** In Indiana, a Lean Six Sigma project resulted in the elimination of a liquid hazardous-waste incinerator, which reduced GHG emissions by 21,750 Metric tons of CO_2 equivalents per year. This project also reduced the amount of waste solvents requiring incineration by 6 million liters per year.

- **Packaging Reduction.** Pharmaceutical packaging is highly regulated and must fulfill many functions—providing information, resisting counterfeiting, and protecting against tampering or access by children. From 2007 to 2008, Lilly's innovative packaging approaches saved more than $14 million by reducing packaging mass and improving productivity.

- **Green Procurement.** Lilly has developed environmental purchasing criteria for specific categories of materials and services, such as energy efficiency and material content. In selected areas, follow-up reviews of a supplier's HSE performance provide validation that the desired capabilities are being realized. This process has encouraged collaboration with suppliers in developing solutions that achieve Lilly's HSE objectives.

Sustainable Packaging for Medical Devices

Lilly markets medical devices that support patient care in the use of its pharmaceuticals. One example is an insulin delivery device commonly known as an "insulin pen." The company's medical device development process emphasizes the use of renewable resources, energy efficiency, and compliance with the evolving product environmental requirements, such as WEEE (see Chapter 3), battery return, and packaging.

Before   After

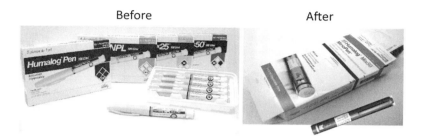

FIGURE 14.4   Packaging improvements in Humalog insulin pen.

For example, the packaging for a new Humalog insulin pen was redesigned to eliminate plastics and reduce overall size (see Figure 14.4). The outcome was a significant reduction in the environmental footprint, including elimination of 146 tons of plastic per year. Package size and mass reduction resulted in a decrease of shipping cases by 106 tons, primary cartons by 18 tons, literature by 76 tons, and number of pallets by 2041 per year. The associated energy savings are 9.8 million megajoules per year—enough to power 260 U.S. homes. In addition, the associated forest products savings for corrugated, fiberboard, and paper are equivalent to saving over 3000 trees per year.

References

1. R. A. Sheldon, "The E Factor: Fifteen Years On," *Green Chem.* **9** (12), 1273–1283 (2007).
2. Personal communication, David Constable, Corporate Environment, Health, and Safety, Glaxo-Smith-Kline.
3. Information about Health Care Without Harm is available at www.noharm.org/us.
4. Information about Johnson & Johnson's environmental programs can be found at www.jnj.com/connect/caring/environment-protection/.
5. The Baxter 2007 Sustainability Report is available at sustainability.baxter.com/documents/sustainability_report_2007.pdf
6. Information about Lilly's EH&S programs is available at www.lilly.com/about/compliance/practices/health/.

CHAPTER 15

Food and Beverage Industries

Overview

Food and beverages represent a product category in which sustainability issues are highly visible to the consumer, perhaps more than any other category. Health consciousness and rising consumer interest in "natural" and "organic" foods have transformed the industry. This has paved the way for a deeper examination of how food consumption impacts the planet, spurred by concerns over global warming. Occasional crises, such as the discovery of *E.coli* in vegetables or melamine contamination in imported commodities, have raised concerns about the safety of the food supply. In opposition to commercial trends toward fast foods and packaged foods, new movements have emerged that promote "slow food" and locally harvested food. Genetically engineered crops have also stirred controversy about health, ecological, and ethical issues. The proximity, freshness, and authenticity of local foods present an increasingly viable alternative to the efficiency and economies of scale that can be achieved by the global food industry.

According to the World Wildlife Fund (WWF), the agricultural sector accounts for over 50% of habitable land around the world, 70% of water use, the highest use of chemicals and the highest pollution among all sectors, and from 25% to 40% of greenhouse gas emissions. Yet global food demand will double in 50 years, and the development of renewable materials and fuels is placing increasing pressure on limited agricultural land. Meanwhile, hundreds of millions of people live in poverty, own no land, and are chronically hungry. WWF has worked extensively with various food and beverage industry sectors to encourage the adoption of more sustainable agricultural practices and to establish certification systems for sustainably harvested commodities, such as soy, sugar cane, cocoa, coffee, palm oil, and seafood. For example, the Marine Stewardship Council is the world's leading certification and eco-labeling program for sustainable seafood.

In early 2007, the U.S. Grocery Manufacturers Association held its first-ever Environmental Sustainability Summit for the Food, Beverage, and Consumer Products Industry. The conference showcased the fast-growing efforts of leading food, beverage, and consumer products companies to contribute to environmental sustainability, and explored available tools for designing sustainable products and packaging, including product life-cycle assessment and carbon footprinting. Much of the credit for this surge of interest goes to Wal-Mart, which has made clear to all of its suppliers that their products and packaging will be evaluated in terms of environmental performance (see Chapter 19). Examples of progressive industry practices include the following:

- Anheuser-Busch has been recycling leftover grain from the brewing process for over 100 years and harnessing renewable energy from the nutrient-rich brewery wastewater for more than two decades. The company has reduced packaging weight and recycles 99% of the solid waste that it generates—over 5 billion pounds of materials per year. From 2003 to 2007, it reduced water consumption per production unit by more than 7% and fuel consumption per production unit by more than 15%.

- Kraft Foods, one of the world's largest food and beverage companies, has implemented an energy management program based on conservation, operating efficiency, and alternative technologies, including conversion of waste biomass, such as whey and meat, to energy. The company was able to decrease absolute energy use by 14% from 2001 to 2006, while production increased by about 10% during that period.

- PepsiCo has worked with the Carbon Trust in the United Kingdom to affix a carbon footprint label to its Walker's Crisps, a snack product. The life-cycle contributions to the footprint are shown in Figure 15.1. Consumer surveys indicated that customers understood the label and responded to it favorably.

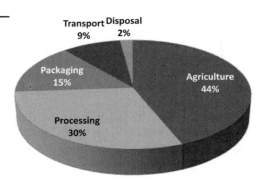

FIGURE 15.1 Proportion of carbon emissions for Walker's Crisps from each life-cycle stage.

- Frito-Lay, a unit of PepsiCo, has collaborated with the water authority in San Antonio, Texas, on a water conservation strategy for its manufacturing plant. Through a series of projects with reasonable pay-back periods, the plant has doubled production while keeping water use below 2003 levels, reducing the water consumption per bag of chips produced by over 50%.

- Cargill has continuously increased its use of renewable energy to more than 10% of energy demand by 2007. The company uses a variety of renewable energy sources including biogas, landfill gas, biofuels such as biodiesel, ethanol and tallow, and biomass such as hulls and bagasse.

- Ben and Jerry's, a division of Unilever, has a long history of progressive environmental actions and was the first public company to endorse the Ceres principles. In recent years, the company has invested in research on thermoacoustic refrigeration technology, which utilizes sound waves to create cooling and, thus, eliminates the use of gases.

- Heinz has established the HeinzSeed program to develop tomato seeds using traditional breeding techniques (no genetic modification) and distribute them globally to farmers. HeinzSeed tomatoes remain ripe longer in the field, are more disease-resistant so they require less pesticide use, and produce a higher yield, thus reducing consumption of water and fertilizer.

The following sections describe DFE programs implemented by several leading companies in the food and beverage industry, namely, Coca-Cola, ConAgra Foods, General Mills, and Unilever. Like many global multinationals, some of these companies have been criticized by activist groups concerned about specific business practices, and no doubt such attacks have heightened their sensitivity to stakeholder interests. The purpose of this book is not to investigate contentious issues, but rather to demonstrate how an earnest commitment to social and environmental responsibility can lead to innovations that benefit both the company and its stakeholders.

Coca-Cola: Global Responsibility
Measuring Environmental Performance

The Coca-Cola Company (Coke) is the world's largest beverage company, with the world's most recognizable brand name. The company employs over 90,000 people in over 200 countries worldwide and distributes more than 2,800 products that account for approximately 1.5 billion consumer servings per day. Coke has more than 300 bottling partners globally, most of which are independent companies, but the company works closely with these partners to measure and improve the overall environmental performance of the system. The system's environmental impacts are primarily due to plant operations and distribution networks, as well as vending machines and coolers. Coke concentrates on three principal areas of environmental responsibility, each with explicit performance metrics [1]:

1. Water stewardship involves both water conservation and operating efficiency initiatives. For example, efficiency is primarily measured by the water use ratio, i.e., liters of water per liter of product. The water use ratio has steadily declined to about 2.5 in 2007, corresponding to a 40% eco-efficiency ratio. In 2008 the company announced a global goal of 20% improvement in water efficiency by 2012 from the baseline year of 2004. Coke has also pledged to achieve "water neutrality" by returning water to nature equivalent to what it uses in its operations.

2. Sustainable packaging is measured through resource use efficiency, including energy and material intensity. These are influenced primarily by packaging design, container recycling, and package material reuse. For example, the company has established a goal in its U.S. operations to recycle or reuse 100% of its PET (polyethylene terephthalate) and aluminum packaging. Coke's innovations in the area of sustainable packaging are described below.

3. Energy and climate protection are primarily measured by the energy use efficiency ratio, i.e., megajoules per liter of product. Vending machines and coolers are the largest contributor to greenhouse gas emissions within the system and produce three times the estimated emissions of beverage manufacturing facilities. Coke's efforts to reduce these emissions through sustainable refrigeration technology are described below.

Neville Isdell, Chairman and former CEO, summarizes Coca-Cola's environmental commitment in simple terms: "We recognize that if the communities we serve are not sustainable, then we do not have a sustainable business for ourselves."

Sustainable Packaging

In food and beverage products, packaging provides convenience as well as safety benefits, such as spoilage reduction and tamper resistance. In Coca-Cola's view, the social value that packaging provides does not diminish the responsibility of the entire packaging chain to reduce waste and consumption of natural resources. In fact, Coke aspires to treat discarded packaging as a resource rather than a waste. Accordingly, the company's DFE efforts include

- Designing packages that use less material without sacrificing quality
- Investing in technologies and recovery systems that enable greater use of recycled materials

To help improve packaging performance, Coke initiated an effort in 2006 to build a global online inventory of primary, secondary, and transport package systems by sales and weight. This extensive database, which also tracks system recycling rates, will enable the system to measure business performance more effectively and assess progress toward longer-term goals. In the first year of reporting, nearly 90% of the Coca-Cola system's global packaging data were captured, accounting for 98% of global sales volume.

Coke continues to make strides in package design efforts focused on improving efficiency, life-cycle effectiveness and eco-innovation. For example, using state-of-the-art computer design software, the company effectively reduced the weight and improved the impact resistance of the widely recognizable glass contour bottle (see Figure 15.2). The innovative "Ultra Glass" bottles, introduced in 2000, are 40% stronger, 20% lighter and 10% less expensive than traditional contour bottles. These redesigned bottles saved 89,000 metric tons of glass in 2006, the CO_2 equivalent of planting more than 13,000 acres of trees.

Coca-Cola has been focused on PET recycling and reuse since introducing the first beverage bottle made with recycled material in 1991. The company has made considerable investments in closed-loop recycling plants around the world. For example, Coke is investing more than $60 million to build the world's largest plastic-bottle-to-bottle recycling plant in Spartanburg, S.C. The plant will produce approximately 100 million pounds of food-grade recycled PET (polyethylene terephthalate) plastic for reuse each year—the equivalent of producing nearly two billion 20-ounce Coca-Cola

---

Figure 15.2 Mass reduction achieved by the Ultra Glass bottle.

Former   Ultra

380 grams   305 grams

bottles. Recycling plastic for reuse yields financial benefits, requires less energy than producing bottles with virgin materials, and reduces waste and greenhouse gases. Over the next ten years, the Spartanburg recycling plant is expected to eliminate the production of one million metric tons of carbon dioxide emissions—the equivalent of removing 215,000 cars from the road.

Sustainable Refrigeration

Since refrigeration equipment is the largest contributor to Coke's climate footprint, the company has established a sustainable refrigeration program as the cornerstone of its energy management and climate protection efforts. Coke's engineers and suppliers have been working to develop technologies that not only improve overall energy efficiency, but also reduce greenhouse gas (GHG) emissions like hydrofluorocarbons (HFCs). The achievements of this program include

- Transitioning to HFC-free insulation foam for new equipment, which will generate 75% fewer direct GHG emissions than traditional equipment
- Deploying more than 8,500 units of HFC-free equipment by 2007 using CO_2 as the refrigerant—a more eco-friendly alternative with 1,300 times less potent GHG emissions than HFCs. Coke is committed to the deployment of 100,000 HFC-free units between 2008 and 2010
- Developing a proprietary energy management system (EMS) for coolers and vending machines, which have been deployed in over one million units of equipment and delivers energy savings of up to 35%

With the support of Greenpeace, Coca-Cola announced that 100% of the coolers and vending machines provided to the 2008 Beijing Olympic Games—more than 5,600 units—would feature an HFC-free natural refrigerant as well as the proprietary EMS technology (see Figure 15.3). This resulted in the reduction of greenhouse gas emissions by approximately 40,000 metric tons—the equivalent of taking more than 194,000 cars off the road for two weeks.

Coke is also working with industry partners and nongovernmental organizations to encourage broader adoption of environmentally-friendly cooling systems. Currently CO_2 equipment is more expensive because of limited demand and high production costs, and many companies have been slow to make the shift. Coke has joined with Unilever, McDonalds and others to create the "Refrigerants, Naturally!" initiative—an industry alliance to promote HFC-free refrigeration.

Taking a life-cycle perspective, Coke is collaborating with suppliers and customers in the practice of DFE. The company works very

FIGURE 15.3 Climate-friendly refrigeration equipment at the 2008 Beijing Olympic Games.

closely with major customers (e.g. Wal-Mart, TESCO) to educate them about the environmental performance of its products, packaging, cold-drink equipment, and distribution network. Likewise, Coke has worked extensively with cold-drink equipment suppliers on the development of sustainable refrigeration and is beginning work on sustainable purchasing, with an initial focus on sugar cane.

Finally, Coke participated in a pilot project with Carbon Trust to evaluate the total emission of GHG from products across their life cycle. Through this project and other similar projects, the company is exploring ways to communicate information related to a product's carbon footprint to consumers. While Coke has no immediate plans to put carbon labels on beverage packages, this information will most likely be shared with interested consumers via Coke's corporate responsibility website.

ConAgra Foods: Packaging Innovation

ConAgra Foods, based in Omaha, Nebraska, is a leading packaged foods company whose brands include Healthy Choice, Hunt's, and Orville Redenbacher's, among others. ConAgra Foods also has a significant presence in commercial food products and is one of the nation's leading specialty potato providers to restaurants and other foodservice establishments. In keeping with its environmental sustainability commitments [2], the company has introduced a number of technological innovations into its packaging.

- ConAgra Foods' packaging team for Hunt's Ketchup worked with its bottle manufacturer, Constar International, to design a single-layer PET bottle that is lightweight and easy to recycle (see Figure 15.4). The technology uses an active oxygen scavenger blended directly into PET in a monolayer structure. It has a number of advantages over multi-layer PET including better integrity (no delamination), lower manufacturing complexity, and reduced cost. The material is impact-resistant and naturally colorless with exceptional clarity. It also provides an excellent barrier against moisture and gases, assuring long shelf life; and the single-layer design reduces both resource consumption and environmental impacts. This innovation was selected by the Institute of Packaging Professionals to receive the 3M Sustainable Packaging Award for 2007.

- ConAgra Foods is the first company in North America to incorporate post-consumer recycled plastic into its frozen meal trays, which will divert approximately 8 million pounds of plastic from landfills into the recycling stream annually. ConAgra Foods has begun to use between 30% and 40% post-consumer recycled plastic in nearly all of its frozen meal trays for Healthy Choice®, Banquet®, Kid Cuisine® and Marie Callender's® products. Most frozen meal trays used by other companies are made of crystallized PET plastic, a material that uses only newly produced plastics and that requires more energy and resources to produce.

FIGURE 15.4   Hunt's ketchup with mono-layer PET bottle wins sustainable packaging award.

- ConAgra Foods introduced a new shrink film that contains more than 50% post-industrial recycled polylactic acid (PLA), a renewable material that is manufactured from corn. This "industrial ecology" innovation avoids the purchase of petroleum-based raw materials and diverts PLA wastes from landfill. The film will be used for tamper evident seals on ConAgra Foods' table spreads—Fleischmann's®, Blue Bonnet® and Parkay®—and for printed shrink labels for multi-packs of Reddi-Wip® topping and PAM® cooking spray. It requires less energy at ConAgra Foods' manufacturing facilities, reducing the temperature necessary to shrink the material by approximately 20 percent. In addition, it provides a higher-quality finished product due to improved shrink performance. The new material will reduce annual greenhouse gas production by approximately 592,000 pounds of carbon dioxide equivalents, the same as taking about 48 cars off the road per year.

General Mills: Nourishing the Future

General Mills is the world's sixth largest food company, with 28,500 employees worldwide and over $13 billion in sales, about $9 billion of which is in the United States. In addition to its popular cereal brands, such as Cheerios, General Mills brands include Betty Crocker, Pillsbury, Yoplait, and Haagen-Dazs ice cream. General Mills has built its entire sustainability program around the theme of nourishment [3]:

- Nourishing lives by providing great food products that meet or exceed expectations for safety, quality, convenience, value and, of course, great taste.

- Nourishing communities by volunteering, donating money and food, supporting minority-owned businesses, and creating healthy and safe workplaces for employees.

- Nourishing the future by establishing sustainable business and manufacturing practices that minimize the company's environmental footprint.

Environmental responsibility has long been an important part of General Mills corporate values and actions. For example, General Mills was able to significantly reduce the environmental impacts of its Green Giant sweet corn products through integrated pest management. Between 1980 and 2005, the company reduced pesticide application volume by 80% and insect control costs by 37%.

Building on its historic commitment to environmental excellence, General Mills established a formal sustainability program in 2003. Later, in 2007, the company appointed Gene Kahn as the new vice president and global sustainability officer. Kahn was an organic foods pioneer who founded Cascadian Farm in 1972 and Small Planet Foods in 1997, both of which were acquired by General Mills in 2000. One of Kahn's first moves at General Mills was to conduct a rigorous assessment of the environmental impacts of the U.S. food industry. His team found that the industry

- Contributes 18% of the country's total greenhouse gas emissions, with U.S. agriculture responsible for half of them
- Consumes 82% of total water use, with U.S. agriculture accounting for 80 of the 82%
- Represents 5% of U.S. energy use, with the food processing sector representing just 1% of the 5%
- Generates 28% of total municipal waste when food service, retail, and consumer food usage are combined

Based on these findings, General Mills is taking steps to improve the sustainability of its product life cycles. For example, by 2010 the company aims to achieve 50% less water consumption in broccoli acreage by using more efficient drip irrigation technology. Across its manufacturing operations, the company has set five-year goals for reducing its key environmental indicators:

- Water usage rate by 5%
- Energy consumption rate by 15%
- Greenhouse gas emission rate by 15%
- Solid waste generation rate by 15%

Manufacturing facilities are constantly seeking innovations to achieve these goals. One plant in Missouri reduced its water usage by more than 14 million gallons per year by reusing water from air compressors for its cooling towers, switching to more dry cleaning than wet cleaning, and installing a high pressure, low volume cleaning system. Another plant in Ohio reduced landfilled waste by about 42% by selling waste as recyclable material; for example, the plant has found buyers for its corrugated cardboard, mixed paper, film, and other trash. Even excess food discarded in the production process is being sent to farms to feed pigs and chickens. The plant has set a goal to convert 100% of its waste into by-products.

Unilever: A Vitality Mentality

Unilever, based in the United Kingdom, employs about 180,000 people and generates over $50 billion in annual revenue from 400 brands in the food, home care, and personal care markets. Recognized as a leader in sustainable business practices, as of 2008 Unilever has been the food industry category leader in the Dow Jones Sustainability World Indexes for a decade. Over 40% of Unilever sales are in emerging markets, and it has established voluntary programs to promote nutrition, health, and hygiene around the world. About 20% of the company's sales are in the United States, including well-known brands, such as Hellmann's, Ragu, and Lipton.

According to Chief Executive Patrick Cescau, "Social responsibility and environmental sustainability are core business competencies, not fringe activities." Unilever has come to believe that today's social and environmental challenges represent opportunities for innovation and product development. For example, Unilever's Indian subsidiary, Hindustan Unilever, was one of the first companies to design products for the "base of the pyramid," distributing affordable products to the hundreds of millions of poor people living in rural villages across India [4].

Unilever's stated mission is to "add vitality to life" by meeting people's needs for nutrition, hygiene, and personal care. Because of its deep roots in local cultures and markets around the world, it has developed strong relationships with consumers, and considers itself a "multi-local" multinational. An example of Unilever's environmental sustainability commitment was its announcement in 2007 that it would purchase all its tea from sustainable, ethical sources. As a first step, the company began working with the international environmental organization, Rainforest Alliance, to certify its tea farms in Africa and audit the company's progress. Unilever is the largest tea company in the world, and tea is the world's most popular beverage, with about five billion cups consumed daily.

One area of environmental responsibility that is common to all food producers is packaging. Like other companies, Unilever has worked on packaging reduction and material selection but has tried to accomplish this within the context of the whole product system. For example, modifying packaging to the extent that there is a risk of compromising the product integrity may have a negative environmental impact. Unilever has formed a Responsible Packaging Steering Team, a global team of senior executives that leads the development and delivery of strategies for responsible packaging, with representation from R&D, Packaging, Marketing, Procurement, Customer Development, and Communications, as well as regional representatives.

An example of a Unilever packaging innovation is the introduction of microwaveable pouches for pasta sauce. Compared to glass jars with metal lids, the total weight of the pouch including primary and secondary packaging is 70% less. Pouches are not subject to breakage during handling and transportation and are more convenient for consumers; moreover, microwave cooking uses less energy than cooking on a stove. Despite such success stories, Unilever acknowledges that challenges remain in achieving sustainable packaging. The company is still grappling with an unintended consequence of its success—how to reduce the litter created by the sale of millions of affordable, single-serving pouches, called "sachets," in developing and emerging markets.

References

1. Information about Coca-Cola's sustainability programs is available at www.thecoca-colacompany.com/citizenship/index.html.
2. ConAgra's corporate responsibility commitments are described at company. conagrafoods.com/phoenix.zhtml?c=202310&p=corp_resp.
3. General Mills' corporate responsibility commitments are described at www.generalmills.com/corporate/commitment/hse.aspx.
4. More information about Unilever's sustainability programs is available at www.unilever.com/sustainability/.

CHAPTER **16**

Consumer Products Industries

Overview

As in the case of the food and beverage industries, consumer products companies are particularly concerned about the brand and reputational implications of environmental performance. Environmental advocacy groups have been fond of targeting well-known brand names to call public attention to the safety and environmental shortcomings of consumer products. However, even in the absence of such attacks, the increasing consumer interest in environmental issues such as carbon footprints, combined with the growing environmental performance emphasis of large retailer chains, has driven a broad range of companies to ramp up their environmental management programs and to discover the benefits of improved eco-efficiency. For example:

- Kodak was a pioneer in the practice of DFE, developing a unique recycling program for its disposable FunSaver cameras. The company has continued to innovate in digital camera design with a careful eye toward environmental impacts. Between 1998 and 2006 Kodak digital cameras have decreased in weight by more than 50%, from over 15 ounces to less than 6 ounces, significantly reducing the consumption of raw materials, packaging, and fuel. Moreover, all Kodak digital camera products were compliant with the European Union's Restriction of Hazardous Substances directive (see Chapter 3), well in advance of the implementation date.

- SC Johnson was one of the first companies to incorporate eco-efficiency into its product development efforts in the early 1990s and has developed the Greenlist™ process that classifies raw materials according to their impact on the environment and human health. In 2005 it became the first major consumer packaged goods company to partner with U.S. EPA's Design for the Environment program. As a result, nearly all of its product lines boast environmental benefits. For example, by reformulating Windex® glass cleaner, the company has avoided 1,800,000 pounds of volatile organic compound (VOC) emissions annually, while achieving 30% more cleaning power.

- Nike has been refining the practice of DFE for over a decade, resulting in a holistic design ethos called "Considered." Nike products are designed from a life-cycle perspective—they are constructed from sustainable materials with minimal use of adhesives in manufacturing and minimal packaging throughout the supply chain, and they are recoverable at the end of their useful life. To assess the product environmental footprint, Nike has developed a Considered Index for both footwear and apparel products, which includes metrics such as the use of solvent-based cleaners, primers, and solvents; the waste generated in cutting and assembly; and sustainable material life cycles, including growing and extraction practices, chemistry, energy intensity, energy source, water intensity, waste, recycled content and end-of-life. For example, Nike's Trash Talk athletic shoe has an upper and midsole made entirely from manufacturing scrap waste, and the outsole also incorporates Nike Grind, which is made from reprocessed post-consumer waste.

- Timberland, a manufacturer of outdoor footwear and clothing, is deeply committed to environmental stewardship and plans to have all of its facilities carbon neutral by 2010. Timberland has taken a unique step in devising a "footprint" label, similar to a nutrition label, which appears on its footwear packaging—a shoebox made from plain brown recycled paper. The label includes a selected set of environmental criteria: the percentage of renewable energy used, the percentage of PVC-free products used, the percentage of eco-conscious materials used, and the number of trees Timberland has planted. Timberland's DFE philosophy is exemplified in the Earthkeepers™ line of premium boots, which boasts recycled, organic, and renewable material content, solvent-free adhesives, and reduced climate impact.

This chapter highlights the application of DFE to a particularly sensitive product category—personal hygiene products such as baby wipes—as practiced by Kimberly-Clark Corporation. Other examples presented below are taken from The Procter & Gamble Company, a leader in global sustainability, Mohawk Industries, a "green" carpet manufacturer, and Patagonia, an early pioneer in eco-design of clothing. But these stories represent only a small part of the vast range of DFE efforts taking place in the diverse arena of consumer products.

Many of the practices described in the previous chapters on electronic, automotive, chemical, pharmaceutical, and food and beverage products are also relevant to consumer products.

Kimberly-Clark: Getting Serious about DFE
Applying DFE to Product Development

Kimberly-Clark Corporation (K-C) is headquartered in Dallas, Texas and employs 53,000 people in 35 countries worldwide. The company's well-known family care and personal care brands include Kleenex, Scott, Huggies, Pull-Ups, and Kotex. K-C has embraced sustainability principles in both product development and manufacturing operations, and has invested considerable effort in reducing the environmental footprint of its products. In 2008, for the fourth consecutive year, K-C was ranked highest in the personal products category of the Dow Jones Sustainability World Index [1].

K-C established a Design for Environment (DFE) initiative in 2007 with the aim of designing products and packaging with more environmentally friendly contents and minimal packaging. This corporate initiative seeks to apply sound environmental design principles in developing processes, products, and packaging. The DFE program was established as part of Vision 2010, a set of overarching, global environmental goals that address key metrics, such as water use reduction, wastewater quality, energy use, greenhouse gas reduction, manufacturing waste reduction, and elimination of landfill waste. These goals flow down to the facility level, where targets are established for each production line.

Led by David Spitzley, Product Sustainability Manager, the DFE team works closely with each of K-C's business sectors to

1. Identify opportunities to improve product design and material sourcing decisions
2. Offer guidance in the environmentally responsible use of virgin and recycled materials
3. Develop environmentally preferred alternative packaging solutions.

In the U.K., K-C launched its first Forest Stewardship Council (FSC) labeled product under the Kleenex brand in 2008. K-C is also working with the Carbon Trust to develop an industry-wide draft methodology for measuring carbon footprints more accurately and consistently. In the United States, K-C Professional offers Kleenex Naturals facial tissue to its customers, and K-C's Scott brand offers Scott Naturals bathroom tissue and paper towels to consumers (see Figure 16.1). All of these innovative products contain a mix of high-quality, post-consumer, recycled fiber and virgin fiber.

Figure 16.1   Scott Naturals products combine post-consumer recycled fiber with virgin fiber [1].

K-C has realized tangible benefits by applying DFE principles. Examples include converting manufacturing wastes into valuable by-products, reducing packaging mass and cost, and increasing the useful life or utility of products. The environmental performance advantages of K-C products are communicated regularly to key customers, such as Wal-Mart or Tesco, who have a keen interest in sustainability. K-C has been an active participant in several of Wal-Mart's Sustainable Value Networks (including Wood and Paper, Sustainable Packaging, and Chemical Intensive Products) and was a beta tester for Wal-Mart's Sustainable Packaging Scorecard (see Chapter 19).

K-C has also extended DFE principles into its supply chain. In 2007, K-C issued *Sustainability at K-C: Guideline for Suppliers*, a document available online that shares K-C's practices and policies with regard to sustainability. K-C is moving to incorporate environmental criteria into procurement decisions, especially with contract manufacturers. In particular, pulp fiber suppliers, who represent a major consumable within K-C's tissue and personal care products, are either in compliance with one of five certification schemes or are in the process of becoming certified.

Using LCA to Develop Greener Baby Wipes

Kimberly-Clark has begun to utilize Life-Cycle Assessment (LCA) as a key tool to support DFE practices. A first step was conducting simple LCA evaluations of material and process options to support technology investment decisions. More recently, as inputs into their sustainability strategies, each business unit has selected three principal products and used LCA to develop preliminary "footprints."

According to David Spitzley, the use of full-scale LCA in conformance with ISO 14040 guidelines, a rigorous and resource-intensive approach, should be undertaken only when there is a need for a high degree of scientific credibility to support major business decisions, public communication, and/or policy making. An example is the trend analysis conducted by EDANA, the European Disposables and Nonwovens Association (see Figure 16.2), showing significant improvement in the environmental performance of baby diapers [2].

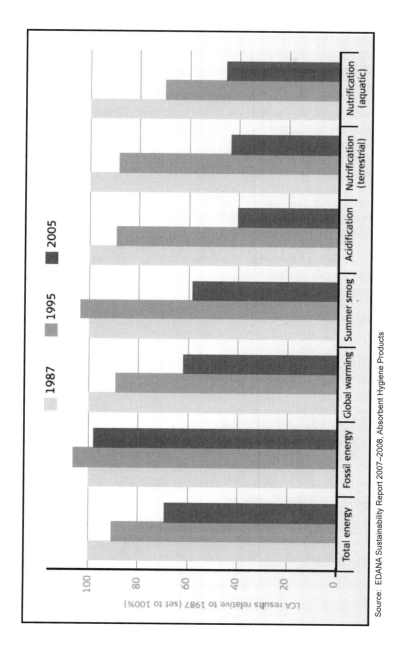

Source: EDANA Sustainability Report 2007–2008, Absorbent Hygiene Products

Figure 16.2 Trends in major environmental impact categories for baby diapers 1987–2006.

Another example is an LCA performed by K-C of its full range of tissue products and associated manufacturing processes to gain insight into the life-cycle environmental implications of the use of virgin versus recycled fibers as raw materials. This study scrupulously followed the ISO process and was validated by an expert review panel. However, Spitzley argues that for routine internal applications of LCA, it is preferable to use less burdensome, more streamlined methods that can provide rapid feedback to product development teams (see Chapter 9).

In the health and hygiene category, disposable baby wipes are a great example of a product that requires careful consideration of sustainability issues. On the one hand, the wipes provide important benefits to families in terms of sanitation and convenience. On the other hand, they are immediately disposed of after use and embody a considerable throughput of water, materials, and energy. To examine these trade-offs more rigorously, the Huggies Baby Wipes team used life-cycle assessment as a means of integrating proactive environmental considerations into new product designs. The results helped to identify the potential impact of design changes, such as the trade-off between basis weight and polymer content changes. Figure 16.3 illustrates the reductions in life-cycle environmental burdens that were achieved by a proposed new design.

Looking to the Future

According to Ken Strassner, Vice President for Global Environmental, Safety, Regulatory, and Scientific Affairs, "Design for Environment will continue to be an important area of activity for Kimberly-Clark because it can be a point of differentiation for us and can help deliver

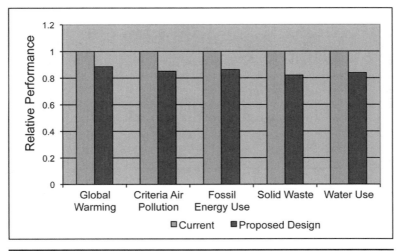

Figure 16.3   Relative life-cycle performance of Baby Wipes proposed design versus current baseline.

value for our business. At K-C we are working to implement a deliberate process for considering environmental concerns in the design of products and packaging. We believe this process will result in reduced costs, increased revenues, reduced risks, and increased brand equity and will position the company as an industry leader."

Specifically, Strassner believes that pursuing DFE can

- Reduce costs—by reducing the consumption of raw materials and energy and by improving operating efficiency
- Increase revenues—by developing innovative solutions for consumers and improved customer partnerships (both price and volume impacts are possible over time)
- Reduce risks—by avoiding negative events that could result in one-time costs and/or market share or stock value loss (temporary or long-term)
- Build brand equity
- Position the company as an industry leader, motivating employees and serving to attract new talent

DFE principles are being integrated into K-C's overall product development process, which includes cost, quality, and safety considerations as well as life-cycle thinking that encompasses the supply chain and the use and disposal of K-C products. As a starting point, K-C introduced environmental questions into its stage-gate-process guidance documents. This is further supported by specific tools to enable environmental evaluations, such as product environmental footprints and life-cycle assessments. In 2008, K-C instituted a DFE training program for engineering and design professionals, and all business units have agreed to incorporate DFE into their sustainability strategies.

Procter & Gamble: Ensuring a Better Quality of Life

The venerable Procter & Gamble Company (P&G) was one of the early pioneers in environmental sustainability and life-cycle assessment. George Carpenter, the now-retired Vice President for Sustainability at P&G, was a prime mover in raising awareness of environmental issues in the business community, and, in 1992, helped to found the Global Environmental Management Initiative (GEMI), a cross-sectoral industry consortium that continues to develop tools and best practices for sustainability (see Chapter 4). P&G has also been an active contributor to the World Business Council for Sustainable Development, and former CEO John Pepper coauthored an influential publication with the CEO of DuPont, arguing that sustainability could be achieved by harnessing market forces [3].

P&G adopted a definition of sustainability first promulgated by the government of the United Kingdom: "sustainable development is about ensuring a better quality of life for everyone, now and for generations to come." In keeping with this notion, the company aspires to use its capacity for innovation to help address the environmental, social, and economic concerns of its consumers and stakeholders [4]. P&G chose to center its sustainable innovation efforts around two key themes that are relevant to many of its businesses namely:

1. Water availability, quality and quantity
2. Health, hygiene and nutritional issues.

For example, P&G collaborated with the United Nations Children's Fund (UNICEF) to develop a nutritional beverage called NutriDelight. This product has the potential to significantly address a common problem in developing nations—micronutrient deficiency in vitamin A, iron, and iodine among school-age children.

As a globally recognized leader in brand development and product innovation, P&G has placed great emphasis on incorporating sustainability awareness into its product development and marketing efforts [5]. In 2007 P&G committed to a five-year goal of generating at least $20 billion in cumulative sales of products with reduced environmental impact. For example, P&G's cold-water detergents reduce energy consumption and greenhouse gas emissions, while saving consumers money and providing better fabric care. Sustainable innovation efforts include the following:

- **Product compaction.** By concentrating its detergent products P&G has significantly reduced the size and weight, thus reducing life-cycle impacts of sourcing, manufacturing, packaging and distribution. Similarly, compacting of household paper products on larger rolls saves on consumption of cardboard cores, reduces fuel costs in distribution, and reduces packaging waste.

- **Sustainable packaging.** To assure that global forest resources are used responsibly, P&G purchases only wood pulp that comes from plantation-grown trees or sustainably managed forests, and is produced using chlorine-free purification processes. P&G also requires suppliers to meet or exceed local and regional laws and gives preference to suppliers that minimize the potential environmental impacts of their operations.

- **Air emission reduction.** None of P&G's products contain chlorofluocarbons (CFCs), which were phased out many years ago. P&G primarily uses hydrocarbons, hydrofluorocarbons, dimethyl ether, and carbon dioxide propellants in its aerosol products. Some P&G products contain VOCs as propellants or to help dissolve ingredients, but they comply with regulatory limits.

- **Life-cycle assessment.** P&G routinely uses LCA to support product and process improvement. In the early 1990s, P&G was one of the first companies to apply LCA, comparing the environmental impacts of disposable diapers versus laundered cloth diapers. The controversy continues to this day, although a recent British government report supported P&G's conclusions that the life-cycle impacts are roughly comparable [6].

- **Eco-efficiency in production.** P&G's program to "design manufacturing waste out" has saved the company over $500 million as well as 2 million metric tons of waste. Today, over 95% of materials that enter P&G plants leave as finished product, and over half of the remaining materials are recycled. In 2007 P&G set a five-year goal to reduce CO_2 emissions, energy and water consumption, and disposed waste per unit of production by an additional 10%, contributing to a 40% reduction for the decade.

Mohawk Industries: Naturally Stain-Resistant Carpet

Mohawk Industries achieved a win-win solution with its Smartstrand™ line of residential carpeting, combining high performance with environmental responsibility. This was accomplished through a partnership with DuPont, which developed Sorona® fiber made from triexta, a renewably sourced polymer. Mohawk Industries' CEO Jeffrey S. Lorberbaum said, "This product will do things that others cannot, making it the biggest fiber innovation we have seen in 20 years. Our confidence in this product is evidenced by Mohawk's first-ever Limited Lifetime Warranty offer for stain resistance. Rarely do consumers get a luxuriously soft carpet with these benefits."

The new technology boasts durability, crush resistance, resilience, and permanent, natural stain resistance that will not wash or wear off like topical stain treatments are prone to do. The stain resistance is an inherent, built-in attribute of the fiber and provides new levels of protection, allowing difficult stains, such as mustard or red wine, to be removed simply with warm water and a mild detergent. The triexta polymer fiber is made from a 1,3- propanediol (PDO),which was originally produced from petrochemical feedstock. However, DuPont found a way to produce PDO through a fermentation process that uses corn sugar, an agricultural feedstock. Based on life-cycle assessment, DuPont has shown that production of renewably sourced Sorona® fiber requires 30% less energy and produces 55% less CO_2 emissions than an equal amount of nylon fiber.

Mohawk has also introduced a variety of other environmentally responsible innovations [7]. For example, Mohawk converts over 3 billion bottles annually—25% of all the bottles collected in North America—into 160 million pounds of recycled carpet fiber. In addition, Mohawk recycles about 10 million pounds per year of crumb rubber from about 720,000 scrap tires into designer door mats. Finally, Mohawk has replaced cardboard carpet cores with those made from recycled carpet edge trim, plastic bottle tops and stretch films. These new cores last longer, are less likely to damage the carpet, and can be recycled into new cores.

Patagonia: Dancing on the Fringe

Patagonia is an acknowledged leader in sustainable product development. The company was founded in 1973 by environmentalist Yvon Chouinard, a mountain climbing enthusiast, whose philosophy and values continue to guide the company. With about $300 million in annual sales and 1400 employees worldwide, the company has achieved extraordinary influence and staying power in the outdoor clothing market. Patagonia's mission statement is to "build the best product, cause no unnecessary harm, and use business to inspire and implement solutions to the environmental crisis."

Unlike larger companies that have injected sustainability concepts into an existing culture, Patagonia has sustainability embedded in its DNA [8]. For example, in 1985 Patagonia formalized its support of environmental activism by committing 10% of pretax profits to grassroots environmental groups; later, the company changed its pledge to at least 1% of sales. Patagonia's human resource policies are family friendly and encourage diversity of thought and behavior. In addition, Patagonia protects the environment through its clothing designs, which are simple, versatile, durable, and in most cases recyclable. The company does not place growth above fidelity to its values.

The Patagonia brand inspires fierce customer loyalty, and vintage garments are prized. The legendary quality of Patagonia products is attributable to its unique design philosophy [9]. Patagonia's designers have developed a list of criteria that define the company mission to make the best product. They pose a series of questions that are universally applicable:

- Is it functional, and does it fit the intended use and market need?
- Is it multifunctional, or versatile, so that it fulfills multiple needs?
- Is it durable, long-lasting, failure-resistant, and repairable if needed?
- Does it fit our customer (especially important for clothing)?
- Is it as simple as possible in design and construction?
- Is the product line simple, offering a few distinct choices?
- Is it an innovation or an invention (the latter is rare but powerful)?

- Is it a global design, yet sensitive to local requirements?
- Is it convenient to care for and clean or maintain?
- Does it have any added value, a tangible benefit to the customer?
- Is it authentic, i.e., rooted in the real world?
- Is it art, reflecting timeless aesthetics rather than current trends?
- Are we just chasing fashion, driven by market cycles?
- Are we designing for our core customer?
- Have we done our homework, including research and testing?
- Is it timely, considering the state of the market and competition?
- Does it cause any unnecessary harm, environmental or otherwise?

The last element summarizes Patagonia's DFE commitment and has led it to adopt innovations such as organic cotton, recyclable polyesters (see Figure 16.4), and Synchilla fabric from soda bottles. However, the sustainability journey has hit a few potholes along the way, and Patagonia is candid in documenting its successes and failures in the "Footprint Chronicles." For example, the company once tried to replace plastic buttons with rainforest tagua nut buttons, supported by an indigenous industry, but the nuts could not survive washers and dryers. Another example is a persistent, bioaccumulative chemical, perfluoro-octanoic acid (PFOA), which is a common constituent of water-repellent membranes and coatings. Patagonia is trying to remove PFOA from its lines without sacrificing performance. It has replaced the membranes with polyester and polyurethane materials but has not yet found a viable alternative to the existing coatings.

FIGURE 16.4 Patagonia's Eco Rain Shell jacket, made from 100% recycled polyester.

In 2005 Patagonia launched its Common Threads Garment Recycling Program: Customers return their worn-out Capilene® Performance Baselayer to headquarters, Distribution Center, or retail stores, and Patagonia will recycle the old into new Capilene® garments. Patagonia has now expanded this program to include cotton T-shirts, Patagonia fleece, and Polartec®-branded fleece from any other manufacturer. Over the last 3 years Patagonia has made progress in increasing the number of styles that can be recycled in the Common Threads program. The percentage of recyclable products in Fall 2008 was 45% and by Fall 2009 it will rise to 65%. The Spring lines have a lower percentage but are still increasing. In addition, the company is exploring alternative approaches, including repurposing old garments by cutting and sewing them into new ones, and "downcycling" garments into filling material or padding.

Not content with compromise, Patagonia will continue to press on in pursuit of environmental solutions, far from the mainstream, like a solitary climber. As founder Chouinard puts it [9]

> Our current landscape is filled with complacency, be it in the corporate world or on the environmental front. Only on the fringes of an ecosystem, those outer rings, do evolution and adaptation occur at a furious pace; the inner center of the system is where the entrenched, nonadapting species die off, doomed to failure by maintaining the status quo. Businesses go through the same cycles…only those businesses operating with a sense of urgency, dancing on the fringe, constantly evolving, open to diversity and new ways of doing things, are going to be here one hundred years from now.

References

1. Information about Kimberly-Clark's sustainability programs is available at www.kimberly-clark.com/aboutus/sustainability.aspx.
2. European Disposables and Nonwovens Association, Sustainability Report: Absorbent Hygiene Products, 2007–2008.
3. C. Holliday and J. Pepper, Sustainability Through the Market—Seven Keys to Success (Geneva: World Business Council for Sustainable Development (WBCSD) Report, April 2001).
4. Information about Procter & Gamble's sustainability programs is available at www.pg.com/company/our_commitment/sustainability.shtml.
5. G. Carpenter and P. White. "Sustainable Development: Finding the Real Business Case," *Corporate Environmental Strategy: International Journal for Sustainable Business*, Volume 11, Issue 2, February 2004.
6. S. Aumonier and M. Collins, "Life Cycle Assessment of Disposable and Reusable Nappies in the UK" (Bristol, UK: Environment Agency, 2005).
7. Information about Mohawk's DFE efforts is available at www.mohawkgreenworks.com/.
8. Information about Patagonia's environmental commitment is available at www.patagonia.com/usa/patagonia.go?assetid=23429.
9. Y. Chouinard, *Let My People Go Surfing: The Education of a Reluctant Businessman* (New York: Penguin Press, 2005).

CHAPTER 17

Materials Production Industries

Overview

Materials are the basic building blocks of our economy, and material selection is an important aspect of every product design. All materials originate from natural resources, whether biological or geological. Materials production industries include agriculture, mining of metal ores, and production of oil and gas, as well as material processing activities, which refine or transform raw materials into commodities that may be incorporated into final products. One of the main principles of DFE is dematerialization (see Chapter 8), since the supply chain activities required to extract, process, and transport materials can generate a significant ecological footprint. Chapter 1 described the enormous amount of waste generated in the flow of materials through our economy. The fundamental challenge of sustainability is to decouple continued economic growth from the throughput of materials [1].

Materials are broadly classified as either organic or inorganic. Organic materials, sometimes called biomass, are derived from living organisms (e.g., cotton, wood, leather), while inorganic materials are typically extracted from the earth's crust. Increasingly, organic materials are being manufactured from biomass that might otherwise be disposed of, such as food waste and agricultural residues. Inorganic materials are considered *nonrenewable*, while organic materials are considered *renewable*, in the sense that they can be rapidly regenerated.* From a DFE perspective, renewable materials are attractive because

---

*One could argue that even fossil fuels are "renewable" except that the time span required is much longer.

they tend to reduce life-cycle carbon emissions and are typically biodegradable or recyclable. However, as shown in Chapter 9, use of renewable materials such as corn may reduce eco-efficiency because of the energy required to harvest and process these materials. Moreover, over-harvesting of renewable materials can threaten the stock of natural capital, such as forests and soils that are needed for replenishment.

Not all materials are created equal in terms of their environmental impact; for example, sand and gravel represent a large proportion of material flow by weight but are much less significant than other materials in terms of their adverse effects. Similarly, while metals and fossil fuels are often lumped into the category of non-renewable resources, the elemental structure of metals means that they are perpetually recyclable and not subject to the degradation that occurs with materials composed of complex molecules [2].

There is a broad range of innovation taking place in the design of new types of materials, and a thorough discussion is beyond the scope of this book. Case studies of such innovations have been discussed in previous chapters, including Xerox's high-yield paper (Chapter 11), Dow's bio-based polyethylene (Chapter 13), Coca-Cola's sustainable packaging (Chapter 15), and numerous examples in Chapter 8. Additional examples of DFE-related innovation in materials production include the following:

- Advanced high-strength steels are being used in the automotive industry to achieve better fuel efficiency and enhanced safety. An international coalition of steel companies has developed an UltraLight Steel Auto Body that achieves a 25% reduction in vehicle mass, establishing steel as a viable lightweight material for the automotive market.

- Rio Tinto, a global mining and minerals giant, launched a "mine of the future" project in Western Australia in 2007. The project is using advanced automation to improve mine safety and decrease the infrastructure environmental footprint, including driverless trains to carry ore, driverless "intelligent" trucks, and remote-control production drills.

- FCB Ciment, a French equipment supplier (now part of Fives Group) introduced the Horomill®, a cement mill using an innovative design that reduces electricity usage for cement grinding by up to 40%. It uses a roller turning at high speed against the concave inner part of the shell, so that particles are compressed repeatedly during a single cycle, with a larger compression surface (see Figure 17.1).

- Georgia Pacific, a leader in the forest products industry, has introduced Greenshield® corrugated paperboard, which is completely recyclable yet provides the moisture resistance

FIGURE 17.1 Horomill high-efficiency cement grinding mill.

needed for refrigerated storage of produce, poultry, and sea-food. Previously, these types of foods had to be shipped in nonrecyclable wax-coated packaging.

• NatureWorks LLC, a joint venture of Cargill and Teijin Limited of Japan, has developed bio-based polylactide resins, manufactured from 100% annually renewable field maize, for food packaging. Used for fruit and vegetable packaging, bottles, deli and bakery containers, egg trays, and film wraps, these materials are available with certification as free of genetically modified organisms.

While materials acquisition, use, and disposal are important issues in every industry, this chapter focuses specifically on producers of inorganic materials. DFE in the context of materials production is mainly concerned with the design of sustainable manufacturing and supply chain processes, although many companies have learned to add value through the development of material products with environmentally friendly characteristics. This chapter highlights three leading global manufacturers in different segments of inorganic material production—Alcoa, an aluminum company, Holcim, a cement company, and Owens Corning, a diversified producer of fiberglass, stone, and asphalt products.

Alcoa: Toward Sustainable Materials Management
Strategic Framework for Sustainability

The aluminum[†] industry has been a leader in the practice of sustainability, and aluminum has advantages for many applications due to its light weight, durability, and recyclability. Increased recycling of aluminum products can save large amounts of energy. For example, according to the International Aluminium Institute (IAI), manufacture of aluminum cans from recycled sources requires only 5% as much energy as manufacture from virgin bauxite ore (see Figure 17.2).The IAI helped to launch an Aluminium for Future Generations initiative in 2003, which comprises thirteen voluntary objectives, covering all phases of the aluminum life cycle, from mining to recycling. These objectives include reductions in energy use, water use, greenhouse gases and other emissions, as well as conversion of spent pot-lining, a residual waste from smelters, into feedstocks for other industries [3].

Alcoa is one of the largest and most fully integrated aluminum companies in the world, with 97,000 employees in 34 countries. Its operations include extraction of raw materials, processing them into aluminum, converting the metal into end-use products, and recycling aluminum products at the end of their useful life. In the year

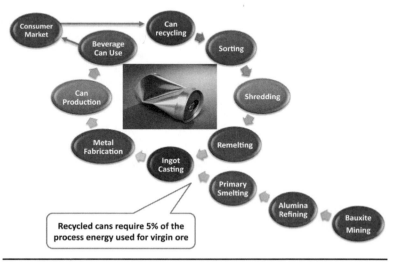

FIGURE 17.2   Life cycle of aluminum can production and recycling.

[†]The North American spelling is "aluminum" although the official worldwide spelling is "aluminium."

2000 a worldwide Alcoa team analyzed environmental and social trends since 1990 and developed a 2020 Strategic Framework for Sustainability, which has evolved into a roadmap for integrating sustainability into all aspects of Alcoa's operations. The framework is organized around six focus areas:

1. Economic benefit
2. Respect and protect people: employees
3. Respect and protect people: communities
4. Safe and sustainable products and processes
5. Meet the needs of current and future generations through efficient resource use
6. Accountability and governance.

Included are long-term targets and short- and long-term metrics for each focus area that were agreed upon with business leaders and technical experts throughout the company. These targets represent aggressive "stretch" goals, intended to drive the organization to think creatively. Table 17.1 shows Alcoa's reported progress in areas 4 and 5 above, as of year-end 2007.

Alcoa's CEO, Klaus Kleinfeld, shortly after taking the helm in 2008, described sustainability as a cornerstone of the company culture:

> Sustainability is about more than "doing good." It is also about smart business. At times like these, when the economy is down, it is important to make that point very clearly. I think too many people have looked at sustainability and issues like climate change as separate from their business. To me, it is critical that sustainability is an integral part of ... all corporate decision making—the products we make; how we manufacture; who we hire; and how we work with the communities where we operate.

Converting Wastes to Products in Brazil

One of Alcoa's key sustainability goals is to have zero landfilled waste by 2015. To help achieve that goal, the company has developed a six-stage process for evaluating and commercializing its industrial waste for reuse in other industries:

1. Develop concept
2. Feasibility and lab work
3. Benchmarks, small-scale testing
4. Pilot plant scale plus demonstration
5. Plant/reuse demonstration phase
6. Plant implementation.

Safe & Sustainable Products & Processes		
Target	**Metric**	**Progress Through 2007**
Increase recycling of aluminum	25% recycled aluminum content in fabricated products by 2010; 50% by 2020	Ratio of purchased scrap to total fabricated product shipments was 30%.[†]
	Increase North American used beverage can recycling rate to 75% by 2015	52%

Meet the Needs of Current & Future Generations through Efficient Resource Use		
Target	**Metric**	**Progress Through 2007**
Improved resource use to reduce environmental footprint	From base year 2005: 10% reduction in selected material use by 2010	Businesses have worked to identify major raw materials for reduction
	From base year 2000: 50% reduction in landfill waste by 2007; 75% by 2010; 100% by 2015	52%
	Reduce energy intensity 10% by 2010	Pursuing several production transformation projects
	60% reduction in process water by 2009; 70% by 2010	26%
	From base year 1990: 25% reduction in GHG emissions by 2010; 50% reduction by 2010[§]	33%
Cleaner production to reduce environmental emissions & impacts	From base year 2000: 60% reduction in sulfur dioxide (SO_2) by 2010	31%
	50% reduction in volatile organic compounds (VOCs) by 2008; 60% reduction by 2010	44%
	30% reduction in nitrogen oxides (NO_x) by 2007; 50% by 2010; 85% by 2015	50%
	80% reduction in mercury emissions by 2008; 90% reduction by 2010; 95% reduction by 2015	10%
	Zero process water discharge by 2020	Initial efforts being made to control process water.

‡The total amount of recycled metal in Alcoa products is somewhat lower since a portion of the purchased scrap would have been included in primary ingot sold to third parties.
§The 2010 target assumes that Alcoa will have success with "inert anode" technology.

TABLE 17.1   Elements of Alcoa's Strategic Sustainability Framework

At each stage, the concept must undergo technical and environmental assessment, commercial analysis, and stakeholder engagement. In addition, any proposed waste reuse must be in accordance with environmental laws and Alcoa standards for waste management. Beginning in 2008, customers must be certified and audited by a third-party company to ensure they are in compliance with the requirements for handling, storing, and using waste approved for commercial use.

As an example, Alcoa's Brazilian refineries and smelters have made significant progress in diverting production wastes from the landfill and converting them into commercially viable products—minimizing their environmental impact and providing a new source of revenue. Examples of successful waste conversion include

- **Carbon Cryolite.** This waste, which had accumulated from results from the Soderberg smelting process, has been developed into an alternative fuel for the cement sector. Alcoa's Poços de Caldas smelter sells approximately 1,000 metric tons each month, which will allow it to eliminate all of this stored waste by mid-2009. The smelter has not produced any carbon cryolite since 2003.

- **Aluminum Oxide Dust.** Since 2004, Poços de Caldas has been selling all of its aluminum oxide dust collected from the calciner department at the refinery as an alumina source for the enrichment of chamot—an inexpensive refractory aggregate. A second application for the dust is being developed for the Alcoa smelter in São Luís, Brazil, which is located too far from most refractory producers.

- **Boiler Coal Dust.** São Luís is currently selling the coal dust generated from its boilers as another alternative fuel source for the cement industry. Almost 12,000 tons of the dust is being diverted from the landfill each year.

Other wastes that are either being sold or provided to other industries include some mineral wastes as well as spent pot lining (SPL), which is generated when the carbon and refractory lining of smelting pots reaches the end of its serviceable life. Currently, Alcoa's two Brazilian smelters are providing 30,000 metric tons of SPL annually to the cement industry to use for fuel and a source of fluoride.

The Brazilian locations are also investigating several potential commercial applications for bauxite residue, the largest volume waste of the refining process. They are working closely in this effort with Alcoa's Australian researchers, who are also developing processes to convert the residue into a raw material for a variety of applications. Primary initiatives including converting the residue for use in the ceramic tile, agriculture, and cement industries.

Holcim: Creating Lasting Value

A Sustainability Mindset

Cement is one of the earliest building materials, based on technology discovered by the ancient Egyptians and further developed by the Greeks and Romans. The process for making "Portland cement" was developed in England in the early nineteenth century and is still the predominant manufacturing method used today. Although the cement industry does not leap to mind as an example of sustainable development, it is actually one of the most progressive industries in the world. Driven by concerns about the resource-intensive nature of cement production and the associated emissions, including greenhouse gases, leading global cement companies began working on environmental responsibility initiatives in the early 1990s.

Holcim Ltd., based in Switzerland, is one of the largest cement companies in the world, with operations in more than 70 countries and approximately 90,000 employees. Its U.S. subsidiary, Holcim (US) Inc., is one of the nation's largest manufacturers and suppliers of cement and mineral components, with 16 manufacturing facilities and more than 76 distribution terminals. Holcim took the innovative step of establishing a Department of Corporate Industrial Ecology in 1991, which continues to oversee its worldwide practices in the management of environmental health, safety, and sustainability. In 2008, for the fourth year in succession, Holcim was named by the Dow Jones Sustainability Indexes as the company with the best sustainability performance in the building materials industry [4].

As illustrated in Figure 17.3, sustainability is tightly woven into Holcim's corporate strategy. Creating value is the company's paramount objective, based on three strategic pillars of product focus, geographic diversification, and local management with global standards. The five mindsets underlying this strategy include environmental performance, social responsibility, cost management, innovation, and human resource excellence. Finally, the foundation on which everything rests is a workforce that gives its best on a daily basis.

Process Improvement Initiatives

The cement production process consists of three major steps: quarrying and raw materials preparation, clinker production in a high-temperature kiln, and cement grinding and distribution. Holcim is conscious of the entire life cycle of cement production, and has deployed a broad range of process improvement programs to reduce its ecological footprint. These include

- **Quarry management.** Other than fuel, the essential raw materials for cement production are limestone, marl, and clay, which are obtained by quarrying. In order to minimize

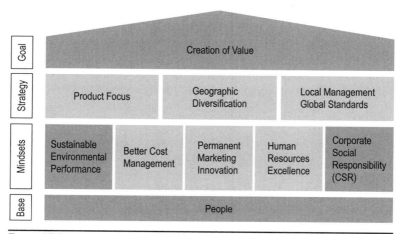

Goal	Creation of Value
Strategy	Product Focus / Geographic Diversification / Local Management Global Standards
Mindsets	Sustainable Environmental Performance / Better Cost Management / Permanent Marketing Innovation / Human Resources Excellence / Corporate Social Responsibility (CSR)
Base	People

FIGURE 17.3   Holcim strategy for value creation incorporating a sustainability mindset.

impacts on local communities, landscape, and ecology, Holcim has a number of programs in place to manage its quarries responsibly. For example, noise and vibration during blasting are mitigated through careful blast design, complete enclosure of processing equipment, and management of truck traffic to and from operating sites. Groundwater effects are managed by design of water handling systems to protect surrounding lands. Dust suppression measures include enclosure of crushing areas and conveyors as well as the installation of water spraying systems. Habitat destruction, wildlife displacement or loss, and visual impact are reduced through quarry design, by berming and tree planting, and through the design and implementation of site reclamation and rehabilitation plans.

- **Greenhouse gas emissions.** Cement manufacturing produces greenhouse gas emissions from both calcination of limestone and from fuel combustion for clinker production. On the average, each ton of cement produces about 0.8 tons of carbon dioxide. Holcim is committed to reducing global average carbon dioxide emissions by 20% by 2010, with 1990 as the reference year. As part of this commitment, Holcim (US) agreed to reduce its carbon dioxide emissions by 12% per ton of product manufactured between 2000 and 2008 as part of the EPA's Climate Leaders program. This is over and above the 15% reduction already achieved between 1990 and 2000.

- **Atmospheric emissions.** In addition to carbon dioxide, cement manufacturing results in the emission of a number of substances to the atmosphere. To assure continuous improvement, Holcim (US) has established a three-pronged global emissions monitoring and reporting standard as follows:

1. Continuous emission monitoring to measure airborne emissions of nitrous gases, sulfur dioxide, carbon dioxide, oxygen, and volatile organic compounds.

2. Annual stack tests for heavy metals, dioxins/furans, hydrogen chloride, benzene, and ammonia.

3. Equipment certification at least once per year by an external, qualified organization. This standard is often more stringent than corresponding federal or state requirements, and allows performance comparisons with other Holcim Group companies.

• **Resource conservation and recycling.** Efficient use of natural resources is an important cornerstone of Holcim's environmental policy. The company promotes eco-efficiency, conservation of nonrenewable natural resources, and recycling of secondary materials (waste and industrial by-products). For example, some Holcim plants use fly ash from electric power plants to help supply natural elements required for cement production. In addition, heat and dust are captured during the production process to make kilns run more efficiently and cleanly. To reduce the environmental footprint of cement production, Holcim uses a number of strategies:

  ○ Substituting natural raw materials with industrial by-products such as fly ash

  ○ Replacing fossil fuels with alternative fuels from industrial waste streams

  ○ Improving process efficiency to reduce energy and material consumption

  ○ Recycling or reuse of solid wastes, primarily cement kiln dust (CKD) and bypass dust.

At selected plants, Holcim has been partially replacing fossil fuels with EPA-approved supplemental fuels, such as used paint thinners, used dry cleaning fluids, and industrial solvents. The highly efficient burning temperatures in cement kilns can eliminate as much as 99.9999% of these organic materials. At some plants, Holcim is using scrap tires as supplemental fuels, which offers several advantages [5]:

• Tires produce the same amount of energy as oil and 25% more energy than coal.

• The ash residues from tire derived fuel may contain lower heavy metals content than some coals.

- Nitrous gas emissions are lower than many U.S. coals, particularly the high-sulfur coals.

To monitor the eco-efficiency of clinker production, Holcim (US) uses four key performance indicators:

1. **Clinker factor**—the percentage of clinker in cement
2. **Thermal energy efficiency**—energy consumed per ton of clinker produced
3. **Thermal substitution rate**—percentage of fossil fuels replaced by alternative, waste-derived fuels
4. **Cement kiln dust disposal rate**—kg of CKD not recycled per ton of clinker

Envirocore™ Family of Products

In addition to process improvements, Holcim has introduced product innovations aimed at improving both the functional and environmental performance of its cement products. Envirocore™ Blended Cements are designed to meet all general construction uses, to incorporate eco-efficiency strategies, and offer additional benefits, such as sulfate attack resistance, reduced heat evolution, alkali-silica reaction resistance, and high early strength. The following are examples of these products.

Holcim Performance Cement (HPC®) is a family of blended cements designed to meet the increasing performance demands, higher standards and specifications, and environmental constraints being placed on today's construction industry. They are manufactured by intergrinding or interblending a pozzolanic substance (typically fly ash from power plants) with conventional clinker. The primary ingredients in HPC® are Portland Cement clinker, fly ash and gypsum ground to a fine powder. Fly ash is primarily silicate glass containing silica, alumina, lime, and iron, and certain fly ashes exhibit both pozzolanic and cementitious properties. A by-product of cement hydration is calcium hydroxide, which is water soluble, and pozzolans react with the calcium hydroxide to form other cementitious compounds, creating a denser, more durable concrete.

Enviroset® is a family of dry blended alkaline reagents for solidification and stabilization, designed specifically for the environmental remediation industry. Alkaline reagents have been generally recognized as the preferred products for remediating sludges containing heavy metals, oils, and solvents. Enviroset® is unusual in that the customer plays a major role in designing the final product. Samples of components are sent to the customer's lab for bench testing with the actual waste stream to be remediated. After evaluation, Holcim will dry blend the reagents in the proportions required by the customer, and in the quantities needed for successful completion of the project.

Envirobase® offers an environmentally responsible, yet economical approach to structural fills, roadbase, slope protection, and flowable fills. Envirobase® products are made of fly ash and cement kiln dust and custom blended with Portland Cement to meet each customer's individual site requirements. The product is versatile enough to be used for multiple tasks on the same project—as a soil cement to stabilize a haul road, as a flowable fill to cover utility cuts, and even as an absorbent to clean up spills.

Holcim Portland-Pozzolan Cement is a high-performance cement that improves the properties of concrete. It is manufactured by inter-grinding or inter-blending a pozzolan (fly ash) with Portland Cement clinker. The resulting product has increased workability, increased durability, decreased permeability, reduced sulfate attack, decreased bleeding and segregation, reduced shrinkage, reduced heat of hydration, increased compressive strength, and increased flexural strength. This product can be used for general construction, as well as the construction of dams, piers, massive mat placements, footings, and similar structures.

Global Cement Sustainability Initiative

Holcim was a founding member of the World Business Council for Sustainable Development (WBCSD) Cement Sustainability Initiative (CSI), a worldwide collaborative program aimed at accelerating the move toward sustainable development. The CSI brought together ten of the world's leading cement companies, and continues to actively engage the broader cement industry and other relevant stakeholders. In addition to Holcim, the member companies are Cemex (Mexico), Cimpor (Portugal), Heidelberg Cement (Germany), Italcementi (Italy), Lafarge (France), RMC (United Kingdom), Siam Cement (Thailand), Taiheiyo (Japan) and Votorantim (Brazil).

In July 2002, at the Johannesburg Summit, CSI launched an "Agenda for Action" in six key areas: climate protection, fuels and raw materials, employee health and safety, emissions reduction, local impacts, and business processes. The foundation for this agenda was a comprehensive report on sustainability in the global cement industry, commissioned by the WBCSD, which included recommendations for sustainability goals and associated key performance indicators, or KPIs [6]. As shown in Table 17.2, specific indicators were chosen for five of the goals, while the other goals required further company-specific exploration. Several of the goals were relatively new for the cement industry, such as "Respect the needs of local communities" and "Support host region economies," so that common indicators had not yet been established. Note that the recommended energy goal uses an eco-efficiency indicator—*production per unit of energy*— which is the inverse of the traditional environmental indicator—*energy consumption per unit produced.*

Goal	Indicator	Range of Values (2002)**
Conserve resources by using less energy	**Metric Tons of cement per MJ (quarry and plant)**	Each metric ton of ordinary Portland cement consumes roughly 3000 MJ of total electrical and thermal energy
Conserve resources by recycling wastes	**Fuel substitution rate (%)** **Raw material substitution rate (%)**	Fuel ranges from 0 to 25% Raw material from 0 to 10%
Reduce environmental waste streams	**Non-product output, i.e., waste (kg) per tonne of cement**	*Airborne and waterborne releases are generally known, but definitions of solid waste vary*
Reduce greenhouse gas emissions	**Net CO_2 (kg) per tonne of cement**	Each metric ton of ordinary Portland cement generates approximately 900 kg of net CO_2 emissions[††]
Assure worker health and safety	**Incident rate (injury, work-related illness) per 200,000 hours**	Ranges from 1 to 5 incidents per 200,000 hours[††]
Reduce adverse impacts of quarrying	*Potential indicators include investments in quarry restoration, overburden waste reduction, water use efficiency, biodiversity action plans, groundwater impact mitigation efforts, etc.*	
Respect the needs of local communities	*Potential indicators include frequency of community meetings, number of hours in volunteer community service, public health initiatives, community complaints, number of advisory panels, community opinion surveys, etc.*	
Support host region economies	*Potential indicators include job creation, local investment, technology transfer, training time, contribution to gross domestic product (GDP), external economic benefits, etc.*	
Create value for shareholders	*Potential indicators include standard financial measures such as return on investment (ROI), return on assets (ROA), return on net assets (RONA), pretax earnings, etc.*	

**Data are for ordinary Portland cement; will vary for differing cement composition.
[††]Blended or composite cements can produce significantly lower CO_2 emissions
[††]Incident measurement may or may not include contractors along with employees.

Italics indicate areas that required further company-specific exploration of indicators.

TABLE 17.2 Sustainability Goals, Indicators, and Baseline Values for the Global Cement Industry

For the five recommended KPIs in Table 17.2, a baseline assessment was performed of the current industry status. Based on a limited sample of company-supplied data, together with literature-based research, a range of values was estimated for each indicator. Note that there was considerable variability in the data due to both differences in company performance and differences in boundary definition for the metrics. While these ranges hardly constitute a precise baseline, they provided a foundation for individual companies to initiate benchmarking, stakeholder dialogue, and establishment of improvement targets.

Owens Corning: A Passion for Environmental Excellence
Organizing for Sustainability

Owens Corning is a leading global producer of residential and commercial building materials including insulation and roofing, as well as glass fiber reinforcements and engineered materials for composite systems. Founded in 1938, Owens Corning is a market-leading innovator of glass fiber technology with sales of $5 billion in 2007 and 18,000 employees in 26 countries on five continents. Owens Corning is committed to driving enterprise sustainability through three major strategies [7]:

1. Greening its operations by reducing their environmental footprint

2. Greening its products by continuously improving their life-cycle impact

3. Accelerating energy efficiency improvements in the built environment.

In 2007, Owens Corning established a Corporate Sustainability group to manage its enterprise-wide sustainability initiatives. According to Chief Sustainability Officer Frank O'Brien-Bernini, "Energy efficiency is a fundamental part of our approach to sustainability." This makes business sense, since rising energy prices will make insulation technologies increasingly important in the construction industry. The amount of energy used in manufacturing insulation is recouped within about four to five weeks of product use in a typical building. That's a pretty impressive payback. Moreover, a study by The Ohio State University estimated that the life-cycle environmental benefits of insulation, in terms of avoided natural resource impacts of energy production, are about 1000 times greater than the impacts of producing the insulation [8].

Today, Owens Corning uses only 9% of the energy needed 50 years ago to melt glass for its insulation and glass reinforcement products, and uses about 55% less raw materials to produce the same insulating characteristics. The company is well on its way to meeting a 10-year goal of 25 percent energy-intensity reduction worldwide by 2012. In addition, Owens Corning's glass melting processes utilize more than 1 billion tons of recycled glass each year. Meanwhile, the company is seeking opportunities to generate business value through sustainable product design in many different markets, such as lightweight fiberglass composites for automobiles and wind turbines.

A Sustainable Factory

One example of Owens Corning's passion for environmental excellence is the Mount Vernon plant, tucked away among the serene rolling hills of Central Ohio. The plant manufactures unbonded loose-fill (ULF) insulation, sold for residential applications under the Atticat® brand. Behind the modest exterior lies an extraordinary example of dedication to high performance in terms of productivity, human development, and environmental sustainability. Originally built for another purpose, the plant was reopened in 2006 with the intent of implementing eco-efficient processes from start to finish. This facility may very well be a model for the sustainable factory of the future.

The manufacturing process involves melting sand together with recycled glass and forming the material into very fine glass fibers. The fibers are then conditioned, compressed, dyed with the familiar

FIGURE 17.4   Owens Corning unbonded loose-fill insulation plant in Mount Vernon, Ohio.

Owens Corning pink color, and bagged for shipment to retail "big box" stores as well as insulating contractors. Over 60% of the raw materials come from recycled "cullet," which is a mixture of crushed post-industrial glass. Since ULF insulation requires no binder material, chemical use and emissions are extremely low. Most of the internal conveyance is accomplished by pneumatic systems, requiring few moving parts. As shown in Figure 17.4, the stack has no visible plume—it emits almost nothing but hot air and carbon dioxide. The plant is permitted through the Ohio E.P.A. as a minor source for waste and air emissions.

While many companies speak of "zero waste" as a stretch goal, Mount Vernon has virtually achieved it. With a production volume of 30,000 tons per year, the plant generates only about 60 tons of solid waste annually, mostly spent filtration media. This is accomplished by dust control systems that recycle every bit of scrap material. The environmental controls include specially designed primary and secondary filtration systems with both visual and electronic monitoring. The plant floor is immaculate, and even stray bits of fiber from packaging operations are quickly vacuumed up and returned to the process. Off-spec product is remixed, and all office trash is recycled. Even waste heat will be recovered using a heat exchange system designed by the plant engineers, which will eliminate the need for natural gas to heat the office areas. With rising energy prices, this system is expected to pay back the initial investment in about 2 years.

The entire plant was designed with material and energy efficiency in mind, and the workforce is constantly coming up with innovations and striving for continuous improvement. Since melting is energy-intensive, the melters have a modular design that can adjust to changes in production volume. Robots are used for packaging the insulation. To improve capital longevity, most of the pipes and tubing have been replaced with stainless steel. Working capital is minimized through "just-in-time" techniques, and there is no finished goods inventory. Packaged insulation is moved directly to trucks waiting at the loading dock, 24 hours a day. Emergency containment systems provide resilience against chemical spills. The facility boasts among the highest productivity and lowest ecological footprint of all Owens Corning insulation plants.

In addition to its environmental leadership, Mount Vernon represents a successful experiment in advanced workplace management. George Bertko, the plant manager, beams with pride as he explains the plant's unique organizational culture. He calls it a "self-directed work environment," where employees are literally business owners and work as a close-knit team. There is no management hierarchy, and only five of the 30 staff members, including George, are not directly involved in production. Every person is a generalist—versatile in their operating and maintenance skills—and has an individualized "business growth plan" that guides their skill development. The facility boasts a record of zero injuries, zero absenteeism, and very low turnover. And of course, everyone shares a commitment to creating a business that operates in harmony with the environment. The plant is surrounded by a pristine wetland preserve, populated by native plants and wildlife.

In keeping with the eco-efficient plant design, the ULF insulation product has strong DFE credentials from a product life-cycle perspective. As mentioned above, 60% of the glass feedstock is recycled material. The product is free of binder chemicals and, therefore, has negligible emissions of the organic compounds that must be reported on U.S. EPA's Toxic Release Inventory. Since the product is extremely compressible, its compact packaging results in higher transportation efficiency than competing products. When applied with a portable Atticat® blowing machine, a 35-lb package of R-19 insulation expands to fill 106 square feet. Finally, if and when buildings are demolished, the ULF is easier to recycle and remelt than conventional bonded insulation products. In 2008 this product line was Cradle to Cradle Certified^{SM} at the Silver level, in recognition of its environmentally intelligent design.

The fanatical elimination of waste at Mount Vernon was perfectly captured when we were being escorted on a plant tour by George Bertko. He stopped in mid-sentence, noticing a small white piece of fluff that resembled glass fiber drifting along the floor. "What is that?" he asked nervously. The plant engineer was quick to reassure him that it was only a bit of cottonweed which had drifted in from the adjacent meadow. George sighed with relief as we continued our tour—another day of environmental harmony.

References

1. J. Fiksel, "A Framework for Sustainable Materials Management," *Journal of Materials*, August 2006.
2. Five Winds International, "Eco-Efficiency and Materials" (Ottawa, Ontario: International Council on Metals and the Environment, 2001).
3. Information about Alcoa's sustainability programs is available at www.alcoa.com/global/en/about_alcoa/sustainability/home.asp.
4. Information about Holcim's sustainability programs is available at www.holcim.com/corp/en/id/1610644016/mod/gnm50/page/channel.html.
5. U.S. EPA, Management of Scrap Tires, www.epa.gov/osw/conserve/materials/tires/.
6. J. Fiksel and T. Brunetti, "Key Performance Indicators," in Battelle Memorial Institute, *Toward a Sustainable Cement Industry*, World Business Council for Sustainable Development, 2002. Available at www.wbcsd.org.
7. Information about Owens-Corning's sustainability programs is available at www.owenscorning.com/sustainability/.
8. Y. Zhang, A. Baral, B. R. Bakshi, G. Jakubcin, J. Fiksel, "Ecologically Based Life Cycle Assessment" (Portland, Ore.: International Life Cycle Assessment and Management, October 2007).

CHAPTER 18

Energy Production Industries

Overview

Sustainable energy production is, quite simply, the paramount challenge facing the global economy. World petroleum reserves are dwindling, and the unprecedented spike in oil prices during 2008—approaching $150 per barrel—was a warning signal to global manufacturers dependent on petroleum fuel for supply chain operations. In the United States, which imports the majority of its oil from abroad, there has been an increasing cry for "energy independence." However, weaning any industrial economy away from petroleum will not be easy. Renewable sources of biofuels, such as corn-based and cellulosic ethanols, are less efficient than gasoline, and are challenged by the life-cycle environmental burdens of harvesting the biomass. Plug-in hybrids are a viable alternative; but electric power is still largely generated by coal, which is confronted with its own environmental challenges, while nuclear power is still controversial.

From a broad sustainability perspective, it is generally accepted that we must drastically reduce global greenhouse gas emissions in order to stabilize atmospheric levels of carbon. Yet, as shown in Figure 18.1, the U.S. economy remains heavily dependent on nonrenewable fossil fuels, which continue to account for about 85% of domestic energy use. Nuclear energy accounts for about 8%, and approximately 6% is derived from renewable sources [1]. The worldwide average consumption data are very similar, and these numbers have not changed appreciably for about a decade.

On a positive note, the greenhouse gas intensity (metric tons per $million) of the U.S. economy has fallen significantly from 1990 to 2005, as shown in Figure 18.2. However, this decrease resulted mainly from increased efficiency in industrial and commercial energy use rather than increased use of low-carbon fuels. The reduced GHG intensity was offset by economic growth, resulting in about a 17% overall increase in GHG emissions during that period. It appears that

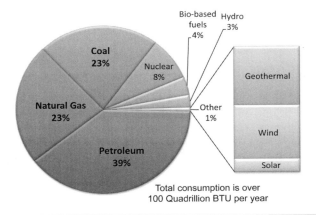

Total consumption is over
100 Quadrillion BTU per year

FIGURE 18.1 U.S. energy consumption by energy source in 2007 [1].

the only practical way to combat the global warming threat is to
increase the proportion of energy supplied by low-carbon energy
sources—renewable, nuclear power, and coal-fired plants with carbon capture—and thus decouple carbon emissions from economic
value creation.

What is the outlook for renewables? Most projections agree that,
even under optimistic investment scenarios, neither biofuels nor renewable electric power sources such as solar, wind, and geothermal

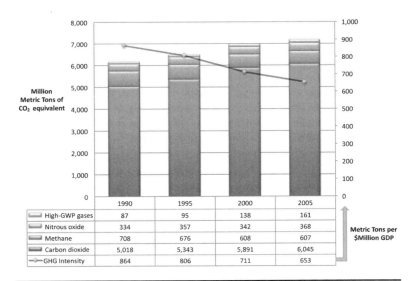

	1990	1995	2000	2005
High-GWP gases	87	95	138	161
Nitrous oxide	334	357	342	368
Methane	708	676	608	607
Carbon dioxide	5,018	5,343	5,891	6,045
GHG Intensity	864	806	711	653

FIGURE 18.2 Growth of U.S. greenhouse gas emissions, 1990–2005, and
change in intensity [1].

energy will be able to fulfill rising energy demand, and that we will continue to rely on fossil fuels for the foreseeable future, well into the twenty-first century. Economic growth, especially in developing nations such as China and India, will most likely overwhelm any gains from conservation. As shown in Figure 18.3, North America has a much higher per capita consumption of energy than the rest of the world, roughly double the consumption rate of Europe or Japan [2]. Reducing that consumption may require a significant change in life-styles—smaller homes, fewer vehicle miles, and less flow-through of consumer merchandise. However, a number of promising technologies have appeared that will contribute to energy conservation, including plug-in hybrid electric vehicles, energy-efficient appliances, and advanced heating and cooling systems.

Ironically, there were warning signals about global energy security several decades earlier. The Arab oil embargo of 1973 prompted a flurry of investigation into alternative technologies and conservation initiatives. Under President Richard Nixon, the U.S. Federal

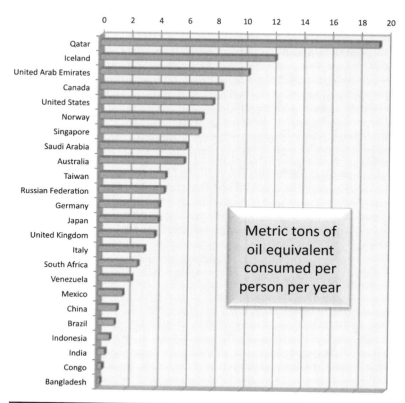

FIGURE 18.3 Per capita energy consumption in 2005 for selected countries [2].

Energy Administration, forerunner of the Department of Energy, launched Project Independence with the goal of achieving energy self-sufficiency by 1980. Although some initial reductions were achieved, the political agenda shifted, and the sense of urgency waned. Instead of foreign oil imports declining from the 1973 level of about 33%, the level of imports has risen to about 67% by 2008.

DFE in Energy Innovation

In response to concerns about energy and climate, there are many companies around the world that have pursued a proactive strategy—grappling with energy innovation rather than clinging to the *status quo*. As described in Chapter 8, energy efficiency, conservation, and carbon emission reduction have become important priorities in the private sector. Companies of every size have become sensitized to the importance of wise energy management, if only for purposes of cost control; and many are implementing more efficient technologies such as combined heat and power. The following are examples of environmentally beneficial innovations that are taking place specifically in the energy-producing industries:

- **Carbon capture and storage.** The coal-fired electric power industry is exploring the use of carbon capture and storage (CCS) technology as a long-term solution for eliminating greenhouse gas emissions. A variety of techniques are being investigated for trapping carbon dioxide (CO_2) from coal combustion before it is released into the atmosphere. While captured CO_2 could be used for applications such as algae farms, the most commonly cited method is pumping the CO_2 into underground storage chambers. For example, a Swedish utility, Vattenfall, has established a pilot 30-megawatt power plant fitted with CCS in Germany. In the United States the Department of Energy was supporting similar pilot studies under the FutureGen program, but funding has been suspended.

- **Advanced biofuels.** Renewable fuels, including next-generation biofuels, are increasingly recognized as a critical component of any U.S. strategy for energy security. Unlike corn ethanol, advanced biofuels will offer both a substantial return on energy and a sufficient use of renewable feedstocks to reduce greenhouse gas emissions per BTU by 50% or more across the full life cycle. Examples of biomass sources being investigated for advanced biofuels production include agricultural residues such as rice straw, woody plants such as poplar, municipal solid waste, and microorganisms such as algae. Figure 18.4 illustrates the many possible pathways for recovering biomass from waste streams and converting it into energy, fuels, or bio-products [3].

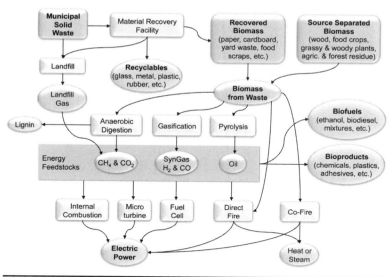

FIGURE 18.4 Potential routes for conversion of waste biomass into energy.

- **Solar energy for rural villagers.** Tata BP Solar, a joint venture of Tata Power and BP Solar, has supplied solar energy systems to more than 7 million people in the province of Uttar Pradesh, India, for applications such as village lighting, water pumping, and telecommunications. Grid power is unavailable for most villages in this region, but Indian banks are offering easy finance schemes that enable villagers to procure solar-home-lighting systems at affordable prices.

- **Solar thermal power.** Unlike solar photovoltaic technology, solar thermal simply concentrates the sun's radiation to power steam turbines. An Australian firm called WorleyParsons is developing the world's largest solar thermal plant, to be completed by 2011; it will deliver 250 megawatts, enough to power the equivalent of 100,000 houses. Another Australian company, Ausra, is creating the manufacturing infrastructure for a 177-megawatt solar thermal power plant in California.

- **Large-scale wind turbines.** As investments increase in wind energy generation, the size and sophistication of wind turbines is growing, including offshore turbines tethered to the ocean floor and advanced technologies for wind power storage. As of 2008, the world's largest wind turbine was the Enercon E-126, with a rotor diameter of 126 meters (see Figure 18.5). Located in Emden, Germany, it is expected to produce more than 7 megawatts, enough to power about 5,000 households in Europe.

FIGURE 18.5   The Enercon E-126 wind turbine.

- **Plug-in hybrids.** One of the major areas of innovation in transportation technologies is the introduction of plug-in hybrid electric vehicles (PHEVs), which utilize batteries that can be recharged by connecting to an electric power source (see Chapter 12). Unlike traditional hybrid vehicles, which use both an electric motor and an internal combustion engine, PHEVs will both draw power from and potentially provide power to the electric grid. PHEVs can substantially reduce fuel costs, air pollution, dependence on petroleum, and greenhouse gas emissions, and can even provide emergency backup power for home-owners.

The traditional gas and electric utility business model in a regulated environment has been based on highly centralized assets, capital investments in infrastructure maintenance, and a focus on meeting demand through efficient and reliable business performance. With the advent of deregulation and distributed energy sources, many have envisioned future utility systems that are more decentralized and interact more closely with customers, with a focus on enhancing revenues while providing innovative services and technology [4]. Whether or not carbon policies are imposed, the electric utility industry will need to invest as much as $2 trillion by 2030 in new generation, transmission, and distribution capacity [5]. Similarly, the traditional liquid fuel industry will likely be transformed as supply chains shift from a focus on petroleum to a variety of bio-based feedstocks.

One emerging technology that will contribute to "greening" of electric power is *smart grid* software. Today's electric grid systems, based on centralized power generation and dispatch, are vulnerable to service disruptions and bottlenecks, and require standby capacity to meet peak demand in local areas. Smart grid systems use Internet connections between power stations, power meters, and appliances to create a two-way flow of information about the state of the grid between utilities and consumers. Thus, utilities can be more responsive to changes in load, and installing smart meters at homes and businesses will enable consumers to make better decisions about their power usage. Smart grid systems enable a more decentralized network model where excess power can be transmitted from one area to another as needed. In addition, energy storage devices such as plug-in hybrids and power generating devices such as residential solar panels can supply energy back to the grid during hours of peak demand. For example, Austin Energy, a Texas power utility, has been working on an initiative to replace all its power meters with smart meters by December 2008.

This chapter highlights the DFE-related strategies of two major U.S. energy companies that are responding proactively to the economic and environmental challenges of the twenty-first century. American Electric Power and Chevron have both demonstrated a willingness to engage in open dialogue about these challenges and to investigate next-generation technologies, while remaining committed to delivering value in their core business.

American Electric Power: Keeping the Lights On
The Challenges of Sustainability

American Electric Power, based in Columbus, Ohio, is one of the oldest and largest electric utilities in the United States, serving more than 5 million customers in 11 states. AEP owns nearly 36,000 megawatts of generating capacity in the U.S., and operates the nation's largest electricity transmission system, a nearly 39,000-mile network that includes more 765 kilovolt extra-high voltage transmission lines than all other U.S. transmission systems combined.

AEP is primarily a coal-burning utility and takes a candid approach toward addressing the sustainability of fossil fuel combustion. It was among the first U.S. electric utilities to acknowledge the problem of global warming. According to Chief Executive Officer, Michael Morris: "Electricity is necessary for a modern society, yet its very production has adverse impacts on society. AEP produces more greenhouse gases than most electric companies in the United States, so we have an increased responsibility to be part of the climate change solution, internationally, nationally, and locally."

Accordingly, the company has been investing in "cleaner" advanced coal technologies, including carbon capture and sequestration, ultrasupercritical pulverized coal, and Integrated Gasification Combined Cycle (IGCC). In addition, AEP is diversifying its energy generation portfolio; in fact, the company by the end of next year will own, operate or have purchased almost 2000 megawatts of wind, or more than 5% of its total generating capacity. At the same time, AEP has been active in promoting sensible legislation for climate change mitigation. This has included public support for the Bingaman-Specter bill in 2006 and the Boucher-Dingell draft bill in 2008—both cap and trade climate bills with sizeable greenhouse gas reductions.

AEP's 2007 sustainability report sets forth a number of key areas that the company will address in the coming years, including climate change; environmental performance; public policy; energy security, reliability and growth; work force issues; stakeholder engagement; and leadership, management, and strategy [6]. For example, the company has set itself the following challenges and goals:

- Achieve 1,000 MW reduction in demand by 2012 with 15% coming from AEP actions (such as operating efficiency) and 85% from customer programs.
- Work with its supply chain partners, including coal suppliers, on improving environmental, safety, and health performance.
- Work constructively to help develop a Federal cap-and-trade program for greenhouse gas emissions that does not unfairly harm the U.S. economy or coal utility customers.
- Implement strategic actions to reduce carbon emissions by about 5 million metric tons per year as new generating plants come online beginning in 2010.
- Develop a diverse portfolio of generating technologies to assure a secure energy future, including addition of 1000 MW of wind power by 2011.

High-Voltage Transmission Line Design

An example of AEP's adherence to sustainability principles was the design of a new 765 kV transmission line, known as the Wyoming-Jacksons Ferry line, which runs through a 90-mile corridor in Virginia and West Virginia. In March 1990, Appalachian Power announced a major transmission reinforcement for this service area to address a growing deficiency in its electric transmission grid. Growing customer demand was creating a situation where the loss of a major line during peak periods could pose a risk of transmission collapse. The last transmission line to serve the area was built in 1973, and peak consumption in that year was 2,720 megawatts. In 2005, the load in this same area was 7,108 megawatts, or more than 161% greater than the 1973 load.

Higher-voltage power lines are more efficient, since they reduce power loss. 765 kV transmission requires less land for rights of way than would be used for the number of lower-voltage transmission lines necessary to carry the same amount of electricity. Specifically, one 765-kV transmission line on a 200-foot wide right of way can transmit electricity equivalent to the capacity of 15 138-kV double circuit transmission lines that would need 1,500 feet of right of way. This new line increased AEP's national 765 kV network to 2,100 miles, more than all other U.S. electric utilities combined. AEP pioneered 765 kV electricity transmission in the early 1960s and put the first 765 kV line in service in 1969.

Completed in 2006 after sixteen years, the $306 million Wyoming-Jacksons Ferry project was the largest ongoing electric transmission infrastructure project in the United States. The design of the new line involved a combination of route selection and technology selection. The power line route was developed by a team of professors from Virginia Tech and University of West Virginia with expertise in biology, anthropology, geographic information systems, and natural resource management. Their goal was to determine the route with the least impact on people and the environment. Technology, ingenuity, public input, and a commitment to people and the environment were combined to create the best alternative possible.

As a result of the careful route planning, only five homes were within the final 200-foot wide right of way, only six eminent domain proceedings were held out of 164 landowners, and only 11 miles of federal lands were crossed. In addition, nonreflective steel and conductor were used to minimize visual impacts of the project. Moreover, the company took extraordinary measures to minimize the environmental impacts. For example:

- In the summer of 2005, a male endangered Indiana bat was found near a county where construction was taking place. To protect the species, construction and tree clearing activities were suspended during the summer for a five-mile radius in the area where the bat was found.
- Helicopter use was prevalent in transporting tower structures or tower steel to construction sites. This helped reduce

wear and tear on roads, and in many cases, reduced the required size of access roads.

- AEP used selective right-of-way maintenance, only removing tall-growing incompatible species while leaving red buds and dogwoods and other low-growing species. In the future, the right-of-way will be maintained with a backpack application of herbicides only to incompatible species.

The final design of the Wyoming-Jacksons Ferry line included 333 transmission towers with an average height of 132 feet. A total of 4,750 tons of conductor were used, spanning 1,620 miles of wire. The line includes 9,876 spacer dampers, which hold the line in a perfect hexagon (see Figure 18.6). The design uses an innovative six-bundle conductor configuration, which reduces audible noise to about half that of earlier four-bundle configurations.

Industrial Ecology in Action

An unusual example of industrial ecology—turning wastes into feedstocks—has taken shape in Moundsville, West Virginia, thanks to a partnership between American Electric Power and CertainTeed, a subsidiary of St. Gobain. Gypsum is a key input to the manufacture of wallboard, and also happens to be a residual from flue gas desulfurization (FGD) in coal-fired electric power plants. To take

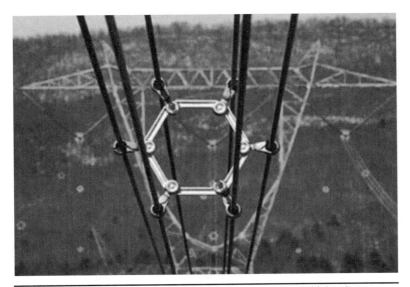

FIGURE 18.6   Hexagonal design of Wyoming-Jacksons Ferry high-voltage power lines.

advantage of this natural synergy, CertainTeed built a new wallboard manufacturing facility in close proximity to AEP's Kammer-Mitchell and Cardinal power plants. The plant began operation in 2008, and is capable of producing up to 800 million square feet of product a year.

AEP's commitment to produce a very consistent, high quality, synthetic gypsum was critical to CertainTeed's decision to locate its newest wallboard plant in Moundsville. The FGD systems at Cardinal and Kammer-Mitchell use both chemical and mechanical methods to remove sulfur dioxide (SO_2) from the flue gas produced during coal combustion. The SO_2 is absorbed into limestone slurry and then reacts with the calcium in the limestone to form a solid compound, calcium sulfate, also known as gypsum. CertainTeed receives the gypsum via a specially designed two-mile conveyor system that spans a four-lane highway to transport the product from Kammer-Mitchell directly into the wallboard plant's gypsum storage facility. Located in Brilliant, Ohio, the Cardinal Plant is nearly 40 miles from the wallboard factory, and the gypsum produced at Cardinal is transported to CertainTeed on the Ohio River by barge.

CertainTeed features state-of-the art technology and makes use of an environmentally sustainable approach in producing its ProRoc® line of gypsum wallboard. This wallboard is manufactured utilizing 96% recycled material, of which gypsum from Mitchell and Cardinal is the main component. An estimated 800,000 tons of gypsum will be used annually to produce wallboard at CertainTeed instead of being sent to landfills. In addition, the wallboard plant will utilize 99% recycled paper in the manufacturing process.

Chevron: The Power of Human Energy
Filling the Pipeline

Chevron is one of the world's largest integrated energy companies, headquartered in San Ramon, California, with annual revenues in 2007 of about $215 billion. The company has more than 65,000 employees in more than 100 countries, and is engaged in every aspect of the oil and natural gas industry, including exploration and production, manufacturing, marketing and transportation, chemical manufacturing and sales, geothermal, and power generation.

Despite negative perceptions about the petroleum industry, Chevron has made an effort to position itself as an environmentally responsible company and to reach out to stakeholders in addressing the challenges of climate change and energy efficiency [7]. Recognizing that worldwide energy consumption is projected to rise more than 50% by 2030, Chevron sees a continuing role for oil and gas companies to find new and better ways to deliver energy, including conventional crude oil and natural gas as well as alternative sources. Chevron's investments are oriented toward four strategic goals:

1. **Finding More Oil and Gas.** As mentioned earlier, fossil fuels account for about 85% of the world's energy supply. In 2007 Chevron invested $20 billion in exploration and production technologies aimed at increasing production of existing fields and drilling deeper to tap into previously unreachable resources.

2. **Using Energy Wisely.** Chevron Energy Solutions, a fast-growing business unit, is devoted to helping schools, government agencies, and businesses use energy more efficiently and reduce energy use. Since 1992, Chevron has increased its own energy efficiency by 27%, through operational improvements ranging from more efficient heat exchangers to new power plants.

3. **Developing Alternative Energy.** Chevron plans to spend $2.5 billion on renewable and alternative energy and energy efficiency projects between 2007 and 2009. For example, Chevron is already the world's largest producer of geothermal energy, powering 7 million homes in Indonesia and the Philippines.

4. **Meeting the Demand.** Chevron sees energy as a fundamental human need that improves quality of life, and is determined to make the right decisions today to meet the growing energy demand, geopolitical pressures, and environmental challenges of the future.

Developing the Next Generation of Biofuels

Chevron is investing across the energy spectrum to develop energy sources for future generations by expanding the capabilities of today's alternative and renewable energy technologies. Focus areas include geothermal power, biofuels, hydrogen and advanced batteries as well as application of wind and solar technologies. Chevron is the largest renewable energy producer among global oil and gas companies, producing 1,152 megawatts of renewable energy primarily from geothermal operations.

To develop alternative sources of energy, Chevron Technology Ventures has entered into a number of research and development partnerships with universities and laboratories. For example, since 2006 Chevron has been working with the U.S. Department of Energy's National Renewable Energy Laboratory (NREL), headquartered in Golden, Colorado, in a strategic research alliance to advance the development of renewable transportation fuels. The objective of the alliance is to collaborate on developing the next generation of process technologies that will convert cellulosic biomass, such as forestry and agricultural wastes, into biofuels, such as ethanol and renewable diesel. The first project initiated under the alliance involves bio-oil reforming, a process by which bio-oils derived from the decomposition of biological feedstocks are then converted into hydrogen and biofuels.

In 2007, Chevron and NREL announced a second collaborative project—investigating technology to produce liquid transportation fuels using algae. This initiative seeks to identify and develop algae strains that can be economically harvested and processed into finished transportation fuels such as jet fuel. Algae are considered a promising potential feedstock for next-generation biofuels because certain species contain high amounts of oil, which could be extracted, processed, and refined into transportation fuels using currently available technology. Other benefits of algae as a potential feedstock are their abundance and fast growth rates. Key technical challenges include identifying the strains with the highest oil content and growth rates and developing cost-effective growing and harvesting methods.

References

1. Energy consumption data are provided by the Energy Information Administration, U.S. Department of Energy.
2. Per capita energy consumption data are provided by the WRI Earthtrends database, earthtrends.wri.org.
3. Personal communication from Mike Long, President, Resource100, Columbus, Ohio.
4. P. Enkvist, T. Nauclér, and J. M. Oppenheim, "Business strategies for climate change," *McKinsey Quarterly*, April 2008.
5. M. W. Chupka, R. Earle, P. Fox-Penner, and R. Hledik, "Transforming America's Power Industry: The Investment Challenge 2010–2030," prepared by The Brattle Group for the Edison Foundation, Nov. 2008.
6. AEP's sustainability report is at www.aep.com/citizenship/crreport/.
7. More information is available at Chevron's website, www.chevron.com.

CHAPTER 19

Service Industries

Environmental Footprint of Services

While the majority of this book has focused on product development and manufacturing industries, service industries are just as important in determining the environmental footprint of our economic system. Service providers utilize products and vehicles, consume materials and energy, occupy space, and generate waste, just like manufacturing companies. According to the U.S. Department of State, services produced by private industry accounted for 67.8% of U.S. gross domestic product in 2006, with real estate and financial services, such as banking, insurance, and investment on top. Other leading categories of services are wholesale and retail sales; transportation; health care; legal, scientific, and management services; education; arts; entertainment; recreation; hotels and other accommodations; restaurants, bars, and other food and beverage services.

Service industries tend to be less natural resource-intensive than manufacturing industries, as shown in Figure 19.1.* From a life-cycle perspective, however, service industries are at the "top of the food chain" because they consume the products of other industry sectors, including material extraction, manufacturing, and power generation [1]. It follows that eco-efficient services can make a substantial difference in a company's environmental footprint. Many companies have begun using electronic communication services, such as telecommuting, teleconferencing, and virtual meetings, as a substitute for physical travel. But according to the Institute for Sustainable Communication those electronic services may also have significant environmental impacts, and companies should be aware of the indirect energy and material burdens associated with their communication supply chain [2].

In addition, service industries rely largely on people, yet most life-cycle studies do not account for the environmental footprint of

---

*This analysis is based on the life-cycle exergy assessment methodology described in Chapter 9, which quantifies natural capital flows in terms of available energy (solar equivalent joules).

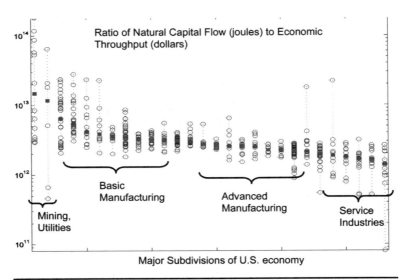

FIGURE 19.1   Average natural capital intensity for 488 sectors of the U.S. economy in 1997 [1].

human resources. Based on a study conducted by the Center for Resilience at The Ohio State University, as shown in Figure 19.2, the life-cycle environmental impacts of a typical household can add up considerably. Although the direct consumption of mass and energy is relatively small, the hidden flows of materials and energy are much larger. A household accounts for about 5000 metric tons per

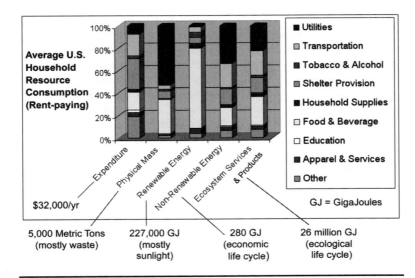

FIGURE 19.2   Household life-cycle footprint assessment.

year of material flow, mainly wastes that are discarded somewhere along the supply chain. Similarly, the energy expended in the economic supply chain to support a household is ten times greater than the direct energy purchased in the form of gasoline and electric power. When one accounts for the total energy derived from ecosystem services, including the large amount of solar energy embedded in biomass, the energy consumption of a household is a million times greater than the direct purchased energy.

DFE in Service Industries

At face value, fulfilling human needs by providing services rather than selling products appears to be less environmentally burdensome. Indeed, one of the largest opportunity areas for DFE, as discussed in Chapter 8, is Design for Servicization. Examples include transportation services such as Zipcar, telephone answering services, and leasing of office equipment. However, as illustrated by the above life-cycle assessment data, care must be taken to assure that the manner in which these services are provided is indeed environmentally efficient. A fair comparison of product vs. service should include the supply chain resources necessary to deliver an equivalent functional unit of value to the customer. As with any business, the same DFE design guidelines and metrics can be applied to improve the environmental performance of these supply chain processes.

One of the most important categories of service industries is transportation, which was addressed in Chapter 12. Note that there are significant differences in environmental performance between different modes of travel; for example, transporting a ton of freight by rail requires only a fraction of the energy of transport by truck. Likewise, electric utilities, discussed in Chapter 18, are effectively service providers, although their generation plants resemble manufacturing sites. Again, there are large differences among generating technologies in terms of cost, reliability, capacity, and environmental footprint.

More generally, many manufacturing businesses offer some form of services, especially when their markets include industrial customers. Depending on the business model, these services may generate additional revenue streams or may be simply bundled with the product to create additional value for the customer. Broadly speaking, there are at least three ways that service providers can incorporate DFE principles into the design of their business:

1. Design innovative services that replace traditional products (i.e., servicization).

2. Design services that improve customer environmental performance, such as energy efficiency consulting and technical assistance.

3. Design the service business model to utilize eco-efficient products and processes, including procurement, operations, and customer interface.

The following sections illustrate how these approaches have been put into practice in a variety of different service industry sectors.

Examples of Industry Practices

Retailing Industry[†]

By far the largest retailer in the world is **Wal-Mart**, with revenues approaching four times its nearest global competitors. Although often maligned for its questionable social impacts, Wal-Mart has embraced environmental excellence with a vengeance and, in doing so, has stimulated a wave of environmental innovation in consumer-oriented industries, such as electronics, clothing, and foods and beverages. The company has engaged its employees and suppliers in various networks aimed at generating value through improving product sustainability [3]. Examples of initiatives being driven by these networks include:

- **Sustainable Packaging**—A web-based packaging scorecard has been developed that utilizes nine sustainability metrics: GHG emissions, material value, product/package ratio, cube utilization, transport distance, recycled content, recovery value, renewable energy, and innovation. Wal-Mart is applying the tool to score all of its suppliers and is using the results as an input to purchasing decisions.

- **GHG Emission Reporting**—Wal-Mart is partnering with the Carbon Disclosure Project in an effort to measure the amount of energy used to create products throughout the supply chain and encourage suppliers to reduce GHG emissions. The partnership includes a pilot project with seven sectors (i.e., DVDs, toothpaste, soap, beer, milk, vacuum cleaners and soda products) to seek new and innovative ways to make the entire supply chain more energy efficient.

- **Chemical Intensive Products**—To help identify and reduce chemical risks, Wal-Mart is developing a list of "chemicals of concern" and working with suppliers to eliminate them from products. The company is also developing a product screening tool to identify potential chemical hazards and risks, and to help buyers and suppliers develop alternative products.

[†]Portions of this section were contributed by Nicole Gullekson, Eco-Nomics LLC.

- **Electronic Products**—An Electronics Scorecard is being developed to rate suppliers' products in terms of energy efficiency, durability, ease and ability to upgrade, compliance with the European RoHS directive (see Chapter 3), end-of-life solutions, and packaging mass. The company is also sponsoring "Take Back" days for electronic wastes, and is working with suppliers to help develop more energy-efficient products.

- **Forest Products**—Wal-Mart is working with several nonprofit organizations to reduce wasted paper and packaging from products, achieve transparency of the wood supply chain, and eliminate the use of illegally sourced wood. The company has also worked with suppliers to develop "extended roll life" products that eliminate plastic wrapping from individual toilet paper or paper towel rolls by selling them as a multiple unit package.

- **Textile Products**—a Textile Scorecard tool has been developed to evaluate the sustainability of textile products, including packaging, GHG emissions, and the impact of dyes. Wal-Mart is the largest buyer of organic cotton goods and is also exploring alternative fibers, such as bamboo and recycled yarn. The company is also developing a closed-loop recycling program for textiles and is working with dye houses to encourage the use of lower toxicity dyes.

- **Global Logistics**—Wal-Mart is an EPA SmartWay Transport Partner and works with major truck manufacturers to help develop diesel hybrid trucks and aerodynamic trucks. Also, Wal-Mart installed diesel-powered Auxiliary Power Units on all trucks that make overnight trips, providing electrical power for heating and communication systems without idling the engine.

Wal-Mart is continuing its sustainability efforts with a new initiative in China, announced at a summit meeting in October 2008 in Beijing and involving more than 1,000 leading suppliers, Chinese officials, and nongovernmental organizations. The company committed to partner with suppliers to improve energy efficiency by 20% in its top 200 supplier factories by 2012, and to hold suppliers accountable for regulatory compliance, transparency, and environmental and social responsibility. Wal-Mart also plans to design and open a new store prototype that uses 40% less energy, to reduce energy use at existing stores 30% by 2010, and to cut water use in all of its stores by 50%.

Other major retailers have launched similar types of programs. For example, in 2007 **JCPenney** established a Corporate Social

Responsibility program under the leadership of Vice President Jim Thomas. The program has five major aspects—Community, Associates, Responsible Sourcing, Environment, and Sustainable Products—summarized by the acronym JCPenney CARES [4]. Among the targets that the company has established are:

- **Responsible Sourcing**
    - Develop an environmental scorecard for private brand suppliers and integrate it into the company sourcing system by 2010
    - Develop and integrate water quality targets for private brand suppliers' mills and laundries into the sourcing system by 2010

- **Environment**
    - Achieve ENERGY STAR certification for at least 200 stores
    - Complete two LEED-certified stores by 2009
    - Ensure 100% of JCPenney facilities are recycling waste in 2008
    - Increase purchase of certified forest products by 5% per year through 2011

- **Sustainable Products**
    - Be a recognized source for eco-friendly products
    - Offer reusable shopping bags for sale in JCPenney stores in 2008 (see below)
    - Reduce packaging weight by 2% by 2010

JCPenney has developed and launched Simply Green™, an exclusive designation that assists customers in making environmentally conscious purchases. A wide range of JCPenney private brand merchandise—from apparel to home accessories—will bear the Simply Green mark, highlighting merchandise that lessens the impact on the environment. To qualify for the designation, merchandise must fall into one of the following three categories:

1. **Organic**—made from at least 70% raw materials, such as organic cotton or linen, which have been grown without chemical fertilizers or pesticides.
2. **Renewable**—made from at least 25% renewable materials, such as bamboo, sorona, ingeo, soy, capiz shells, or wood that comes from certified, well-managed forests. These materials are replenished by natural processes at a rate comparable to, or faster than, their rate of consumption.

3. **Recycled**—containing at least 25% recycled materials, such as recycled cotton, recycled glass (home products), or recycled polyester made from plastic bottles. These materials help reduce the amount of waste sent to landfills.

In August 2008, JCPenney introduced a reusable shopping bag bearing the Simply Green mark called the EcoBag and sold out of the initial order of 350,000 in about 2 months (see Figure 19.3). At the same time, the company introduced recycling bins for plastic bags, no matter what their origin, in all its stores.

Other retail chains have developed similar "green designations"; for example, **Home Depot** launched its Eco Options mark in 2007. The program identifies over 3000 environmentally preferred products in five categories: Sustainable Forestry, Energy Efficiency, Water Conservation, Clean Air, and Healthy Home. That same year, the British-based retail and grocery chain Tesco launched a Greener Living brand and a website that not only identifies green products but also includes a host of sustainability initiatives, as well as information for consumers about sustainable lifestyles. It should be noted that these in-house marks are different from the eco-labels discussed in Chapter 3; the latter generally require independent third-party certification.

Food and Beverage Services

McDonald's, the global fast food restaurant chain, was one of the first food service companies to explore environmental sustainability, starting with their engagement with the Environmental Defense Fund (EDF) in 1990. The company has adopted a life-cycle approach to waste minimization, stressing avoidance, recycling, and proper disposal in that order. The company's waste avoidance efforts focus

Figure 19.3 JCPenney has introduced a variety of merchandise bearing the simply green mark, as well as a reusable "eco-bag."

on minimizing the environmental footprint in the sourcing and design of food packaging [5]. For example:

- McDonald's strives to source raw materials for paper-based consumer packaging from well-managed forests and is developing a comprehensive forestry policy. For example, cartonboard for sandwich clamshells in Australia is sourced from Forest Stewardship Council accredited suppliers. In Germany, France, and the United Kingdom, nearly 57% of the paper fiber for McDonald's packaging comes from certified forests.

- Approximately 82% of the consumer packaging used in McDonald's nine largest markets is made from renewable materials (paper or wood-fiber), and approximately 30% of the material comes from recycled fiber. Recycled paper is used in trayliners, napkins, bags, sandwich containers (see Figure 19.4), and other restaurant items such as shipping containers. In 2007, McDonald's purchased almost $530 million in recycled content in the United States alone.

- In 2007, McDonald's voluntarily phased out the coating on some food packaging that could produce perfluorooctanoic acid (PFOA), a chemical shown to persist in the environment. By the end of 2007, the phase-out was completed for the majority of packaging items in all four major geographic areas of the world, and it was completed in the first quarter of 2008.

- McDonald's Europe achieved a nearly 2,000 ton per year reduction in the consumption of nonrenewable materials through the launch of a paper salad bowl and wooden coffee stirrer. Redesign of the McFlurry spoon eliminated 286 metric tons of polypropylene plastic, and 423 metric tons of paper materials were saved through enhancements to the Big Mac carton.

- McDonald's Australia has incorporated 35% post-consumer recycled PET plastic in cold beverage and dessert cups, re-

FIGURE 19.4   Over 80% of McDonald's consumer packaging comes from renewable materials, with about 30% recycled content.

ducing the amount of virgin plastic resin needed. A 915,000 pound reduction in packaging weight was achieved through implementation of the company's carton board reduction strategy.

- McDonald's USA reduced the weight of the 32-ounce polypropylene cup, saving 650 tons of resin per year. The F-Flute containers (see Figure 19.4) for the Big Mac, Filet-O-Fish, and Quarter Pounder with Cheese now weigh 25% less and incorporate a minimum of 46% post-consumer recycled content and 71% unbleached fiber. The recycled content in these containers saves more than 23,000 tons of wood, equivalent to 161,000 trees, according to the EDF Paper Calculator.

Similarly, **Starbucks** introduced hot beverage paper cups made with 10% post-consumer recycled content into its U.S. and Canadian stores in 2006. These cups were the first direct contact food packaging containing post-consumer recycled content to receive a favorable safety review by the U.S. Food and Drug Administration. The company estimates that this will yield annual savings of 11,300 tons of wood (about 78,000 trees), 58 billion BTUs of energy (enough to supply 640 homes for a year), 47 million gallons of wastewater, and 3 million pounds of solid waste.

Package Delivery Services

In an economy that emphasizes speed, package delivery has become an indispensable service. The availability of rapid, affordable delivery combined with electronic commerce has enabled innovative business models in a number of industries, such as computers (e.g., Dell) and books (e.g., Amazon).

UPS, formerly United Parcel Service, has evolved unimaginably from its origins in 1907 as a messenger service, with deliveries made on foot or on bicycles. Today, UPS is a global powerhouse employing 425,000 people operating airplanes, ground vehicles, logistics facilities, call centers, data centers, and retail stores. UPS believes in "connecting the world responsibly" and has made sustainability a corporate priority [6]. Examples of environmental sustainability initiatives include:

- **Ground and Air Fleets**
  - UPS continues to develop and use sophisticated aircraft routing technology to improve both fuel efficiency and environmental performance.
  - UPS operates the largest private "green" fleet in the transportation industry, and continues to enhance the fleet by testing and deploying hybrid, compressed natural gas and propane-powered delivery vehicles.

- ○ UPS employs telematics, i.e., technology to collect and analyze data from delivery trucks, to help identify ways to reduce energy and emissions while improving driver safety and customer service.

- **Infrastructure**
  - ○ UPS has introduced energy-saving innovations in vehicle dispatching, ranging from simply avoiding left turns to sophisticated "package flow technology" that uses geographic tools to analyze and edit dispatch plans. During 2007 UPS reduced U.S. delivery routes by 30 million miles, which saved 3 million gallons of fuel and reduced CO_2 emissions by 32,000 metric tons.
  - ○ The company is exploring alternative and renewable energy sources such as solar energy. For example, use of roof-top solar panels in the Palm Springs, Calif., sorting facility has reduced CO_2 emissions by 1 million pounds since deployment in 2003.

- **Recycling**
  - ○ The company has recycled electronic equipment since 2000, totaling 22.3 million pounds to date, with 2.65 million pounds in 2007.
  - ○ In 2007, UPS recycled 45,400 tons of solid waste materials, including metals, plastics, paper, corrugated materials, pallets, and wood waste. Additionally, UPS purchased 159,100 tons of materials with recycled content.

FedEx, formerly Federal Express, was founded in 1971 as an airborne delivery service, grew rapidly through acquisitions, and in 1998 acquired RPS, a major ground package carrier now designated as FedEx Ground. Today, its global network of companies employs about 290,000 people. FedEx strives to integrate sustainability into every aspect of its operations [7]. The following are highlights of its accomplishments:

- **Fuel Efficiency**
  - ○ By rebalancing the airplane fleet and optimizing routes, FedEx Express improved total fleet miles per gallon within the United States by 13.7% from 2005 to 2007, saving 45 million gallons of fuel and 452,573 metrics tons of CO_2 emissions.
  - ○ FedEx is upgrading its fleet by replacing 90 narrow-body Boeing 727 aircraft with Boeing 757 planes, which reduce fuel consumption up to 36% while providing 20% more

capacity. This aircraft replacement program is expected to eliminate more than 350,000 metrics tons of CO_2 emissions annually.

o FedEx wide-body planes with flight management systems use continuous approach descent, which keeps the plane in idle during the descent, reducing engine thrust and fuel. Also, ground support equipment at FedEx operations at select airports has been converted from internal combustion engine models to electric units, which saves almost 1 million gallons of jet fuel per month.

o The company has set a goal to reduce greenhouse gas emissions from FedEx Express global air operations by 20% per available ton mile by 2020.

o As a result of route optimization, more than one-fourth of the FedEx fleet has been converted to smaller more fuel-efficient vehicles, saving more than 50 million gallons of fuel.

- **Material Efficiency**

  o FedEx has extensive recycling programs at its operating companies, and, in 2008, recycled a total of 17.6 million pounds of waste materials including paper, metal, plastic, wood, and electronic waste.

  o FedEx has worked to minimize the environmental impact of packaging by using recycled content and maximizing recyclability; for example, most FedEx envelopes are made from 100% recycled content.

  o The FedEx Packaging Lab came up with an innovative alternative for a commonly used Chinese cushioning material—expanded polystyrene foam. Using environmentally preferable honeycomb-style packaging with corrugated pads achieved more effective packaging at the same cost.

- **Alternative Energy**

  o FedEx worked with Environmental Defense Fund to design and specify a hybrid-electric truck that improves fuel economy by 42%, reduces greenhouse gas emissions by 25% and cuts particulate pollution by 96%. FedEx currently operates the largest fleet of commercial hybrid trucks in North America, consisting of more than 172 vehicles.

  o In August 2005, FedEx Express, an express transportation subsidiary of FedEx, activated California's then-largest corporate solar power installation at its Oakland hub.

This 904 kilowatt system produces the equivalent of the power required to supply more than 900 homes, and meets up to 80% of the hub facility's peak energy demand. The 81,000 square feet of roof space at the facility is covered with more than 5,700 solar electric panels, which also help insulate the buildings.

Financial Services

Apart from the environmental footprint of their physical operations, financial service industries have a broad impact on society and the environment through their investment, insurance, and lending practices. Financial institutions have the power to take into account environmental factors such as climate change risks in their lending or investment strategies, and thus influence decision making on the part of consumers and all types of businesses, large and small. During the 2008 U.S. economic collapse, many financial services firms experienced severe losses, and some did not survive. Nevertheless, the fundamental principles of environmental responsibility will continue to have a strong influence on business practices in this sector.

Citigroup was one of the companies damaged by the 2008 mortgage foreclosure crisis, but its approach to corporate citizenship and sustainability has been extremely progressive. An international financial conglomerate with operations in consumer, corporate, and investment banking, as well as insurance, Citigroup employs about 300,000 people around the world. Its environmental efforts include several major thrusts [8]:

- Reducing its environmental footprint through sustainable buildings and consolidation of data centers. For example, Citigroup's Frankfurt data center uses up to 25% less energy, saving more than 16,000 MWh per year, as well as 11,750 metric tons of CO_2, 46.5 million liters of water, and 129 metric tons of water treatment chemicals. Citigroup's 15-story office building in Long Island, New York (see Figure 19.5) includes eco-design features such as storm-water recycling, energy-efficient fixtures, and emphasis on natural lighting.

- Reducing overall GHG emissions by 10% from 2005 to 2011 through adoption of "best practice" energy saving measures across its global operations. The company hopes to save as much as $1 per square foot in energy costs, yielding savings of almost $100 million annually.

- Implementing sustainable procurement policies that emphasize recycled content; for example, 33% of the 2007 expenditures were for recycled products, and 63% of the 2007 paper purchased contained recycled content.

FIGURE **19.5** Citigroup's green office tower in Queens, New York.

With respect to its business operations, Citigroup has developed an overarching Environmental and Social Risk Management Policy to help address sustainability considerations from a credit risk as well as reputational risk perspective. The policy draws upon a variety of codes and standards, including the Equator Principles for project financing, in which Citigroup played a leadership role. These principles call for projects to undergo a Social and Environmental Assessment, that covers a broad variety of issues, including: sustainable development and use of renewable natural resources; protection of human health and cultural properties; biodiversity; endangered species and sensitive ecosystems; use of dangerous substances; major hazards; occupational health and safety; fire prevention and life safety; socio-economic impacts; land acquisition and land use; involuntary resettlement; impacts on indigenous peoples and communities; cumulative impacts of existing and anticipated future projects; participation of affected parties in the design, review and implementation of the project; consideration of feasible environmentally and socially preferable alternatives; efficient production, delivery and use of energy; pollution prevention and waste minimization; pollution controls; and solid and chemical waste management.

The financial service industries have introduced a number of other sets of sustainability principles. For example, the Carbon Principles is a set of best practices to help banks manage the lending risks associated with coal projects. **Bank of America**, which adopted these principles, announced in 2008 that it would phase out loans to coal mining companies that use mountaintop extraction as their primary means of production. Another example is the Climate Principles, a global framework introduced in December 2008 to aid banks and insurers in managing climate change-related risks in their products and services.

Tourism and Recreational Services

The travel, tourism, leisure, and entertainment industries have responded to both customer interest and broader social responsibility concerns by adopting a host of environmentally friendly policies and practices. These range from energy and water conservation programs at hotels to specially designed eco-tourism adventure trips in exotic settings. According to the World Tourism Organization, "sustainable tourism" involves management of all resources in such a way that economic, social, and aesthetic needs can be fulfilled while maintaining cultural integrity, essential ecological processes, biological diversity, and life support systems [9]. In other words, tourism can benefit local communities by creating employment and income while remaining ecologically and culturally sensitive.

One entertainment company that was an early adopter of sustainability principles is the **Walt Disney Company**. In 1990, the company formed an Environmental Policy Division, which focuses on the education and maintenance of six key priorities: climate protection, energy conservation, green purchasing, waste minimization, water conservation and wildlife conservation. In addition, Disney introduced their environmental brand: Environmentality™, representing a fundamental ethic that blends business growth with the preservation of nature [10]. Examples of Disney initiatives include

- **Green Standard.** The Green Standard is a global employee program rolled out in 2008 that encourages every Cast Member and employee to reduce Disney's operational impact on the environment by adopting specific environmentally friendly behaviors at work, in meetings, while planning events, during travel, and while dining during work hours. The standard is divided into five phases: Green Workspace, Green Meetings, Green Events, Green Travel, and Green Dining. Employees are encouraged to recycle wastes, eliminate use of individual plastic water bottles, use double-sided copying and recycled-content paper for everyday printing and copying, turn off unnecessary lights, and minimize driving alone. They are also encouraged to go "Above and Beyond" the standard, and have access to an extensive, online resource center that provides guidance on environmentally beneficial actions. To support employee efforts, Disney is partnering with Hewlett-Packard® to launch the HP Document Output

Management program, replacing printers, copiers, and fax machines with multifunctional units that automatically default to duplex printing, resulting in potential paper savings. The new devices are also energy efficient, with the potential to save the Company nearly $800,000 annually. In addition, business units have access to bulk water dispensers and water filtration systems, are installing light sensor technology in buildings, and are able to use 30% post-consumer recycled content paper at a comparable price to virgin paper.

- **Waste reduction.** Disney's worldwide operations have recycled more than 925,000 tons of materials since 1991. Tokyo Disneyland Resort continues to lead the Company in overall waste diversion totaling 64%. In the United States, Disney earned the EPA WasteWise 2007 Gold Achievement Award for Paper Reduction, largely as a result of transitioning the entire Disney Catalog to an online shopping experience, which eliminated 1.8 billion sheets of paper. Disney has also established electronic waste recycling channels to provide cost-effective services to business units nationwide, so that central processing units, monitors, keyboards, stereos, and other devices can be collected and safely recycled. Electronics that cannot be refurbished or salvaged for repair parts are dismantled via industrial grinders that shred materials and separate metal and plastic components, which are then sold to industry. Meanwhile, Disney Music Group is using more recycled materials and a lot less plastic in media packaging; some compact discs are being released in the new CDVU+ format, which replaces traditional, physical packaging and booklets with an on-disc multimedia experience including photos, videos, and lyrics.

- **Energy Savings.** Disney is pursuing a wide variety of initiatives worldwide to cut electricity use and to support alternative fuels and renewable energy markets. For example, Walt Disney Parks and Resorts has a number of energy conservation initiatives in place:

  - Walt Disney World Resort has been working to reduce energy consumption related to electricity, natural gas, and hot and chilled water by 5% property-wide. The program, called "Strive for Five," has saved the company millions of dollars and reduced related greenhouse gas emissions. For example, an Energy Management System controls the heating and air conditioning systems and monitors dampers, humidity, and temperature set points, in order to minimize energy consumption while maintaining desired comfort levels. Also, Cast Members replaced neon lights with light emitting diodes (LEDs) in many signs and decorations including numerous holiday installations.

o In 2006 Disneyland Resort invested in a Central Energy Plant and permanently eliminated eight combustion sources, all of which produced nitrogen oxide, a major contributor to smog. The Hotel also replaced two old boilers with new state-of-the-art, high efficiency water heaters that burn natural gas. This renovation is projected to reduce annual natural gas usage by as much as 200 million cubic feet, equivalent to a full year's usage of approximately 1,500 homes, and will reduce annual CO_2, CO, NO_x emissions by 11,000, 240, and 30 tons respectively.

o At Tokyo Disney Resort, a Central Energy Plant provides energy to all facilities of Tokyo Disney Sea Park, and collects and reuses waste heat to produce steam with help from a boiler. Power is generated using a gas turbine system fueled by municipal gas, providing approximately 16% of the power required. Likewise, the Tokyo Disneyland Central Energy Plant was rebuilt to increase energy-efficiency, eliminating the need for local air compressors and leading to a 10% decrease in air-conditioning-related energy consumption.

A more recent adopter is **Hilton Hotels Corporation**, which in 2008 announced short and long term goals for building sustainability into the core fabric of its businesses. By 2014, the worldwide hotel chain aims to reduce direct energy consumption, CO_2 emissions, and solid waste by 20%, and water consumption by 10%. In addition, the company will focus on high-impact areas, such as sustainable building design, construction, and operations and renewable energy infrastructure for hotels and corporate offices. Already, in the European region, energy and water consumption were reduced by 10% from 2005 to 2007, and in the United Kingdom and Ireland, introduction of carbon-free electricity has reduced CO_2 emissions in participating Hilton hotels by more than 56%. In the United States, Hilton was the first in the hotel industry to install a commercial fuel cell system atop the Hilton New York.

References

1. N. U. Ukidwe and B. R. Bakshi, "Flow of Natural versus Economic Capital in Industrial Supply Networks and Its Implications to Sustainability" *Environmental Science & Technology*, 2005, 39, 24, 9759–9769.
2. For more information about sustainable communication, see www.sustainable-communication.org/.
3. Information about Wal-Mart's sustainability programs is available at walmart-stores.com/Sustainability/.
4. Information about JCPenney's corporate social responsibility programs is available at www.jcpenney.net/social_resp/default.aspx.
5. More information about McDonald's environmental programs is available at www.mcdonalds.com/usa/good/environment.html.
6. Information about UPS's sustainability programs is available at www.sustainability.ups.m/.
7. Information about FedEx's DFE programs is available at about.fedex.designcdt.com/corporate_responsibility/the_environment.
8. Information about Citigroup's environmental programs is available at www.citigroup.com/citi/environment/.
9. For more information ab.out sustainable tourism, see www.unwto.org/sdt/.
10. Information about Disney's environmental programs is available at disney.go.com/disneyhand/environmentality/environment/index.html.

PART 4

Conclusion:
The Road Ahead

CHAPTER 20

Sustainability and Resilience

In a world that is getting hot, flat, and crowded, the task of creating the tools, systems, energy sources, and ethics that will allow the planet to grow in cleaner more sustainable ways is going to be the biggest challenge of our lifetime.

THOMAS FRIEDMAN [1]

The Outlook for Sustainable Growth

This book began with a simple premise—that environmental sustainability is compatible with economic growth. Experience has shown that environmental excellence actually can be *synergistic* with corporate goals of profitability and shareholder value creation. It is clear that the changing business landscape has made environmental performance an important value driver, and that many companies are retooling their business processes to incorporate life-cycle thinking. In particular, Design for Environment is becoming an essential component of product and process development. It is fair to say that environmentally sustainable business practices will increasingly become a prerequisite for competitiveness in most industry sectors.

The positive response of the business community is encouraging. In fact, if all companies around the world were to adopt current best practices with regard to corporate sustainability, it might be possible to gradually reduce the enormous flows of material and energy that are required to support the expanding global economy. However, this optimistic scenario does not seem to be unfolding. Rather, it appears that global atmospheric concentrations of greenhouse gases will continue to rise, and that environmental resources will continue to be depleted or degraded, with potentially catastrophic consequences for future generations (see Chapter 3). For

357

example, one worrisome side effect of increasing atmospheric levels of carbon dioxide is ocean acidification, which could cause corals to go extinct. If corals cannot adapt, the cascading effects in reef ecosystems will reduce global biodiversity and threaten food security for hundreds of millions of people dependent on reef fish [2].

A 2007 study, published by the National Academy of Sciences, showed that global CO_2 emissions from fossil-fuel burning and industrial processes have been accelerating, with their growth rate increasing from 1.1% per year for the decade 1990–1999 to more than 3% per year for the period 2000–2004 [3].The observed rise in worldwide greenhouse gas emissions since 2000 can be attributed to increases in both the energy intensity of production as well as the carbon intensity of energy generation (see Chapter 18), coupled with continuing increases in population and per-capita gross domestic product (GDP). Not surprisingly, the growth rate in emissions has been strongest in rapidly developing economies, particularly China. Although the economic recession of 2008–2009 caused a temporary lull, the long-term pattern is alarming.

The overall ecological burdens of global economic growth can be understood from the following equation, which is a generalization of the well-known Kaya identity [4].

Total burden = population × (\$GDP/capita) × (resources/\$GDP) × (burden/resource unit)

The same equation holds whether the resources are fossil fuels and the burdens are greenhouse gas emissions, or whether the resources are material flows and the burdens are ecosystem service degradation. The first two factors are inexorably rising; and even if population growth slows, the GDP per capita will most likely continue to rise in developing nations. Yet scientific projections indicate that we need to sharply reduce our overall emissions and waste in order to stabilize atmospheric CO_2 concentrations and protect natural capital. Therefore, the focus of sustainability strategies needs to be on the latter two factors:

1. **Resource intensity** (resources/\$GDP) can potentially be reduced by decoupling material and energy throughput from economic growth. This is as much a behavioral challenge as it is a technological challenge. Despite improvements from 1970 to 2000, resource intensity seems to be flattening out, and could even begin to rise again as personal wealth increases in developing nations. Dematerialization strategies (see Chapter 8) are the best avenue for achieving further reductions in this factor.

2. **Burden intensity** (burden/resource unit) can potentially be reduced through process innovations; for example, the

carbon intensity of energy consumption will decline if we can scale up the use of biofuels or carbon sequestration (see Chapter 18). Likewise, the waste generated per unit of material throughput will decline if we can achieve greater eco-efficiencies through product life-cycle management (see Chapter 10).

The implication is that we need to seek disruptive innovations in both production and consumption of goods and services in order to drastically reduce our material and energy requirements. For developed countries, this could mean significant lifestyle changes, but not necessarily diminished quality of life. It is conceivable that a shift toward smaller-scale distributed production, reduced reliance on motorized transportation, denser living communities, more modest consumption patterns, and reduced waste generation might actually result in a less stressful and more healthful lifestyle. For developing countries, sustainable growth would imply a nontraditional pattern of growth that favors highly efficient "clean" technologies, with an emphasis on social equity and inclusion. Arguably, much can be accomplished simply by scaling up existing technologies, such as alternative energy sources and green buildings [5].

Achieving sustainable growth will require global collaboration on an unprecedented scale aimed at public education, environmental policy, and innovation. Companies will need to push the boundaries of their DFE efforts beyond the individual enterprise, working with customers, suppliers, competitors, and other interested parties. Governments will need to become more innovative in developing policies and strategies for large-scale infrastructure systems—urban systems, water resource systems, regional transportation systems, and energy distribution systems. To avoid the paralysis of parochial debate and traditional lobbying, we will need to form joint industry-government task forces and public-private partnerships. Actually, a President's Council for Sustainable Development was convened in the United States during the 1990s with high-profile industry participation, but it accomplished little in terms of genuine change. Perhaps a greater sense of urgency will yield more meaningful results in the future.

To address these challenges, the United Nations has set in motion the Marrakech Process, a series of global initiatives through which countries are working towards a 10-Year Framework of Programmes on Sustainable Consumption and Production. Seven government-led Task Forces have been created to carry out activities at national or regional levels to help accelerate a shift to more sustainable consumption and production patterns throughout the world. These voluntary teams of experts from government ministries, regional organizations, academic research institutes, technical agencies and UN bodies are attempting to tackle pressing problems in innovative ways. They focus their work in seven specific areas: sustainable products, life-styles, education, building and construction, tourism, public procurement, and cooperation with Africa.

The Need for Systems Thinking

The greatest challenge to achieving environmental sustainability may be the tendency of business and government leaders to deal with issues piecemeal rather than striving for a more holistic perspective. The Kaya logic above, while it provides helpful insights, is a perfect illustration of the prevalent linear, reductionist, and incremental approach toward analyzing sustainability opportunities. As a result, sustainability policies and practices have focused mainly on *reducing unsustainability* rather than strengthening the systemic underpinnings of sustainability [6]. Indeed, most of the company programs discussed in Part 3 of this book are directed largely at reducing environmental burdens, measured in terms of resource consumption and waste emissions. Little is understood about the broader impacts of these material and energy flows, or about the qualitative differences among sustainability conditions in different social and economic settings. Organization learning expert Peter Senge is a strong advocate of systems thinking because considering the whole system may reveal breakthrough opportunities that are not evident when one is busy optimizing the individual parts of the system [7]. One example is Dow AgroSciences' innovative Sentricon™ system, which achieved a 10,000-fold reduction in pesticide volume by rethinking how termites could be detected and controlled (see Chapter 18).

Systems thinking is being practiced in many parts of the world by ecologists and city planners who are teaming up to develop sustainable communities. Hammarby Sjöstad, a suburb of Stockholm, Sweden, is an example. The target of this development was to build affordable homes that use half the energy and water that a conventional Swedish home did in the early nineties. Rather than homes operating independently, heat and power are provided by a central plant. Combustible waste is sucked through a system of tubes, rather than being collected by trucks, and burned in a combined heat and power plant to provide electricity and heat. A dedicated wastewater treatment plant generates biogas from sewage and uses it to power local buses. Even warm wastewater is made to yield its energy, which is then used for space heating. A similar project in Abu Dhabi is developing Masdar City, a "green" community in the desert with 50,000 residents, which will utilize advanced energy and transportation technologies to achieve zero net carbon emissions and zero waste.

While "systems thinking" sounds good in principle, it is not easy. Many analysts are tempted to model complex systems from a static perspective, as if they were in "equilibrium." In truth, the ecosystems and industrial systems that we try to "manage" are dynamic, open systems operating far from equilibrium, exhibiting nonlinear and sometimes chaotic behavior. To better understand these phenomena, scientists in many disciplines have been pursuing research in the field of *biocomplexity*, which is concerned with characterizing the interdependence of human and biophysical systems [8]. As illustrated in Figure 20.1, such studies investigate the flows of information, wealth, materials, energy, labor, and waste among industrial systems (energy, transportation, manufacturing, food production, etc.), societal systems (urbanization, mobility, communication, etc.) and natural systems (soil, atmospheric, aquatic, biotic, etc.) [9]. The complexity, dynamics, and nonlinear nature of these interdependent systems imply that the notion of "sustainability" as a steady-state equilibrium is not realistic. Forces of change, such as technological, geopolitical, or climatic shifts will inevitably disrupt the cycles of material and energy flows, sometimes leading to unintended consequences. For example, few people foresaw that corn-based ethanol production in the United States might drive up food prices in Mexico, or that floods in the Mississippi basin might cause fuel shortages.

While ecosystems can be investigated on a local or regional basis, the connectedness of the global economy makes it difficult, and

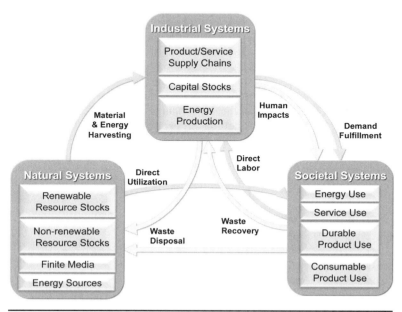

Figure 20.1 Interdependence among natural, industrial, and societal systems [9].

perhaps meaningless, to study industrial systems in isolation. The need to draw appropriate system boundaries is a barrier to complex systems modeling; for example, one impediment to the Kyoto Protocol is the potential for carbon "leakage" across borders due to economic imbalances between participating countries. Apart from connectivity and complexity, another barrier is the high degree of uncertainty that is inherent in large-scale complex systems. The turbulence of natural, political, and economic forces exceeds our ability to predict the outcomes of our actions with any confidence. Who could have anticipated the sequence of events that swept through the United States in the early twenty-first century, including the September 11, 2001 attack on the World Trade Center, the devastation of Hurricane Katrina, and the economic collapse of 2008? Plans for sustainable growth need to be based on assumptions about the future, but assumptions are often dead on arrival. In a time of turbulence and discontinuity, old business models based on precision, stability, and repeatable processes are no longer viable.

Toward Resilient Enterprise Systems

A number of multinational companies are trying to incorporate a broader systems perspective into their enterprise risk management programs. However, the traditional risk management approach is predicated on identifying threats and reducing their likelihoods or consequences. It is not effective for dealing with obscure threats that are difficult to anticipate and quantify. Faced with a dynamic and unpredictable business environment, some companies have abandoned traditional approaches to risk management and strategic planning in favor of a more flexible, resilient approach [10]. Enterprise resilience can be defined as the *capacity for an enterprise to survive, adapt, and grow in the face of turbulent change.* According to the Council on Competitiveness, innovation, enterprise resilience, and sustainability are the three cornerstones of economic competitiveness and value creation [11].

Enterprises need to grow, just as natural organisms do, and the concept of a static, no-growth enterprise is unrealistic in the business world. Enterprises can learn much from natural ecosystems, in which individual creatures and entire species are engaged in a constant struggle for food, security, survival, and growth. In a complex, connected, and uncertain world, resilience is a fundamental attribute that enables both biological and human systems to cope successfully with continual waves of change. Resilience can be seen in many contexts—in forests and wetlands that recover from devastation, in cities that

RESILIENCE WILL ENABLE ENTERPRISES TO REMAIN COMPETITIVE AND ACHIEVE THEIR LONG-TERM SUSTAINABILITY GOALS.

survive disasters and achieve renewed vitality, in companies that overcome competitive pressures by leveraging new technologies, and hopefully in industrial societies that ensure their future prosperity by learning to use natural resources more wisely.

Human enterprises have some advantages over natural systems—they are capable of foresight and planning, and they can transform very rapidly, if necessary. However, in contrast, most engineered systems—software, machines, buildings, and infrastructure—tend to be brittle and vulnerable to sudden failure or gradual decay. Therefore, the established principles of systems engineering are not sufficient to "engineer" human enterprises. The challenge of enterprise resilience is not merely to recover from disruptions and return to business as usual; rather, it is to continually reexamine the world with fresh eyes and be prepared to transform the organization in response to the emerging needs of customers and other stakeholders. Resilience may just be the missing ingredient that will enable sustainable growth.

Nowhere is the importance of resilience more evident than in the field of supply chain management, where rapid globalization and outsourcing have created massive interdependencies. All enterprises rely on both their suppliers and their customers for business continuity; therefore, an enterprise is only as resilient as its supply chain. Numerous studies have shown that supply chain disruptions can cause an immediate sharp decline in shareholder value, and some companies never fully recover [12]. Enterprise resilience management augments traditional risk management by designing supply

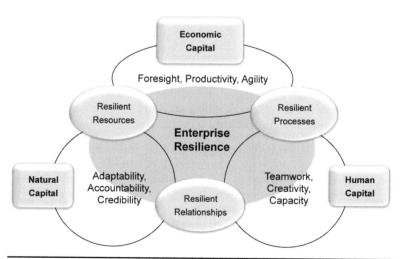

Figure 20.2 Conceptual framework for enterprise resilience.

chains that are *inherently resilient*. A classic example is Nokia's success in overcoming a March 2000 supply interruption that crippled its competitor, Ericsson, enabling Nokia to increase its market share in cellular phones. In the wake of disruptions, such as natural disasters and power failures, resilient enterprises quickly recover and sometimes are able to gain a lasting advantage over their less nimble competitors.

There is a fundamental linkage between resilience and sustainability—resilience will enable enterprises to remain competitive, overcome unforeseen obstacles, and achieve their long-term sustainability goals. Figure 20.2 depicts a strategic framework that shows how the successful operation and growth of an enterprise is linked to the vitality of external systems. The framework is built around the three sources of capital defined in Chapter 8—human capital, natural capital, and economic capital. A resilient enterprise seeks to identify vulnerabilities that may threaten these sources of capital, and to develop capabilities that offset, mitigate, or eliminate those vulnerabilities. To assure the protection and renewal of its capital, an enterprise can improve its resilience as follows:

- Improve the foresight, productivity, agility, and effectiveness of its business **processes**, from order fulfillment to knowledge management. A DFE-related example would be developing a life-cycle management process as part of new product development.

- Improve the quality, reliability, productivity, capacity, and adaptability of its available **resources**, including human, ecological, structural, and technological assets. An example would be designing closed-loop recycling technologies that reduce both waste and resource use.

- Improve collaboration, creativity, communication, and credibility in the context of its key stakeholder **relationships**, including employees, suppliers, contractors, customers, investors, regulators, communities, and advocacy groups. An example would be pursuing stakeholder engagement around climate change mitigation initiatives.

Thus, a resilience strategy will lead an enterprise to strengthen its position with respect to the network of interdependent systems in which it operates. As observed by strategy expert Michael Porter, enterprise growth and prosperity are linked to the health of the "competitive context," the social and environmental assets that provide employee talent, market demand, and a reliable supply of materials and energy [13]. Any type of product, process, or service innovation can influence these linkages in numerous ways. From this perspective, "design" is more than just creating an artifact; it is a deliberate intervention within a complex set of relationships. Companies that wish to ensure their long-term resilience must reach beyond their own boundaries, develop an understanding of the intricate systems in which they participate, and strive for continuous innovation and renewal. In this broader playing field, the rules are different: strategic adaptation becomes more important than strategic planning, and decision makers need to embrace uncertainty rather than try to eliminate it [14].

Finally, enterprise resilience operates at different time scales. *Short-term* resilience involves coping with sudden disruptions in real time to assure safety, security, and business continuity. Because of competitive pressures, redundant capacity is not a viable approach, so resilient enterprises need to develop operations that are both lean and agile. In contrast, *long-term* resilience involves pursuing competitive strategies that anticipate emerging changes in technologies and markets. Such foresight may be the key to assuring enterprise sustainability in the face of external pressures such as global climate change. In fact, while much attention has been give to climate change mitigation through carbon emission reduction, an equally important issue is *adaptation* to the emerging impacts of climate change, including sea level rise and habitat alteration, which are already being felt around the globe [15].

Returning to the practical challenges of DFE, it will be difficult for product development teams to incorporate systems thinking and resilience concepts without appropriate metrics and analytical tools. However, anticipating the implications of a design change in terms of secondary impacts on natural and social systems will require a level of analysis that is not readily available today. Integrated assessment of sustainable systems cannot be accomplished by simply linking together a collection of domain-specific models. To understand the higher-order interactions among interdependent systems will require new tools to capture emergent behaviors and dynamic relationships in complex systems. As discussed in Chapter 9, a number of multidisciplinary groups around the world are developing such tools, and some companies have begun pilot applications. The next phase of the journey is just beginning.

References

1. T. L. Friedman, *Hot Flat and Crowded: Why We Need a Green Revolution—and How It Can Renew America* (New York: Farrar, Straus, and Giroux, 2008).
2. K. E. Carpenter *et al*, "One-Third of Reef-Building Corals Face Elevated Extinction Risk from Climate Change and Local Impacts," *Science* July 25, 2008: Vol. 321, No. 5888, pp. 560–563
3. M. R. Raupach, G. Marland, P. Ciais, C. Le Quéré, J. G. Canadell, G. Klepper, and C.B. Field, "Global and regional drivers of accelerating CO_2 emissions," *Proceedings National Academy of Sciences*, June 12, 2007, Vol. 104, No. 24, 10288–10293.
4. K. Yamaji, R. Matsuhashi, Y. Nagata, and Y. Kaya, "An Integrated System for CO_2/Energy/GNP Analysis: Case Studies on Economic Measures for CO_2 Reduction in Japan. Workshop on CO_2 Reduction and Removal: Measures for the Next Century, March 19, 1991" (Laxenburg, Austria: International Institute for Applied Systems Analysis).
5. R. Sokolow and S. Pacala, "A plan to keep carbon in check," *Scientific American*, Sept. 2006, pp. 50–57.
6. J. R. Ehrenfeld, "The Roots of Sustainability," *Sloan Management Review*, 2005, 46(2): pp. 23–25.
7. P. Senge, B. Smith, N. Kruschwitz, and J. Laur, *The Necessary Revolution: How Individuals and Organizations Are Working Together to Create a Sustainable World* (New York: Doubleday, 2008).
8. R. Colwell, "Balancing the biocomplexity of the planet's living systems: A twenty-first century task for science," *BioScience*, 1998, Vol. 48, No. 10: pp. 786–787.
9. J. Fiksel, "Sustainability and Resilience: Toward a Systems Approach," *Sustainability: Science, Practice, and Policy*, September 2006.
10. G. Hamel, and L. Valikangas, The Quest for Resilience. *Harvard Business Review*, September, 2003: 52–57.
11. D. van Opstal, *Transform. The Resilient Economy: Integrating Competitiveness and Security* (Washington, DC: Council on Competitiveness, 2007).
12. T. J. Pettit, J. Fiksel and K. L. Croxton, "Can You Measure Your Supply Chain Resilience?," *Supply Chain and Logistics Journal*, 2008, Vol. 10, No. 1, pp. 21–22.
13. M. Porter and M. Kramer, "Strategy and Society," *Harvard Business Review*, Dec. 2006.
14. J. Fiksel, "Designing Resilient, Sustainable Systems," *Environmental Science & Technology*, December 2003.
15. A. D. Hecht, "Resolving the Climate Wars," *Sustainable Development Law and Policy*, Vol. IX, No. 2, Winter 2009, pp. 4–14.

CHAPTER 21

Summary

The following is a summary of the key points presented in each chapter of this book.

Chapter 1: Introduction. Environmental awareness is pervasive nowadays, due to anxieties over energy security and global warming. Many leading companies have embraced sustainability and corporate citizenship. But we have not fully acknowledged the magnitude of the environmental challenges that we face. The increasing throughput of materials in developed economies generates a hidden mountain of waste, depleting natural resources and threatening ecosystem integrity. Finding a path to sustainable growth will require global collaboration and innovation on an unprecedented scale. The business practice called Design for Environment (DFE) enables companies to address the twin goals of sustainable development and enterprise integration, assuring that new products are developed with a full understanding of life-cycle environmental considerations.

Part 1 Answering the Call: The Green Movement

Chapter 2: Motivating Forces. Recent authoritative studies have confirmed the reality of climate change and the degradation of ecosystem services. The business community has found common cause with the environmental movement, which began as a protest against industrial pollution motivated by reverence for nature. A major turning point was the Earth Summit of 1992 in Rio de Janeiro, where the fundamental principles of sustainable development were first articulated, and the groundwork was laid for the Kyoto Protocol. The World Business Council for Sustainable Development has been instrumental in the widespread adoption of sustainable business practices.

Chapter 3: External Drivers: The Voice of Society. Companies have been motivated to embrace sustainability principles because of a number of external driving forces. These include the growing "green" consciousness among customers and many other stakeholder groups; the emergence of carbon management as a key aspect of corporate strategy; enactment of a number of environmental directives by the European Union, as well as partnership initiatives by the U.S. government; establishment of international standards for environmental management systems; proliferation of sustainability rating schemes, eco-labeling programs, and voluntary codes and principles; and the blossoming of relationships between businesses and environmental advocacy groups.

Chapter 4: Business Value Drivers. Apart from their basic values and beliefs, companies have identified key drivers of business value that justify the adoption of corporate citizenship, sustainability, and associated business practices such as DFE. As a result, environmental strategy has moved beyond compliance and risk management toward pollution prevention, product stewardship, and eventually supply chain sustainability. There are several different pathways whereby environmental excellence creates shareholder value—tangible financial returns, enhancement of intangible assets such as reputation and human capital, and delivering value to stakeholders, which indirectly strengthens intangible assets.

Part 2 Charting the Course: The Art and Science of Design for Environment

Chapter 5: Managing Environmental Innovation. Responding to market expectations, many companies have incorporated DFE practices into their innovation processes. DFE should be viewed as a standard component of integrated product development and concurrent engineering. Successful implementation of DFE within a product development organization requires appropriate communications, enabling tools, and reward systems. There are three main elements of DFE practice that need to be established—performance indicators and metrics, design rules and guidelines, and analysis methods. It is important to understand the full spectrum of product life-cycle concerns, including stakeholder perceptions.

Chapter 6: Principles of Design for Environment. The following are seven basic principles of DFE:

1. Embed life-cycle thinking into the product development process.

2. Evaluate the resource efficiency and effectiveness of the overall system.

3. Select appropriate metrics to represent product life-cycle performance.

4. Maintain and apply a portfolio of systematic design strategies.

5. Use analysis methods to evaluate design performance and trade-offs.

6. Provide software capabilities to facilitate the application of DFE practices.

7. Seek inspiration from nature for the design of products and systems.

Chapter 7: Performance Indicators and Metrics. An environmental performance measurement process is essential for establishing product requirements, evaluating design improvements, communicating with stakeholders, and benchmarking of performance results. Ideally, environmental performance should be integrated with a company's existing measurement and reward systems. There are many dimensions of environmental performance, and companies need to identify the most important aspects that fit their business. Performance indicators and metrics should be chosen carefully with the intended audience in mind. Aggregation and weighting schemes should be constructed with care to avoid bias and oversimplification.

Chapter 8: Design Rules and Guidelines. DFE guidelines can be organized into four major strategies.

A. **Design for Dematerialization**—Minimize material throughput as well as the associated energy and resource consumption at every stage of the life cycle.

B. **Design for Detoxification**—Minimize the potential for adverse human or ecological effects due to waste and emissions at every stage of the life cycle.

C. **Design for Revalorization**—Recover residual value from materials and resources that have already been utilized in the economy, thus reducing the need for virgin resources.

D. **Design for Capital Protection and Renewal**—Ensure the availability and integrity of human, natural, and economic capital that are the basis of future prosperity.

Chapter 9: Analysis Methods for Design Decisions. Design teams need analysis methods to support screening of design alternatives, assessment of design performance, and trade-off comparison. There are a variety of analysis methods available, including tangible evaluation; qualitative assessment using checklists or matrices; environmental analysis using footprint indicators, life-cycle assessment methods, or predictive simulation; risk analysis methods; and financial analysis methods including life-cycle accounting and cost-benefit analysis. Choosing the right methods and level of detail to support decision making is a challenge.

Chapter 10: Product Life-Cycle Management. DFE should be embedded within a broader cross-functional process that monitors performance over the product life cycle and provides feedback for purposes of product improvement. For example, Caterpillar has established a separate Remanufacturing Division that oversees the worldwide take-back and refurbishment of engines and components, making extended producer responsibility a profitable business. 3M has developed a Life Cycle Management process that is routinely applied to the design of new products across all divisions. Companies are extending their life-cycle management and decision processes to consider stakeholder interests throughout their global supply chains.

Part 3   Walking the Talk: The Real-World Practice of Design for Environment

Chapter 11: Electronic Equipment Industries. Many electronics companies were early adopters of DFE. The industry has formed a collaborative global initiative to assess the environmental performance of suppliers, and most companies have emphasized equipment recycling programs.

> **Examples:** Xerox has taken a systems approach in applying DFE to its products, including development of "greener" high-yield paper and designing machines for end-of-life recovery. Hewlett Packard applies DFE as part of global product stewardship and has developed technologies for reducing energy use in data centers. Sony has taken a life-cycle approach to technology development and product design.

Chapter 12: Transportation Industries. Transportation accounts for a large share of global greenhouse gas emissions, and there has been a great deal of research into environmentally sustainable alternatives for enabling human mobility. Companies in the rail, air transport, and automotive industries have been working on greener technologies.

> **Examples:** General Motors, despite its financial difficulties, has introduced the electric-powered Chevrolet Volt and is converting all its manufacturing wastes to by-products. Toyota is developing new types of hybrid propulsion systems,

building on the success of the Prius. DuPont pioneered a water-based automotive paint system that has been adopted by Volkswagen.

Chapter 13: Chemical Industries. Chemical manufacturers have adopted the rigorous Responsible Care® code and are emphasizing the practice of "green chemistry" for sustainable product and process development.

> **Examples:** Dow Chemical is planning to manufacture polyethylene from sugar cane in Brazil and has developed an award-winning, bio-derived "green" insecticide. DuPont has integrated sustainable growth into its business models and corporate goals. BASF developed an eco-efficiency analysis methodology to assess the ecological and economic benefits of new products.

Chapter 14: Medical and Pharmaceutical Industries. Medical and pharmaceutical companies have also adopted green chemistry practices. Health Care Without Harm, a nonprofit, has been influential in promoting environmentally conscious practices, such as green hospital buildings.

> **Examples:** Johnson & Johnson established a Healthy Planet Initiative that includes application of DFE across all its divisions. Baxter International performs systematic product sustainability reviews and has implemented product take-back programs. Eli Lilly uses Lean Six Sigma to improve environmental performance, and has developed sustainable packaging for insulin pen devices.

Chapter 15: Food and Beverage Industries. Consumers are particularly sensitive about food and beverage products, and the industry is working on reducing the material and energy intensity of its products and packaging.

> **Examples:** Coca-Cola has been pursuing major initiatives in water stewardship, sustainable packaging, closed-loop recycling, and climate-friendly refrigeration. ConAgra has introduced a number of packaging innovations. General Mills has adopted aggressive environmental performance goals. Unilever has worked on responsible sourcing and packaging in developing economies.

Chapter 16: Consumer Products Industries. Companies with strong consumer brands have integrated DFE into product development and frequently incorporate environmental claims into their marketing efforts.

> **Examples:** Kimberly-Clark has established a DFE initiative and has used life-cycle assessment to improve the design of products such as baby wipes. Procter & Gamble has focused

its sustainability programs on water, health, and hygiene. Mohawk Industries developed a carpet made with renewable polymers. Patagonia has embraced DFE since its founding and is striving to make all of its products recyclable.

Chapter 17: Materials Production Industries. Materials are the building blocks of product design, and material producers are increasingly emphasizing recyclability and other environmentally friendly characteristics.

> **Examples:** Alcoa stresses recycling of aluminum and is converting wastes to products in its Brazilian manufacturing operations. Holcim, a global leader in the cement industry, is reducing the environmental footprint of its operations and blending waste into its products. Owens Corning is reducing the energy intensity of glass products and operates an eco-efficient insulation plant that is a model of sustainable manufacturing.

Chapter 18: Energy Production Industries. Energy production is in the spotlight due to concerns over global warming and petroleum consumption. Many alternative technologies are being developed to supplement conventional fuel sources and enable more efficient power generation and distribution.

> **Examples:** American Electric Power, a major coal-burning utility, demonstrates a commitment to sustainability principles in carbon management, design of transmission lines, and recovery of by-products such as gypsum. Chevron is helping to provide energy conservation solutions and is developing new biofuels in partnership with the National Renewable Energy Laboratory.

Chapter 19: Service Industries. Service providers can generate a significant footprint based on their utilization of energy, products, and human labor. Global companies in a variety of service industries are pursuing environmental excellence.

> **Examples:** Wal-Mart has launched a score of environmental initiatives that are driving supplier performance; JCPenney and other retailers are also introducing greener products and supply chain operations. McDonald's, Starbucks, and other food and beverage service chains are redesigning their packaging. UPS and FedEx, the leading delivery service providers, have pursued a variety of energy efficiency and recycling initiatives. The financial industries have adopted sustainability principles, and major companies such as Citigroup are greening their facilities and operations. Tourism and recreational service companies, such as Disney and Hilton, are also reducing their environmental footprint.

Part 4 Conclusion: The Road Ahead

Chapter 20: Sustainability and Resilience. Despite the efforts of industry leaders, carbon emissions and resource consumption continue to rise with economic growth. It is not enough to reduce unsustainability; we must change the fundamental systems that give rise to these environmental impacts. Systems thinking is difficult because industrial and ecological systems are highly interconnected and exhibit nonlinear behavior. Enterprises will be increasingly challenged by the complexity and turbulence of the global business landscape. Traditional risk management must be supplemented by enterprise resilience— the ability to anticipate surprises, recover from disruptions, and adapt to changing conditions. The journey to global sustainability will require development of new tools, as well as partnerships among industry, government, and nongovernmental organizations.

Glossary

Acute exposure/effect   Acute exposures refer to short duration, high intensity human exposures to chemical, physical, or biological hazards. Acute effects refer to severe symptoms that are generally of short duration.

Air pollutant   A substance released into the air from industrial emissions or other sources such as vehicles and product use. Air pollutants can have adverse effects on human health or the environment.

Alternative fuels   Energy-containing wastes used as substitutes for conventional thermal energy sources.

Aspect   A broad, measurable characteristic of an enterprise or activity. An example of an "environmental aspect" is energy intensity.

Balanced scorecard   A framework for measuring corporate performance in terms of both financial outcomes and non-financial drivers, including learning and growth, internal business process excellence, and customer relationships.

Benchmark   A standard against which something is measured; a reference point.

Bioaccumulative substance   A substance that tends to accumulate in living tissues and therefore is found at high concentrations in the food chain.

Bio-based product   A product that is partly or wholly composed of agricultural or biological materials; also refers to products manufactured using biological processes regardless of the feedstock.

Biodegradable material   A material that can be decomposed by microorganisms into organic constituents, and therefore does not accumulate in the environment.

Biodiversity   Variability among living organisms within species, between species, and between ecosystems. Biodiversity helps to assure the resilience of ecosystems.

Biofuel   A fuel made in whole or in part from biomass, i.e., renewably sourced biological materials, such as crops, agricultural residues, or organic waste materials.

Biomimicry   The practice of adapting designs and technologies found in nature to solve human problems.

Biotechnology   A set of biological techniques, including but not limited to genetic modification, which utilize living organisms for industrial, agricultural and pharmaceutical applications.

Brand equity   An intangible value-added aspect of particular goods otherwise not considered unique.

Brominated compounds   Bromine-based compounds often used as fire retardants, particularly in textiles and polymers. Some brominated compounds have been identified as persistent, bioaccumulative, and toxic.

Business case   A rationale for making a business decision, usually involving quantitative analysis of costs, benefits and trade-offs.

By-product   Secondary or residual product of an industrial process, with potential economic value.

By-product synergy   The conversion of industrial process wastes into by-products that can become feedstocks for other processes, organized through collaboration between waste generators and waste utilizers.

Capital   Economic capital is something owned that provides ongoing services, including land, durable investment goods, physical structures, and equipment. (See also Natural capital, Human capital.)

Carbon footprint   Estimate of the total greenhouse gas emissions associated with an enterprise or activity, including direct and indirect emissions. The boundaries of indirect emissions included can vary considerably.

Carcinogenic   Capacity to cause cancer in humans. Many substances are considered "suspect" carcinogens based on limited evidence from animal or in vitro studies.

Cash flow   Earnings before depreciation, amortization and non-cash charges (sometimes called cash earnings).

Cellulosic ethanol   An advanced biofuel produced by converting the starch extracted from cellulose (derived from plant materials) into sugars and fermenting the sugars into ethanol.

Certified   Approved or qualified by an independent third party according to predefined criteria. Products, services, or management systems can be certified.

Chlorinated compounds   Substances that contain chlorine molecules, such as carbon tetrachloride. Some chlorinated compounds have been identified as persistent, bioaccumulative, and toxic.

Chronic   Chronic exposures refer to long-duration, low-intensity human exposures to chemical, physical, or biological hazards. Chronic disease effects generally have a delayed onset and may last for years.

Climate change   See Global warming.

Code of conduct   A set of principles governing the behavior and business practices of a company or other organization, often addressing issues such as business ethics and responsible governance.

Compost   Material formed by decomposition of organic wastes that can be used as a soil conditioner or mulch.

Concurrent engineering   A product development practice in which different engineering disciplines work in a parallel, coordinated fashion to address life-cycle requirements, including quality, manufacturability, reliability, maintainability, safety, and sustainability. (Also known as simultaneous engineering.)

Continuous improvement   An ongoing effort to improve the measurable performance of products, services or processes, usually based on the "Plan, Do, Check, Act" cycle.

Corporate citizenship   Company activities concerned with treating the stakeholders of the firm ethically and in a socially responsible manner.

Corporate governance   The system of oversight, including a Board of Directors, whereby a corporation maintains accountability for protecting the interests of shareholders and other stakeholders.

Corporate social responsibility   Commitment to uphold the rights of citizens and communities, behave according to accepted ethical standards, and contribute to socio-economic development and quality of life.

Cost-benefit analysis   Comparison of the costs and benefits associated with alternative plans, designs, or investments, for purposes of decision making.

Cradle-to-cradle   Scope of a product life cycle that extends from raw material acquisition through recovery and renewal of the product residuals at end-of-life.

Cradle-to-gate   Scope of a product life cycle that extends from raw material acquisition through final manufacturing (the factory gate).

Cradle-to-grave   Scope of a product life cycle that extends from raw material acquisition through disposition of the product at end-of-life.

Design for environment   Systematic consideration of design performance with respect to environmental, health, safety, and sustainability objectives over the full product and process life cycle.

Dioxins   A family of chlorinated compounds known as halogenated aromatic hydrocarbons. Dioxins are by-products of certain combustion and manufacturing processes, and are generally considered to pose risks of chronic effects.

Discounted cash flow   Analysis of future cash flows multiplied by discount factors to obtain present values.

Earnings before interest, taxes, depreciation and amortization (EBITDA)   A financial indicator used to analyze profitability of operations.

Eco-effectiveness   Characteristic of life-cycle product/service systems that leads to progressive improvements in environmental and economic well being.

Eco-efficiency   Delivery of valuable goods and services that satisfy human needs while progressively reducing life-cycle environmental impacts and resource intensity. Eco-efficiency is often measured by the ratio of value delivered to resource inputs, such as materials, natural resources, and energy.

Eco-label Designation of a product or service with a label or mark, indicating that it has met specified environmental criteria. Eco-labels can be self-declared or based on third-party certification.

Ecological economics An interdisciplinary field of academic research that studies the interdependence between human economies and natural ecosystems.

Economic value added (EVA) A financial indicator of the shareholder wealth or "economic profit" created by a particular activity within an organization. EVA is calculated as the difference between after-tax operating profit and capital charge.

Ecosystem A dynamic complex of plant, animal, and micro-organism communities and their nonliving environment that interact as a functional unit. Examples of ecosystems include deserts, coral reefs, wetlands, rainforests, boreal forests, grasslands, urban parks, and cultivated farmlands.

Ecosystem services Beneficial services that human communities obtain from ecosystems. Examples include fresh water, timber, climate regulation, erosion control, and recreation.

Emissions Airborne releases of gaseous pollutants.

Endangered species Species that are in imminent danger of extinction, due to loss of habitats, loss of food sources, pollution, increases in predators, or dramatic reductions in population.

Energy intensity A measure of environmental efficiency in production, calculated by dividing the net energy consumption by the quantity or monetary value of the output.

Enterprise integration Re-engineering of business processes and information systems to improve teamwork and coordination across organizational boundaries, thereby increasing the effectiveness of the enterprise as a whole.

Environmental footprint A quantitative measure of the impacts that a product, process or activity has upon the environment. Footprint measures may include energy use, water use, material consumption, waste and emissions, or productive land area required, and may extend over all or part of the life cycle.

Environmental impacts Adverse changes in ecosystems, habitat conditions, flora, fauna, etc. due to human activities such as resource consumption, pollution and land use.

Environmental performance The performance of a product, process, activity, or business entity according to selected indicators of environmental impact.

Environmental, health, and safety (EH&S) A professional discipline concerned with protection of the environment, human health, and safety through the application of scientific, engineering, and management methods.

Environmentally preferable   Superior to comparable products or services in terms of environmental footprint. Analogous terms include environmentally friendly, environmentally benign, and green.

Ethanol   The most widely used biofuel today; ethanol is an alcohol that can be produced by converting starch from crops into sugars and fermenting the sugars with microbes.

Exergy   A thermodynamic measure of available energy that takes into account energy quality.

Financial performance   The performance of a corporation over time as measured by selected financial indicators, typically measuring the return.

Fossil fuel   A general term for combustible hydrocarbon deposits of biological origin, including coal, oil, natural gas, and oil shale.

Genetic engineering   Modification of organisms, either plants or animals, by transfer of genes from other species into their genetic material. Concerns about genetic engineering include the possibility of uncontrolled gene transfer to non-target species, emergence of unintended side effects, as well as ethical concerns about gene manipulation.

Global warming   Gradual increase in average temperatures at the earth's surface, believed to result from the "greenhouse effect" due to increased atmospheric concentrations of carbon dioxide and other gases.

Green chemistry   Development of chemical products and chemical reactions that are inherently benign in terms of environmental impacts. Also known as "sustainable chemistry."

Green marketing   The analysis of market opportunities related to product environmental performance and positioning of products to address those opportunities.

Green purchasing   A business practice whereby purchasing agents in business or government evaluate products and services based upon selected environmental performance attributes.

Greenhouse gas (GHG)   A gaseous substance that contributes to the greenhouse effect, i.e., global warming. The most abundant greenhouse gas is carbon dioxide, but other gases released by human activities have significant global warming potential, including methane, nitrous oxide, and fluorinated gases.

Habitat   The environment or ecosystem where a plant or animal naturally or normally lives and grows.

Hazard   A material or condition whose presence may cause harm to humans, wildlife, or property. Examples of hazards include flammable, explosive, corrosive, or toxic chemicals, as well as heat, noise, radiation, and biological agents. The effects of exposure to hazards may be acute or chronic.

Hazardous waste   A material which is potentially harmful to people or the environment because of its toxic, poisonous, explosive, corrosive, flammable, infectious characteristics.

Heavy metals   Metals that are relatively dense and toxic at low concentrations. For example, cadmium, chromium, nickel and certain of their compounds are known to be carcinogenic; lead and mercury can cause irreversible functional impairment.

Human capital   The set of skills which employees acquire on the job, through training and experience, and which increase their value in the marketplace.

Indicator   A quantifiable performance aspect of a product, process, service, facility, or enterprise.

Industrial ecology   Framework for improvement in the efficiency of industrial systems by imitating aspects of natural ecosystems, including the cyclical transformation of wastes into input materials. Alternative terms include "industrial metabolism" and "industrial symbiosis."

Intangible asset   A non-monetary asset, including people, ideas, networks, relationships, and processes, which is not traditionally accounted for on the balance sheet.

Integrated product development   A cross-functional design process that considers the entire spectrum of quality factors, including safety, testability, manufacturability, reliability, maintainability, and sustainability throughout the product life cycle.

Intellectual capital   Knowledge that can be exploited for business purposes, including the skills, knowledge, and documents that a company or its employees have accumulated about the business.

Key performance indicator (KPI)   One of a small number of indicators that correspond to critical corporate goals and are reflected in compensation or recognition systems.

Lagging indicator   An indicator of performance outcomes that can be observed after the period of performance.

Leading indicator   A predictive indicator of anticipated performance that can be observed prior to the period of performance.

License to operate   The ability of a corporation or business to continue operations based on ongoing acceptance by external stakeholder groups.

Life-cycle accounting   Quantification of direct and indirect costs and benefits across the life cycle of a facility, product, or process. Cost/benefit categories include hidden, contingent, good will, and external.

Life-cycle assessment (LCA)   A systematic technique for identifying and evaluating the potential environmental benefits and impacts associated with products or processes throughout their life cycle. LCA is a standardized method and is documented in the ISO standards 14040 series. (See cradle-to-cradle, etc.)

Life-cycle impacts   The impacts of a product on the environment from extraction of raw materials to production, transportation, use, recycling and final disposal. Potential impacts include energy and water consumption, liquid discharges, gaseous emissions, solid wastes, etc.

Management system   A management approach that enables an organization to identify, monitor and continuously improve its performance, including financial, environmental, and/or social aspects.

Material intensity   A measure of environmental efficiency in production, calculated by dividing the net material consumption by the quantity or monetary value of the output.

Materiality   Importance of information about company performance to decision making on the part of stakeholders, including company shareholders and management.

Metric   A specific unit of measure used to quantify a performance indicator. An indicator (e.g., energy efficiency) can have many different metrics.

Natural capital   Ecological resources and services that make possible all economic activity. Ecological resource flows include edible organisms, sand, wood, grass, metals, and minerals, while ecological services include various forms of energy provided by the water cycle, wind, tides, soil, and pollination.

Net present value   The amount of cash today that is equivalent in value to a payment, or to a stream expected future cash flows minus the cost.

Non-governmental organization (NGO)   An organization, typically not-for-profit, which is independent of both industry and government, e.g., charitable foundations, advocacy groups.

Non-financial performance   The performance of a business measured in terms of non-financial aspects such as environmental and social responsibility.

Non-renewable energy   Energy derived from sources that cannot be replenished in a short period of time relative to a human life span. Examples include fossil fuels, such as oil, natural gas, and coal, and nuclear fuels.

Non-renewable resource   A natural resource that cannot be replaced within the same time scale that it is consumed for industrial purposes, e.g., fossil fuels.

Normalization   The use of common denominators to adjust metrics for purposes of comparison, e.g., emissions per pound of product is a measure of emissions normalized by product mass.

Organic   A product of organic agriculture, based on minimal use of synthetic inputs as well as management practices that maintain and enhance ecological harmony.

Persistent substance   A chemical substance that persists in the environment, and is not easily biodegradable. Examples include industrial by-products such as dioxins and pesticides such as DDT.

Pollution prevention (P2)   Modification of production processes and technologies so that they generate less waste and pollution. Principal P2 techniques include source reduction, toxics avoidance, and recycling.

Polymers Complex, chain-like molecules produced by uniting simpler molecules called monomers through chemical bonding. Polymers may be persistent and bioaccumulative, though not necessarily toxic.

Post-consumer waste Waste materials generated by a business or consumer that have served their intended end uses, and can be separated for the purposes of collection, recycling and disposition.

Post-industrial waste Waste materials generated in a manufacturing process and can be recycled back into the process or used for another purpose.

Precautionary principle A principle advanced at the 1992 UN Summit, stating that scientific uncertainty should not be a basis for postponing cost-effective measures to prevent serious or irreversible environmental degradation.

Product differentiation A competitive business strategy that seeks to offer products with distinctive features in order to differentiate them from those of competitors.

Product life cycle (1) A series of stages in the physical life of a product, including resource extraction, procurement, transportation, manufacturing, product use, service, and end-of-life disposition or recovery. (2) A series of stages in the commercial life of a product, including research and development, design, introduction, growth, extension, phase-out, and discontinuance.

Product stewardship Consideration of health, safety and environmental protection as an integral part of designing, manufacturing, marketing, distributing, using, recycling and disposing of products.

Product take-back A program, either voluntary or regulated, whereby manufacturers take responsibility for recovering obsolete products at the end of their useful lives, and seek to re-use or recycle their components.

Rebound effect A phenomenon that occurs when increased efficiency lowers the cost of consuming an economic good or service (e.g., gasoline), resulting in greater consumption of that or other goods and services.

Recycled content Percentage of recycled material incorporated into a product or raw material. Generally, the higher the recycled content, the lower the life-cycle energy use and environmental impacts.

Renewable resource An energy or material resource (e.g., wood) that can be renewed or regenerated by natural ecological cycles or sound management practices within a short time relative to a human life span.

Requirements management A process for assuring that product requirements are met, consisting of three main functions that are performed iteratively: requirements analysis, requirements tracking, and requirements verification

Resilience The capacity of a system, such as an enterprise, a community, or an ecosystem, to survive, adapt and flourish in the face of turbulent change.

Return on net assets (RONA) A measure of a corporation's profitability determined by dividing net income for the past 12 months by total average assets minus total liabilities, i.e., net worth.

Risk (1) The presence of uncertainty in a business or other activity. (2) The possibility of an adverse incident due to the presence of hazards.

Risk management The process of identifying and evaluating risks and selecting and managing techniques to adapt to risk exposures.

Screened investing The application of social and environmental criteria to evaluate conventional investments, such as stocks, bonds, and mutual funds.

Servicization An innovation strategy that shifts the business focus from designing and selling physical products to selling a system of products and services that fulfill customer needs.

Shareholder resolution A recommendation or requirement, proposed by a shareholder, that a company and/or its board of directors take action presented for a vote at the company's general shareholders' meeting.

Shareholder value The economic market value that a shareholder would realize from liquidation of their equity.

Socially responsible investing The incorporation of an investor's social, ethical, or religious criteria in the investment decision-making process.

Solvents Solvents are liquids (e.g., acetone) capable of dissolving or dispersing other substances, used in manufacturing and consumer cleaning applications. Solvents can have adverse effects on human health and atmospheric pollution (see VOCs).

Stakeholder (1) A person or group that has an investment, share, or interest in something, as a business or industry. (2) Any party that has an interest, financial or otherwise, in a firm, including shareholders, creditors, employees, customers, suppliers, communities, interest groups, and the government.

Stakeholder engagement Deliberate outreach on the part of a corporation in order to engage in dialogue and develop relationships with stakeholders.

Strategy A set of goals and aspirations combined with a comprehensive action plan for achieving those goals.

Supply chain A sequence of suppliers and customers that add value in the form of materials, components, or services, ultimately resulting in a final product. For products with multiple inputs, supply chains are actually networks.

Sustainability Conformance with principles of sustainable development, encompassing the environmental, health and safety, social, economic, and ethical aspects of a corporation or other entity.

Sustainable business A business that is able to anticipate and meet the economic, environmental, and social needs of present and future generations of customers and other stakeholders.

Sustainable business practice   An activity or process implemented by a corporation in order to enhance its sustainability. An example is incorporating Design for Environment into new product development.

Sustainable consumption   The consideration of sustainability issues in managing the consumption patterns of an organization or community.

Sustainable development   Economic development that meets the needs of the present without compromising the ability of future generations to meet their own needs.

Sustainable production   The consideration of sustainability issues in managing the production processes of a company or industry, e.g., computer manufacturing.

Synthetic   Chemical or material substance manufactured by humans from materials of non-biological origin.

Tangible asset   An asset whose value depends on particular physical properties, including reproducible assets such as buildings or machinery and non-reproducible assets such as land, a mine, or a work of art.

Time to market   The time interval or cycle time between the launch of a new product development effort and the market introduction of the new product.

Total quality management   A management approach, centered on quality, based on the participation of all employees and aimed at long-term success through customer satisfaction.

Transparency   Openness of a company or organization with regard to disclosing information about its policies, principles, and decision-making processes.

Triple bottom line   A framework for sustainable development that defines three fundamental aspects of corporate performance—economic, environmental, and social.

Upgradeable design   A design for a durable product that allows the product to be upgraded by the replacement of outdated components.

Value chain   See Supply chain.

Value driver   A fundamental and persistent characteristic of a business enterprise that influences its market value positively.

Volatile organic compounds (VOCs)   Carbon compounds that contribute to smog formation through atmospheric photochemical reactions. Examples include: formaldehyde, benzene, and toluene.

Waste material   A by-product of a process or activity having no or minimal economic value. Waste materials can be solids or liquids; gases are considered air pollutants.

Index